T0389639

Statistical Complexity

K.D. Sen
Editor

Statistical Complexity

Applications in Electronic Structure

Editor
Professor Dr. K.D. Sen
School of Chemistry
University of Hyderabad
Hyderabad 500046
India
kalidas.sen@gmail.com

ISBN 978-90-481-3889-0 e-ISBN 978-90-481-3890-6
DOI 10.1007/978-90-481-3890-6
Springer Dordrecht Heidelberg London New York

Library of Congress Control Number: 2011936030

Art work by S.L. Parasher: Prism Light, 1979 [Bela and Vinay Sethi Collection]

Printed on acid-free paper

Springer is part of Springer Science+Business Media (www.springer.com)

Dedicated to Robert G. Parr on his 90th Birthday

Foreword

"I think the next century will be the century of complexity", Stephen Hawking in San Jose Mercury News, Morning Final Edition, January 23 (2000)

During the recent years, several different measures defining *complexity* have been proposed within the scientific disciplines of physics, biology, mathematics and computer science. Such measures of complexity are generally context dependent. In chemical physics, a set of statistical complexity measures have been introduced using the information theoretical tools centering around the electron probability density of N-electron system and its other characteristics. The monograph presents a detailed description of *such* measures of complexity. Starting with the information theoretical foundations this monograph discusses at length their applications in the electronic structure of atoms and molecules including reactivity. The subject matter of the title is covered by the leading research scientists in the field who present an up-to-date account of their contributions, including future projections.

The following topics are dealt with by leading research groups from 9 countries who are currently active in this area of research: Entropic Uncertainty Relationships in Quantum Physics; Scaling properties of Information Theoretical Uncertainty-like Measures; Derivation of Generalized Weizsäcker Kinetic Energy Functionals from Quasiprobability Distribution Functions; Complexity in Atoms; Statistical Complexity and Fisher-Shannon Information: Applications; Rényi Entropy and Complexity; Entropy and Complexity Analyses of D-dimensional Quantum Systems; Information Uncertainty, Similarity and Relative Complexity Measures; Atomic and Molecular Complexities: Their Physical and Chemical Interpretations including Relativistic Electron Density.

The monograph is intended for the practicing physical, theoretical and computational chemists, material scientists, bio physicists and mathematical physicists.

C.R. Rao, Sc.D., F.R.S.
National Medal of Science Laureate, USA
Eberly Professor Emeritus of Statistics, Penn State
Distinguished Professor Emeritus, CR RAO AIMACS
University of Hyderabad Campus, Prof CR RAO Road
Hyderabad 50046, India

Preface

Information theoretical tools are being increasingly applied in several disciplines of scientific enquiry ranging from chemistry, physics, biology, medicine, genetics, electrical engineering, computer science, to linguistics, economics and other social sciences. Recently, the useful intuitive concept of complexity has attracted considerable attention. The system specific context and scale dependence, spread across disciplines, has given rise to several a host of different measures of complexity. In chemical physics, several structural and reactivity related properties and processes can be analyzed in terms of the probability distribution function which describes the uncertainty, randomness, disorder or delocalization, within an information theoretical framework. The present monograph is a collective endeavor of nine research groups in physics and chemistry actively engaged in the development and applications in this rapidly growing and truly interdisciplinary field of research. Each chapter is planned as being self contained. The opening contribution of Iwo Bialynicki-Birula and Łukasz Rudnicki entitled *Entropic Uncertainty Relations in Quantum Physics* presents a critical study of the information theoretical uncertainty-like relations covering various generalizations and extensions of the uncertainty relations in quantum theory that involve the Rényi and the Shannon entropies. More direct connection of such relationships to the observed phenomena is emphasized. The contribution by Debajit Chakraborty and Paul W. Ayers entitled *Derivation of Generalized von Weizsäcker Kinetic Energies from Quasiprobability Distribution Functions* focuses on the Fisher Information of the electronic distribution functions and describes how generalizations of the Weizsäcker kinetic energy density functional can be derived from the canonical momentum-space expression for the kinetic energy with further extension to higher-order electron distribution functions. In the chapter written by C.P. Panos and coworkers entitled *Atomic Statistical Complexity* a review of applications of the Shannon information entropy, Fisher measure, Onicescu energy, and two of the simple statistical complexity measures on atoms are presented using the Hartree-Fock density in position and momentum space. Interesting correlations with the experimental properties such as the ionization potential and static dipole polarizability are reported using such measures. The chapter entitled *Statistical Complexity and Fisher-Shannon Information Applications* by Ricardo López-Ruiz and coworkers defines the currently most widely used statistical measure of complexity, and is known as the LMC measure named after its main developers. In it, a variety of applications in discrete and continuous standard model quantum systems are presented. The statistical indicators discern and highlight some conformational properties of the model systems. J.S. Dehesa and coworkers review the present status of the analytic information theory of quantum systems with non-standard dimensionality in the position and momentum spaces in their chapter entitled *Entropy and Complexity Analyses of D-dimensional Quantum Systems*. Here the power and entropic moments are used to study the most relevant information-theoretic one-element Fisher, Shannon, Rényi, Tsallis as well as some composite two-elements Fisher-Shannon, LMC shape and Cramér-Rao complexities measures which describe the spreading measures of the position and momentum probability densities well beyond the standard deviation. The chapter entitled *Atomic and Molecular Complexities—Their Physical and Chemical Interpretations* by J.C. Angulo and coworkers further develops on the meaning, interpretation and applications of the complexity measures by utilizing different order-uncertainty planes in the position and momentum electron densities of several atomic (neutrals, singly-charged ions, isoelectronic series) and molecular (closed shells, radicals, isomers) systems. The quantities sustaining such planes are the exponential and the power Shannon entropies, the disequilibrium, the Fisher information and the variance. Each plane gives rise to a measure of complexity, determined by the product of its components. By means of extensive numerical examples an emphasis is placed on the necessity to consider both position and momentum space probability density. In the chapter entitled *Rényi Entropy and Complexity* by Á. Nagy and E. Romera the relationship between the statistical complexity and the Rényi entropy is studied. A detailed study on the uncertainty-like relations for the Rényi entropy including uncertainty relations for single particle densities of many particle systems in position and momentum spaces is presented. The maximum Rényi entropy principle is used to generalize the Thomas-Fermi model. A relationship with the atomic radius and quantum capacitance is discussed. The chapter entitled *Scaling Properties of Net Information Measures for Bound States of Spherical Model Potentials* by K.D. Sen and S.H. Patil reviews the results obtained on the scaling properties from dimensional analysis of the position and momentum variances defining the quantum mechanical Heisenberg uncertainty product as well as several *composite* information theoretical measures. A test set of free and spherically confined model potentials has been included and the main dimensionless variables on which the dynamical variables depend have been listed. In their chapter entitled *Chemical Information from Information Discrimination and Relative Complexity* Paul Geerlings and coworkers have presented a review of their work which derives heavily from the Kullback-Leibler relative information measure in which a suitable choice of the reference prior density reveals useful information. Interesting chemical insights, for example, on the periodicity in Mendeleev's table and the interesting reaction profiles can be extracted in this manner. The investigation of atomic complexity measures is reported in terms of the non-relativistic and relativistic shape electron densities so as to quantify the relativistic effects on the complexity measure. It is shown that a relative complexity measure can be defined so as to reflect the *diversity* of electron density functions with respect to a reference atom as the sub shells are filled across the periodic table.

I thank Professor C.R. Rao for his constant support and encouragement. This project has been carried out during the period of award of J.C. Bose national fellowship by Department of Science and Technology, New Delhi, which is gratefully acknowledged. I remain forever indebted to Dr. Bela Sethi, Vinay Sethi and Dr. Aloka Parasher-Sen for their generous permission to use the painting of Shri S.L. Parasher for the cover design.

Hyderabad, India K.D. Sen

Contents

Contributors

J.C. Angulo Departamento de Física Atómica, Molecular y Nuclear, Universidad de Granada, 18071 Granada, Spain; Instituto Carlos I de Física Teórica y Computacional, Universidad de Granada, 18071 Granada, Spain, angulo@ugr.es

J. Antolín Instituto Carlos I de Física Teórica y Computacional, Universidad de Granada, 18071 Granada, Spain; Departamento de Física Aplicada, EUITIZ, Universidad de Zaragoza, 50018 Zaragoza, Spain, antolin@unizar.es

Paul W. Ayers McMaster University, Hamilton, Ontario, Canada, L8S 4M1, ayers@mcmaster.ca

Iwo Bialynicki-Birula Center for Theoretical Physics, Polish Academy of Sciences, Al. Lotników 32/46, 02-668 Warszawa, Poland, birula@cft.edu.pl

Alex Borgoo Vrije Universiteit Brussel, Brussels, Belgium, aborgoo@vub.ac.be

Xavier Calbet BIFI, University of Zaragoza, Zaragoza 50018, Spain, xcalbet@googlemail.es

Debajit Chakraborty McMaster University, Hamilton, Ontario, Canada, L8S 4M1, chakrd2@mcmaster.ca

K.C. Chatzisavvas Department of Theoretical Physics, Aristotle University of Thessaloniki, 54124 Thessaloniki, Greece; Department of Informatics and Telecommunications Engineering, University of Western Macedonia, 50100 Kozani, Greece, kchatz@auth.gr

J.S. Dehesa Departamento de Física Atómica, Molecular y Nuclear and Instituto Carlos I de Física Teórica y Computacional, Universidad de Granada, 18071 Granada, Spain, dehesa@ugr.es

R.O. Esquivel Instituto Carlos I de Física Teórica y Computacional, Universidad de Granada, 18071 Granada, Spain; Departamento de Química, Universidad Autónoma Metropolitana, 09340 México D.F., México, esquivel@xanum.uam.mx

Paul Geerlings Vrije Universiteit Brussel, Brussels, Belgium, pgeerlin@vub.ac.be

S. López-Rosa Departamento de Física Atómica, Molecular y Nuclear and Instituto Carlos I de Física Teórica y Computacional, Universidad de Granada, 18071 Granada, Spain, slopez@ugr.es

Ricardo López-Ruiz Department of Computer Science, Faculty of Science and BIFI, University of Zaragoza, Zaragoza 50009, Spain, rilopez@unizar.es

D. Manzano Departamento de Física Atómica, Molecular y Nuclear and Instituto Carlos I de Física Teórica y Computacional, Universidad de Granada, 18071 Granada, Spain, manzano@ugr.es

S.E. Massen Department of Theoretical Physics, Aristotle University of Thessaloniki, 54124 Thessaloniki, Greece, massen@auth.gr

C.C. Moustakidis Department of Theoretical Physics, Aristotle University of Thessaloniki, 54124 Thessaloniki, Greece, moustaki@auth.gr

Á. Nagy Department of Theoretical Physics, University of Debrecen, 4010 Debrecen, Hungary, anagy@madget.atomki.hu

N. Nikolaidis Department of Theoretical Physics, Aristotle University of Thessaloniki, 54124 Thessaloniki, Greece; Department of Automation, Faculty of Applied Technology, Alexander Technological Educational Institute (ATEI) of Thessaloniki, 57400 Thessaloniki, Greece, niknik@teithe.gr

C.P. Panos Department of Theoretical Physics, Aristotle University of Thessaloniki, 54124 Thessaloniki, Greece, chpanos@auth.gr

S.H. Patil Department of Physics, Indian Institute of Technology, Mumbai 400 076, India, sharad@phy.iitb.ac.in

E. Romera Departamento de Física Atómica, Molecular y Nuclear, Universidad de Granada, 18071 Granada, Spain, eromera@ugr.es

Łukasz Rudnicki Center for Theoretical Physics, Polish Academy of Sciences, Al. Lotników 32/46, 02-668 Warszawa, Poland, rudnicki@cft.edu.pl

Jaime Sañudo Department of Physics, Faculty of Science, University of Extremadura, Badajoz 06071, Spain, jsr@unex.es

K.D. Sen School of Chemistry, University of Hyderabad, Hyderabad 500 046, India, kalidas.sen@gmail.com

Chapter 1
Entropic Uncertainty Relations in Quantum Physics

Iwo Bialynicki-Birula and Łukasz Rudnicki

Abstract Uncertainty relations have become the trademark of quantum theory since they were formulated by Bohr and Heisenberg. This review covers various generalizations and extensions of the uncertainty relations in quantum theory that involve the Rényi and the Shannon entropies. The advantages of these entropic uncertainty relations are pointed out and their more direct connection to the observed phenomena is emphasized. Several remaining open problems are mentioned.

1.1 Introduction

In recent years we have seen many applications of the Rényi and Shannon entropies in many fields from biology, medicine, genetics, linguistics, and economics to electrical engineering, computer science, geophysics, chemistry, and physics. In particular, the Rényi entropy has been widely used in the study of quantum systems. It was used in the analysis of quantum entanglement [1–5], quantum communication protocols [6, 7], quantum correlations [8], quantum measurement [9], and decoherence [10], multiparticle production in high-energy collisions [11–13], quantum statistical mechanics [14], pattern formation [15, 16], localization properties of Rydberg states [17] and spin systems [18, 19], in the study of the quantum-classical correspondence [20], in electromagnetic beams [21], and the localization in phase space [22, 23].

Our aim in this review is to use the Shannon and Rényi entropies to describe the limitations on the available information that characterizes the states of quantum systems. These limitations in the form of mathematical inequalities have the physical interpretation of the uncertainty relations. We will not enter here into the discussion (cf. [24, 25]) of a fundamental problem: which (if any) entropic measure of uncertainty is really most adequate in the analysis of quantum mechanical measurements.

I. Bialynicki-Birula (✉)
Center for Theoretical Physics, Polish Academy of Sciences, Al. Lotników 32/46, 02-668 Warszawa, Poland
e-mail: birula@cft.edu.pl

K.D. Sen (ed.), *Statistical Complexity*,
DOI 10.1007/978-90-481-3890-6_1, © Springer Science+Business Media B.V. 2011

The uncertainty relations discussed in this paper are valid as mathematical inequalities, regardless of their physical interpretation. Since this is a review, we felt free to use the results of our previous investigations. Of course, such self-plagiarism would be highly unethical in an original publication.

1.2 Information Entropy as a Measure of Uncertainty

Statistical complexity, a tool presented and explored in this book is based on information entropy. This concept was introduced by Claude Shannon [26] in 1948 and grew up to have many applications. There are also several generalizations and extensions which we will discuss in our contribution. We shall focus on the uncertainty, a notion closely connected with the information entropy but a little bit wider and having different meanings depending on a context. We shall use here the term uncertainty as a measure of missing information. The use of the information entropy as a measure of uncertainty becomes then very natural. All we have to do is to reverse the sign. The lack of information—negative information—is the uncertainty. In simple terms, we can measure the capacity of an empty tank by measuring the volume of water filling this tank. Therefore, the uncertainty or missing information can be measured in exactly the same manner as the information is measured. We shall exploit this point of view showing that the measure of uncertainty based on the information entropy may be used to replace famous quantum mechanical uncertainty relations. Moreover, we shall argue now that this method of expressing the uncertainty relations is much closer to the spirit of quantum theory.

1.2.1 Standard Deviation

In classical physics the positions and momenta of every particles can be determined without any fundamental limitations. In that case, when we talk about an uncertainty of position or momentum of a particle, we mean an uncertainty caused by a lack of precision of our measuring instruments. When we measure some observable Q we usually repeat this measurement many times obtaining a distribution of values. We apply then various standard procedures, well known to experimentalists, to extract the "best value" of the measured quantity. A crude, but often a sufficient procedure is to evaluate an average value $\langle Q\rangle$ of Q. We can also calculate the standard deviation:

$$\sigma_Q = \sqrt{\langle (Q - \langle Q\rangle)^2\rangle}, \tag{1.1}$$

which we treat as a measure of uncertainty of $\langle Q\rangle$. It is the uncertainty connected directly with a given experimental setup and has no fundamental significance. A more skilled experimentalist will be able to reduce this uncertainty.

In quantum physics we face a dramatic change because we have to deal with a spread of measured values which is of fundamental nature. The uncertainty in most measurements (there are some exceptions) cannot be indefinitely decreased. This is due, at least at this stage of our theoretical understanding, to the probabilistic nature of the quantum world. For example, the state of a single nonrelativistic quantum particle can be described by a wave function $\psi(\boldsymbol{r})$. The square of the modulus of this wave function $\rho(\boldsymbol{r}) = |\psi(\boldsymbol{r})|^2$ determines the probability distribution of the position of the particle. Since the probability to find the particle anywhere must be equal to 1, the function $\psi(\boldsymbol{r})$ must be normalized according to:

$$\int_{\mathbb{R}^3} d^3r |\psi(\boldsymbol{r})|^2 = 1. \tag{1.2}$$

This probabilistic distribution of values at our disposal is not connected with some faults of our measuring procedure but it is an intrinsic property—the spread of values of the position $\boldsymbol{r}$ cannot be avoided. The classical position $\boldsymbol{r}_{cl}$ can at best be associated with the average value

$$\boldsymbol{r}_{cl} = \langle \boldsymbol{r} \rangle = \int_{\mathbb{R}^3} d^3r \boldsymbol{r} \rho(\boldsymbol{r}). \tag{1.3}$$

Having a probability distribution $\rho(\boldsymbol{r})$ we can proceed according to the rules of statistics and calculate the standard deviation, say σ_x, that characterizes the spread of the values of the coordinate x,

$$\sigma_x = \left[\int_{\mathbb{R}^3} d^3r (x - \langle x \rangle)^2 \rho(\boldsymbol{r}) \right]^{1/2}. \tag{1.4}$$

After the Fourier transformation, we can obtain another description of *the same state*

$$\tilde{\psi}(\boldsymbol{p}) = \int_{\mathbb{R}^3} \frac{d^3r}{(2\pi\hbar)^{3/2}} e^{-i\boldsymbol{p}\cdot\boldsymbol{r}/\hbar} \psi(\boldsymbol{r}). \tag{1.5}$$

The Fourier transform $\tilde{\psi}(\boldsymbol{p})$ of a wave function, according to the rules of quantum mechanics, gives the probability distribution in momentum space $\tilde{\rho}(\boldsymbol{p}) = |\tilde{\psi}(\boldsymbol{p})|^2$. Note, that $\tilde{\rho}(\boldsymbol{p})$ is not the Fourier transform of $\rho(\boldsymbol{r})$. Due to the Plancherel theorem for the Fourier transform this probability distribution is normalized since the original function was normalized as expressed by (1.2). Using this probability distribution we can calculate σ_{p_x} the standard deviation of the p_x component of the momentum,

$$\sigma_{p_x} = \left[\int_{\mathbb{R}^3} d^3p (p_x - \langle p_x \rangle)^2 \tilde{\rho}(\boldsymbol{p}) \right]^{1/2}. \tag{1.6}$$

Even though for different states both standard deviations σ_x and σ_{p_x} can be arbitrarily small when treated separately, they become correlated when calculated for the same state. This correlation is usually expressed by the Heisenberg uncertainty relation,

$$\sigma_x \sigma_{p_x} \geq \frac{\hbar}{2}. \tag{1.7}$$

The bound in the inequality (1.7) is saturated by Gaussian wave functions. Heisenberg explained this relation in the following words [27]:

> "the more accurately the position is known, the less accurately is the momentum determined and *vice versa*".

Quantum mechanics gave an important reason to study the inherent incompleteness (or uncertainty) of our information about quantum objects. However, in our opinion, the expression of the uncertainty in terms of standard deviations is too much ingrained in classical physics to be of importance at a more fundamental level. We shall argue in the next section that the measure of uncertainty in terms of information entropy is much more appropriate in the quantum world. We shall prove the uncertainty relations in quantum mechanics expressed in terms of the information entropy. Finally, in the last section we introduce further generalizations of these ideas. We begin with a critical analysis of the standard deviations as measures of uncertainty.

1.2.2 Limitations on the Use of Standard Deviations

As we have already noticed, the standard deviation is a well known and widely accepted measure of uncertainty. However it is rarely mentioned that it is not an ideal tool, failing even in very simple situations. The reason, why we should be careful while using the standard deviation, is explained with the help of the following examples [28].

1.2.2.1 Example I

Let us specify two regions on the real axis (see Fig. 1.1). The first region is the line segment $A = [L(N+1/N), L(N+1)]$ with its length equal to $L(1-1/N)$ where L is the unit length and N is a large number. The second region B is the line segment with its length equal to L/N separated from the first region by the large distance NL. In the next step let us assume that we have the probability distribution of the

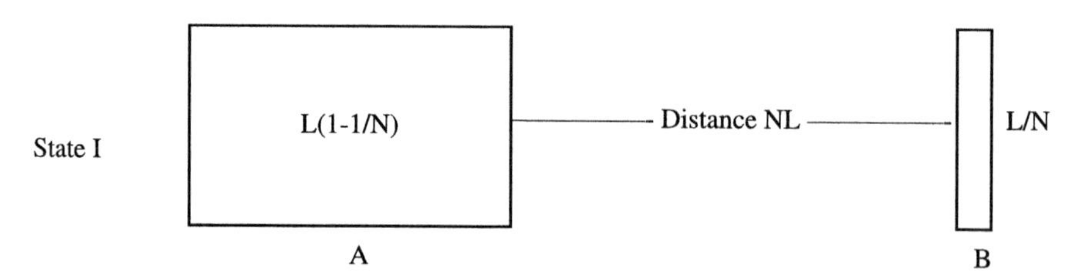

Fig. 1.1 The particle is to be found mostly in a central region. In addition, there is a tiny probability that the particle can be found far away

form (the boxes in Fig. 1.1 represent the probability distribution):

$$\rho(x) = \begin{cases} 1/L, & x \in A, \\ 1/L, & x \in B, \\ 0, & \text{elsewhere.} \end{cases} \tag{1.8}$$

The standard deviation of the variable x calculated in this case is:

$$\sigma_x^2(\text{I}) = L^2\left(N - \frac{1}{N} + \frac{1}{12}\right), \tag{1.9}$$

and for sufficiently large N we obtain:

$$\sigma_x(\text{I}) \approx L\sqrt{N}. \tag{1.10}$$

Therefore, σ_x tends to infinity with $N \to \infty$, while common sense predicts that the uncertainty should remain finite. So, why the uncertainty measured by the standard deviation grows with N? It is simply the effect of a growing distance between the two regions. Standard deviation is based on the second moment which is very sensitive to the values of an observable lying far away from the average. This is the first flaw of the standard deviation. Another one appears in the second example [28].

1.2.2.2 Example II

In the previous example the size of the space where the probability distribution is constant is preserved. Now we omit this condition to present, one more time, inadequateness of the standard deviation and prepare the reader for the solution of this problem. Let us assume two different cases (see Figs. 1.2 and 1.3). In the case A we take:

$$\rho_A(x) = \begin{cases} 1/L, & x \in [0, L], \\ 0, & \text{elsewhere,} \end{cases} \tag{1.11}$$

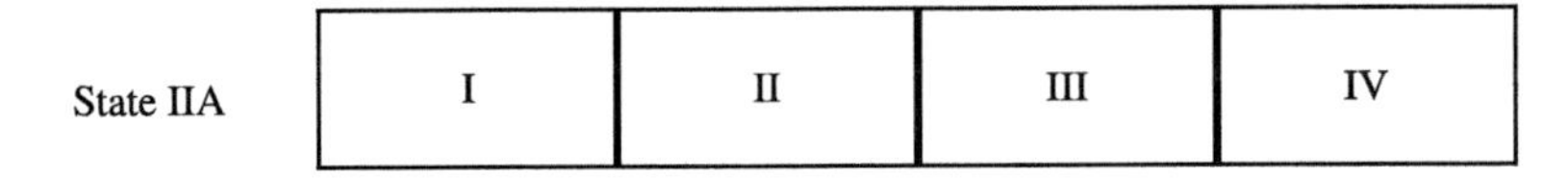

Fig. 1.2 The particle is localized with a uniformly distributed probability in a box of length L

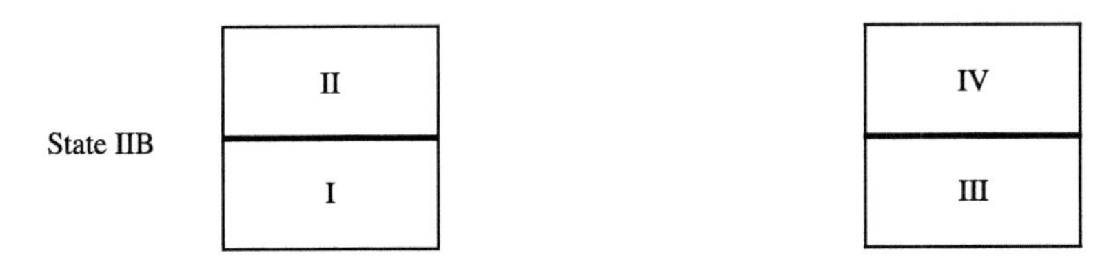

Fig. 1.3 The particle is localized with equal probabilities in two smaller boxes each of length $L/4$

in the case B we have:

$$\rho_B(x) = \begin{cases} 2/L, & x \in [0, L/4], \\ 2/L, & x \in [3L/4, L], \\ 0, & \text{elsewhere}. \end{cases} \tag{1.12}$$

It is easy to see that in the case A the size of the space where the particle can be found is L, while in the case B this size is $L/2$. In the case B we know more about position than in the case A. According to this obvious argument it is rather disappointing that the uncertainty of the position is greater in the case B,

$$\sigma_x(\text{IIA}) = \frac{L}{\sqrt{12}}, \tag{1.13a}$$

$$\sigma_x(\text{IIB}) = \sqrt{\frac{7}{4}}\frac{L}{\sqrt{12}}. \tag{1.13b}$$

This is another manifestation of the fact that the gap between two regions (line segment $[L/4, 3L/4]$) where the particle cannot be found contributes to the standard deviation.

1.2.3 Uncertainty of Measurements

In all natural sciences measurements play a fundamental role. Every measurement has some uncertainty, so we want our theoretical tools to be directly correlated with our knowledge about that uncertainty. On the other hand, as it has been aptly stressed by Peres [29], "The uncertainty relation such as $\sigma_x\sigma_p \geq \hbar/2$ is not a statement about the accuracy of our measuring instruments." The Heisenberg uncertainty relations in their standard form (1.7) completely ignore these problems. In what follows we show how to incorporate the quantum mechanical restrictions on the information obtained in an experiment. We shall consider, however, only the ideal situation when the restrictions on the information gathered in the process follow from the statistical nature of quantum mechanics. Other sources of information loss will not be taken into account. Still, the real experimental errors will be, to some extent, represented by the assumed precision of the measurements. To start, we shall rephrase our quotation from Heisenberg by using the notion of information:

> "the *more information* we have about the position, the *less information* we can acquire about the momentum and vice versa".

1.2.4 Shannon Entropy

Since we know precisely how to measure information, we may try to give a rigorous mathematical relation that will reflect this statement. Shannon connected the measure of the information content with a probability distribution. The use of the information entropy fits perfectly the statistical nature of quantum mechanical measurements. All we have to do is to insert the set of probabilities obtained from quantum mechanics into the famous Shannon formula for the information entropy. Considering our mathematical applications, we shall use this formula with natural logarithms, not with logarithms to the base 2,

$$H = -\sum_k p_k \ln p_k. \tag{1.14}$$

We shall now associate information about the experimental errors with the probabilities p_k. The proposal to associate the Shannon entropy with the partitioning of the spectrum of a physical observable was made by Partovi [30]. As an illustration, let us consider the measurements of the position. For simplicity, we shall consider a one dimensional system and assume that the experimental accuracy δx is the same over the whole range of measurements. In other words, we divide the whole region of interest into bins—equal segments of size δx. The bin size δx will be viewed as a measure of the experimental error.

Each state of a particle whose position is being measured can be associated with a histogram generated by a set of probabilities calculated according to the rules of quantum mechanics. For a pure state, described by a wave function $\psi(x)$, the probability to find the particle in the i-th bin is

$$q_i = \int_{(i-1/2)\delta x}^{(i+1/2)\delta x} dx\, \rho(x). \tag{1.15}$$

The uncertainty in position for the bin size δx is, therefore, given by the formula

$$H^{(x)} = -\sum_{i=-\infty}^{\infty} q_i \ln q_i. \tag{1.16}$$

For the experimental setup characterized by the bin size δx the uncertainty of the measurement is the lowest when the particle is localized in just one bin. The probability corresponding to this bin is equal to 1 and the uncertainty is zero. In all other cases, we obtain some positive number—a measure of uncertainty. When the number of bins is finite, say N, the maximal value of the uncertainty is $\ln N$—the probability to find the particle in each bin is the same, equal to $1/N$.

The uncertainty in momentum is described by the formula (1.14) in which we substitute now the probabilities associated with momentum measurements. Since most of our considerations will have a purely mathematical character, from now on we shall replace momentum by the wave vector $\boldsymbol{k} = \boldsymbol{p}/\hbar$. In this way we get rid of the Planck constant in our formulas. In physical terminology this means that we use the system of units in which $\hbar = 1$. According to quantum mechanics, the probability to find the momentum k_x in the j-th bin is

$$p_j = \int_{(j-1/2)\delta k}^{(j+1/2)\delta k} dk\, \tilde{\rho}(k), \tag{1.17}$$

where $\tilde{\rho}(k)$ is the modulus squared of the one-dimensional Fourier transform $\tilde{\psi}(k)$ of the wave function,

$$\tilde{\psi}(k) = \int_{-\infty}^{\infty} \frac{dx}{\sqrt{2\pi}} e^{-ikx} \psi(x), \tag{1.18}$$

and δk is the resolution of the instruments measuring the momentum in units of $\hbar$. The uncertainty in momentum measured with the use of the Shannon formula is constructed in the same way as the uncertainty in position,

$$H^{(k)} = -\sum_{j=-\infty}^{\infty} p_j \ln p_j. \tag{1.19}$$

1.2.4.1 Examples

The uncertainties measured with the use of the formulas (1.16) and (1.19) do not suffer from the deficiencies described by our examples I and II. One can easily check that in the case I the uncertainty caused by the smaller region does not contribute in the limit when $N \to \infty$. In turn, in the case II when $\delta x = L/4$ the uncertainty in the state A is equal to $2\ln 2$ while in the state B it is equal to $\ln 2$. Thus, as expected, the position uncertainty in the state A is greater (we have less information) than in the state B.

Now, we shall give one more very simple example that clearly shows the merits of entropic definition of uncertainty. Let us consider a particle in one dimension localized on the line segment $[-a, a]$. Assuming a homogeneous distribution, we obtain the following wave function:

$$\psi_a(x) = \begin{cases} 1/\sqrt{2a}, & x \in [-a, a], \\ 0, & \text{elsewhere}. \end{cases} \tag{1.20}$$

The Fourier transform of $\psi_a(x)$ is:

$$\tilde{\psi}_a(k) = \sqrt{\frac{1}{\pi a}} \frac{\sin(ak)}{k}. \tag{1.21}$$

This leads to the following probability distribution in momentum space:

$$\tilde{\rho}_a(k) = \frac{1}{\pi a k^2} \sin^2(ak). \tag{1.22}$$

The standard uncertainty relation (1.7) for this state is meaningless because the second moment of $\tilde{\rho}_a(k)$ is infinite. However the uncertainty in momentum measured with the Shannon formula is finite. Taking for simplicity the bin size in momentum as $\delta k = 2\pi/a$, we obtain the following expression for the probability to find the momentum in the j-th bin:

$$
\begin{aligned}
p_j &= \frac{1}{\pi a}\int_{(j-1/2)\delta k}^{(j+1/2)\delta k} dk\,\frac{\sin^2(ak)}{k^2} \\
&= \frac{1}{\pi}\int_{2\pi(j-1/2)}^{2\pi(j+1/2)} d\eta\,\frac{\sin^2(\eta)}{\eta^2} \\
&= \frac{1}{\pi}\int_{2\pi(2j-1)}^{2\pi(2j+1)} d\eta\,\frac{\sin(\eta)}{\eta} \\
&= \frac{\mathrm{Si}[(4j+2)\pi]}{\pi} - \frac{\mathrm{Si}[(4j-2)\pi]}{\pi},
\end{aligned}
\tag{1.23}
$$

where Si is the integral sine function. The uncertainty in momentum is obtained by evaluating numerically the sum (1.19) which gives $H^{(k)} = 0.530$. To obtain the position uncertainty, we take just two bins ($\delta x = a$), so that $H^{(x)} = \ln 2$. The sum of these two uncertainties is about 1.223.

1.2.5 Entropic Uncertainty Relations

We have shown that the use of the Shannon formula gives a very sensible measure of uncertainties that takes into account the resolution of measuring devices. The situation becomes even more interesting when we consider measurements on the same quantum state of two "incompatible" properties of a particle. The term incompatible has a precise meaning in terms of the experimental setup and is reflected in the mathematical framework of quantum mechanics. In this study, we shall treat not only the most important example of such incompatible properties—the position and the momentum of a particle—but also angle and angular momentum.

As has been argued by Bohr and Heisenberg [31], it is impossible to measure the position without loosing information about the momentum and vice versa. This property is embodied in the standard Heisenberg uncertainty relation and we will express it now in terms of Shannon measures of uncertainty.

$$
H^{(x)} + H^{(p)} > 1 - \ln 2 - \ln\left(\frac{\delta x\delta p}{h}\right). \tag{1.24}
$$

To stress the physical meaning of this relation, we reinserted the Planck constant. It is clear that this inequality is not sharp since for large $\delta x\delta p$ the right hand side becomes negative. However, in the quantum regime, when the volume of phase space $\delta x\delta p$ does not exceed the Planck constant h, we obtain a meaningful limitation on the sum of the uncertainties in position and in momentum. When one uncertainty tends to zero, the other must stay above the limit. In the last example considered in the previous section the right hand side in (1.24) is equal to $1 - \ln 2 = 0.307$, while the calculated value of the sum of two uncertainties was equal 1.223.

The proof of this inequality was given in [32] and it proceeds as follows. First, we use the integral form of the Jensen inequality for convex functions [33, 34]. The function $\rho \ln \rho$ is a convex function. Therefore, the value of a function evaluated at the mean argument cannot exceed the mean value of the function so that the following inequality must hold:

$$\langle \rho \ln \rho \rangle \geq \langle \rho \rangle \ln \langle \rho \rangle . \tag{1.25}$$

Upon substituting here the probability distribution, we obtain:

$$\frac{1}{\delta x} \int_{(i-1/2)\delta x}^{(i+1/2)\delta x} dx\, \rho(x) \ln \rho(x)$$
$$\geq \left[\frac{1}{\delta x} \int_{(i-1/2)\delta x}^{(i+1/2)\delta x} dx\, \rho(x) \right] \ln \left[\frac{1}{\delta x} \int_{(i-1/2)\delta x}^{(i+1/2)\delta x} dx\, \rho(x) \right], \tag{1.26}$$

or after some straightforward rearrangements and with the use of (1.15)

$$-q_i \ln q_i \geq - \int_{(i-1/2)\delta x}^{(i+1/2)\delta x} dx\, \rho(x) \ln[\rho(x)\delta x]. \tag{1.27}$$

This form of the inequality is more satisfactory from the physical point of view since $\rho(x)$ has the dimension of inverse length and the dimensional quantities should not appear under the logarithm. Adding the contributions from all the bins, we obtain:

$$H^{(x)} \geq - \int_{-\infty}^{\infty} dx\, \rho(x) \ln[\rho(x)\delta x]. \tag{1.28}$$

Applying the same reasoning to the momentum distribution, we arrive at:

$$H^{(k)} \geq - \int_{-\infty}^{\infty} dk\, \tilde{\rho}(k) \ln[\tilde{\rho}(k)\delta k]. \tag{1.29}$$

Adding these two inequalities, we get on the left hand side the sum of the uncertainties in position and in momentum as in (1.24). What remains is to establish a bound on the sum of the integrals appearing on the right hand side. This problem has a long history. More than half a century ago a bound has been conjectured by Hirschman [35] and Everett [36, 37]. Actually, Hirschman proved only a weaker form of the inequality and Everett showed that the left hand side in (1.24) is stationary for Gaussian wave functions. The inequality was later proved by Bialynicki-Birula and Mycielski [38] and independently by Beckner [39]. In this way we arrive at the entropic uncertainty relation (1.24). We shall give a detailed derivation of this inequality and its extensions in the next section.

The measure of uncertainties with the use of standard deviations requires a certain measure of the distance between events. Sometimes there is no sensible definition of a distance. The simplest example is the original application of the Shannon information entropy to text messages. The value of H can be calculated for each text but there is no sensible measure of a distance between the letters in the alphabet. Important examples are also found in physics. The simplest case is a quantum particle moving on a circle. The configuration space is labeled by the angle φ and the canonically conjugate variable is the angular momentum represented in quantum mechanics by the operator $\hat{M} = -i\hbar\partial/\partial\varphi$. Wave functions describing this system are integrable *periodic* functions $\psi(\varphi)$ or alternatively the coefficients c_m in the Fourier expansion of $\psi(\varphi)$,

$$\psi(\varphi) = \frac{1}{\sqrt{2\pi}} \sum_{m=-\infty}^{\infty} c_m\, e^{im\varphi}. \tag{1.30}$$

Even though formally φ and $\hat{M}$ obey canonical commutation relations, these relations are not mathematically consistent because φ cannot be treated as an operator—multiplication by φ produces a function that is not periodic. Therefore, the notion of a standard deviation of φ can only be used approximately for states for which $\Delta\varphi$ is very small. Several, more or less natural methods, were introduced to deal with this problem. One of them is to use periodic functions of φ instead of the angle itself (cf. [40] for a list of references). This leads to some uncertainty relations which are mathematically correct but physically less transparent. The use of entropic measures of uncertainties solves this problem. The uncertainty in position on the circle is defined as for position on an infinite line, except that now there is finite number of bins $N = 2\pi/\delta\varphi$. Thus, the Shannon entropy of the angle is:

$$H^{(\varphi)} = -\sum_{n=0}^{N-1} q_n \ln q_n, \tag{1.31}$$

where

$$q_n = \int_{n\delta\varphi}^{(n+1)\delta\varphi} d\varphi\, |\psi(\varphi)|^2. \tag{1.32}$$

According to the rules of quantum mechanics, the probability to find the value $\hbar m$ of angular momentum is $p_m = |c_m|^2$. Therefore, the Shannon entropy for the angular momentum is:

$$H^{(M)} = -\sum_{m=-\infty}^{\infty} p_m \ln p_m. \tag{1.33}$$

The proof of the uncertainty relation for the angle and angular momentum entropies is much simpler than for the position and momentum. It will be given later in the more general case of the Rényi entropies. The uncertainty relation has the following form:

$$H^{(\varphi)} + H^{(M)} \geq -\ln \frac{\delta\varphi}{2\pi}, \tag{1.34}$$

or

$$H^{(\varphi)} + H^{(M)} \geq \ln N. \tag{1.35}$$

This inequality is saturated for every eigenstate of angular momentum. Then, the uncertainty in angular momentum vanishes and the uncertainty in position is exactly $\ln N$ because the probability to find the particle in a given bin is $1/N$.

1.3 Rényi Entropy

The Shannon information entropy has been generalized by Rényi [41–43]. The Rényi entropy is a one-parameter family of entropic measures that share with the Shannon entropy several important properties. Even though the Rényi entropy was introduced in 1960, its substantial use in physics is more recent—it took place during the last decade (cf. references listed in the Introduction).

Rényi entropy H_α is defined by the following formula:

$$H_\alpha = \frac{1}{1-\alpha} \ln\left[\sum_k p_k^\alpha\right], \tag{1.36}$$

where p_k is a set of probabilities and α is a positive number. Rényi in [41–43] restricted α to positive values but we may consider, in principle, all values. Sometimes it is useful to consider only the values $\alpha > 1$ since we may then introduce a conjugate positive parameter β satisfying the relation

$$\frac{1}{\alpha} + \frac{1}{\beta} = 2. \tag{1.37}$$

In the limit, when $\alpha \to 1$, the Rényi entropy becomes the Shannon entropy. To see this, we have to apply the L'Hôpital rule to the definition (1.36) at the singular point $\alpha = 1$. Since the Shannon entropy is a special case of the Rényi entropy, we shall proceed, whenever possible, with proving various properties of the Rényi entropy, taking at the end the limit $\alpha \to 1$. This approach is particularly fruitful in the case of entropic uncertainty relations because the proofs are in a way more natural for the Rényi entropies than for the Shannon entropies.

Shannon entropy can be defined as a function $H(p_k)$ of the set of probabilities, obeying the following axioms:

1. $H(p_1, p_2, \ldots, p_n)$ is a symmetric function of its variables for $n = 2, 3, \ldots$
2. $H(p, 1-p)$ is a continuous function of p for $0 \leq p \leq 1$
3. $H(tp_1, (1-t)p_1, p_2, \ldots, p_n) = H(p_1, p_2, \ldots, p_n) + p_1 H(t, 1-t)$ for $0 \leq t \leq 1$

We presented the axioms in the form taken from Rényi [41–43] which we liked more than the original axioms given by Shannon [26]. Sometimes an additional axiom is added to fix the scale (the base of the logarithm) of the function H. For example by

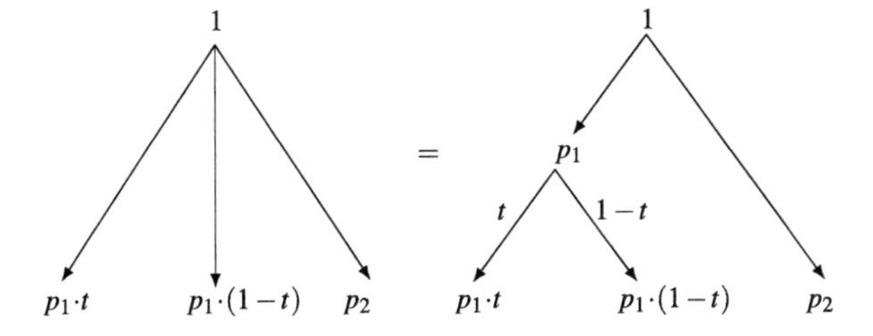

Fig. 1.4 Probability trees representing the main idea of the third axiom

requiring that $H(1/2, 1/2) = 1$ we obtain the entropy measured in bits (logarithm to the base 2).

The first two axioms are of a simple mathematical nature but the third axiom is of crucial importance. This axiom may be figuratively expressed as the invariance of the entropy with respect to breaking down the probability tree. This invariance rule is illustrated in Fig. 1.4. It leads, in particular, to the very important property that characterizes the physical entropy. The entropy is an extensive quantity—for a composed system it must satisfy the additivity law. This law in terms of probabilities can be expressed as follows. Given two independent probability distributions $(p_1, p_2, \ldots, p_n)$ and $(q_1, q_2, \ldots, q_m)$, the entropy of the composed system must be equal to the sum of the entropies for the subsystems,

$$\begin{aligned} &H(p_1q_1, p_1q_2, \ldots, p_1q_m, p_2q_1, \ldots, p_2q_m, \ldots, p_nq_1, \ldots, p_nq_m) \\ &\quad = H(p_1, p_2, \ldots, p_n) + H(q_1, q_2, \ldots, q_m). \end{aligned} \tag{1.38}$$

It turns out that by replacing the third axiom by the additivity law we substantially increase the set of allowed functions. The most prominent of these generalized entropies is the Rényi entropy. It is simple task to check that the Rényi entropy obeys the additivity law. Indeed, from the distributive property of the logarithm we obtain

$$\begin{aligned} \frac{1}{1-\alpha}\ln\Big[\sum_{kl}(p_kq_l)^\alpha\Big] &= \frac{1}{1-\alpha}\ln\Big[\sum_k p_k^\alpha \sum_l q_l^\alpha\Big] \\ &= \frac{1}{1-\alpha}\ln\Big[\sum_k p_k^\alpha\Big] + \frac{1}{1-\alpha}\ln\Big[\sum_l q_l^\alpha\Big]. \end{aligned} \tag{1.39}$$

Note that any linear combination (discrete or continuous) of Rényi entropies with different values of α would obey the same additivity law. For example, we may take a sum of two terms

$$\frac{1}{1-\alpha}\ln\Big[\sum_k p_k^\alpha\Big] + \frac{1}{1-\beta}\ln\Big[\sum_k p_k^\beta\Big]. \tag{1.40}$$

A particular expression of this type was introduced in [44] and is very useful in the context of entropic uncertainty relation. It is obtained by requiring that the two parameters α and β are constrained by the relation (1.37). In this case, they can be expressed in terms of a single variable: $\alpha = 1/(1-s)$, $\beta = 1/(1+s)$. The new symmetrized entropy $\mathcal{H}_s$ is a one parameter average of two Rényi entropies,

$$\mathcal{H}_s = \frac{1}{2}\left(1-\frac{1}{s}\right)\ln\left[\sum_k p_k^{1/(1-s)}\right] + \frac{1}{2}\left(1+\frac{1}{s}\right)\ln\left[\sum_k p_k^{1/(1+s)}\right]. \tag{1.41}$$

Since $\mathcal{H}_s$ is a symmetric function of s, it is sufficient to consider only the positive values $0 \le s \le 1$. The Shannon entropy is recovered in the limit when $s \to 0$.

The Rényi entropy is a decreasing function of α. For a given set of probabilities it attains its maximal value at $\alpha = 0$. The symmetrized Rényi entropy attains its maximal value at $s = 0$. Starting from this value, as a symmetric function of s, it drops down in both directions.

1.3.1 Rényi Entropy and the Uncertainty Relations

Our goal is now to generalize the uncertainty relations to the Rényi entropies. In this section we shall consider only the measurements of position and momentum. We have already defined the Shannon information entropies that contain information about the precision of measurements. The definitions of the uncertainty in position and momentum, as measured by the Rényi entropy, are obvious:

$$H_\alpha^{(x)} = \frac{1}{1-\alpha}\ln\left[\sum_{i=-\infty}^{\infty} q_i^\alpha\right], \tag{1.42}$$

$$H_\beta^{(k)} = \frac{1}{1-\beta}\ln\left[\sum_{j=-\infty}^{\infty} p_j^\beta\right], \tag{1.43}$$

where the probabilities q_i and p_j are given by the formulas (1.15) and (1.17). The reason, why we have chosen the Rényi entropies with different parameters will become clear later. For definiteness, we shall assume at this point that $\alpha \ge \beta$ (later this restriction will be lifted). This means that according to (1.37) $1 \le \alpha < \infty$ and $1/2 < \beta \le 1$. Therefore, z^α is a convex function, while z^β is concave. These properties enable us to use the Jensen inequality in the same way as we did before, and arrive at:

$$\left(\frac{1}{\delta x}\int_{(i-1/2)\delta x}^{(i+1/2)\delta x} dx\,\rho(x)\right)^\alpha \le \frac{1}{\delta x}\int_{(i-1/2)\delta x}^{(i+1/2)\delta x} dx[\rho(x)]^\alpha, \tag{1.44a}$$

$$\frac{1}{\delta k}\int_{(j-1/2)\delta k}^{(j+1/2)\delta k} dk\left[\tilde{\rho}(k)\right]^\beta \le \left(\frac{1}{\delta k}\int_{(j-1/2)\delta k}^{(j+1/2)\delta k} dk\tilde{\rho}(k)\right)^\beta. \tag{1.44b}$$

Summing (1.44a) and (1.44b) over the indices i and j we obtain:

$$(\delta x)^{1-\alpha} \sum_{i=-\infty}^{\infty} q_i^{\alpha} \leq \int_{\mathbb{R}} dx[\rho(x)]^{\alpha}, \tag{1.45a}$$

$$\int_{\mathbb{R}} dk[\tilde{\rho}(k)]^{\beta} \leq (\delta k)^{1-\beta} \sum_{j=-\infty}^{\infty} p_j^{\beta}. \tag{1.45b}$$

At this point we invoke a powerful Babenko-Beckner inequality for (p, q) norms of a function and its Fourier transform. This inequality was proved for restricted values of α and β by Babenko [45] and for all values by Beckner [39]. In our notation, this inequality reads:

$$\left(\int_{\mathbb{R}} dx\,[\rho(x)]^{\alpha}\right)^{1/\alpha} \leq n\,(\alpha, \beta) \left(\int_{\mathbb{R}} dk\left[\tilde{\rho}(k)\right]^{\beta}\right)^{1/\beta}, \tag{1.46}$$

where

$$n\,(\alpha, \beta) = \left(\frac{\alpha}{\pi}\right)^{-1/2\alpha} \left(\frac{\beta}{\pi}\right)^{1/2\beta}. \tag{1.47}$$

This inequality enables us to put together the formulas (1.45a) and (1.45b). To this end, we raise to the power $1/\alpha$ both sides of (1.45a). Next, we raise to the power of $1/\beta$ both sides of (1.45b) and multiply the resulting inequality by $n(\alpha, \beta)$. In this way we can obtain the following chain of inequalities:

$$\begin{aligned}\left[(\delta x)^{1-\alpha} \sum_{i=-\infty}^{\infty} q_i^{\alpha}\right]^{1/\alpha} &\leq \left(\int_{\mathbb{R}} dx[\rho(x)]^{\alpha}\right)^{1/\alpha} \\ &\leq n(\alpha, \beta) \left(\int_{\mathbb{R}} dk[\tilde{\rho}(k)]^{\beta}\right)^{1/\beta} \\ &\leq n(\alpha, \beta) \left[(\delta k)^{1-\beta} \sum_{j=-\infty}^{\infty} p_j^{\beta}\right]^{1/\beta}.\end{aligned} \tag{1.48}$$

Now, we take only the first and the last term of this chain and multiply both sides of the resulting inequality by $(\alpha/\pi)^{1/2\alpha}$ to obtain:

$$\left[(\delta x)^{1-\alpha} \sqrt{\frac{\alpha}{\pi}} \sum_{i=-\infty}^{\infty} q_i^{\alpha}\right]^{1/\alpha} \leq \left[(\delta k)^{1-\beta} \sqrt{\frac{\beta}{\pi}} \sum_{j=-\infty}^{\infty} p_j^{\beta}\right]^{1/\beta}. \tag{1.49}$$

Next, we evaluate the logarithms of both sides and multiply the result by a positive number $\alpha/(\alpha - 1)$. After a rearrangement of terms, with the use of the relationship

(1.37), we arrive at the final form of the uncertainty relation (with Planck's constant reinserted) between position and momentum in terms of the Rényi entropies [44]:

$$H_\alpha^{(x)} + H_\beta^{(p)} \geq -\frac{1}{2}\left(\frac{\ln\alpha}{1-\alpha} + \frac{\ln\beta}{1-\beta}\right) - \ln 2 - \ln\frac{\delta x\delta p}{h}. \tag{1.50}$$

We can lift now the restriction that $\alpha \geq \beta$. Owing to a symmetry between a function and its Fourier transform we can write the same inequality (1.50) but with x and p interchanged. After all, it does not make any difference from the mathematical point of view, what is called x and what is called p, as long as the corresponding wave functions are connected by the Fourier transformation. Therefore, we can take the average of (1.50) and its counterpart with x and p interchanged and obtain the following uncertainty relation for symmetrized entropies:

$$\mathcal{H}_s^{(x)} + \mathcal{H}_s^{(p)} \geq \frac{1}{2}\ln(1-s^2) + \frac{1}{2s}\ln\frac{1+s}{1-s} - \ln 2 - \ln\frac{\delta x\delta p}{h}. \tag{1.51}$$

This relation is more satisfactory than (1.50) because the uncertainties of both quantities x and p are measured in the same way. In the limit, when $s \to 0$ we obtain the uncertainty relation (1.24) for the Shannon entropies. The inequality (1.51) (or its asymmetric counterpart (1.50)) is the best known uncertainty relation for the entropies that takes into account finite resolution of measurements. However, this inequality is obviously not a sharp one. For $\delta x\delta p/h$ only slightly larger than 1 the right hand side of (1.50) becomes negative while the left hand side is by definition positive. For finite values of $\delta x\delta p/h$ there must exist a better lower bound. Since the Babenko-Beckner inequality (1.46) cannot be improved (it is saturated by Gaussian functions), we conclude that Jensen inequalities (1.44a, 1.44b) used together are not optimal. Of course, each of them can be separately saturated but when they are used jointly one should somewhere include the information about the Fourier relationship between the functions. An improvement of the uncertainty relations (1.50) or (1.51) is highly desired but it does not seem to be easy. A recent attempt [46] failed completely (see [47]).

1.4 Uncertainty Relations: From Discrete to Continuous

In this section we describe various types of uncertainty relations that depart from the original Heisenberg relations for position and momentum much further than those described so far. We shall also comment on different mathematical tools that are used in proving these relations.

1.4.1 The Uncertainty Relations for N-level Systems

Pure states of an N-level system are described in quantum mechanics by normalized vectors in the N-dimensional Hilbert space $\mathcal{H}_N$. We shall use the Dirac notation denoting vectors by $|\psi\rangle$ and their scalar products by $\langle\psi_1|\psi_2\rangle$. In this notation, the normalization condition reads $\langle\psi|\psi\rangle = 1$. Let us choose two orthonormal bases $\{|a_i\rangle\}$ and $\{|b_j\rangle\}$, where $i, j = 1, \ldots, N$. We can then represent each vector $|\psi\rangle$ in two ways:

$$|\psi\rangle = \sum_{i=1}^{N} \langle a_i|\psi\rangle|a_i\rangle = \sum_{j=1}^{N} \langle b_j|\psi\rangle|b_j\rangle. \tag{1.52}$$

The squares of the absolute values of the complex expansion coefficients are interpreted in quantum mechanics as the probabilities, say q_i and p_j, to find the states represented by the vectors $|a_i\rangle$ and $|b_j\rangle$, when the system is in the state $|\psi\rangle$,

$$q_i = |\langle a_i|\psi\rangle|^2, \qquad p_j = |\langle b_j|\psi\rangle|^2. \tag{1.53}$$

The normalization of vectors and the orthonormality of the bases guarantees that the probabilities q_i and p_j sum up to 1.

1.4.1.1 Deutsch Inequality

Having at our disposal two sets of probabilities $\{q_i\}$ and $\{p_j\}$ we can construct two Shannon entropies:

$$H^{(a)} = -\sum_{i=1}^{N} q_i \ln q_i, \qquad H^{(b)} = -\sum_{j=1}^{N} p_j \ln p_j. \tag{1.54}$$

Since we aim at finding some uncertainty relation we consider the sum of these entropies:

$$H^{(a)} + H^{(b)} = \sum_{i=1}^{N}\sum_{j=1}^{N} q_i p_j Q_{ij}, \tag{1.55}$$

where

$$Q_{ij} = -\ln(\langle a_i|\psi\rangle\langle\psi|a_i\rangle) - \ln(\langle b_j|\psi\rangle\langle\psi|b_j\rangle). \tag{1.56}$$

In 1983 Deutsch [48] found a lower bound on the sum (1.55). He observed that for a normalized vector $|\psi\rangle$ each Q_{ij} is nonnegative. Therefore, by finding the minimal value of all Q_{ij}'s we determine a lower bound on the sum (1.55). We shall present here an expanded version of the Deutsch analysis. For each pair (i, j), the minimal value of Q_{ij} is attained for some function ψ_{ij}. This function is found by varying Q_{ij} with respect to the wave function ψ with the normalization constraint $\langle\psi|\psi\rangle = 1$.

$$\frac{\delta}{\delta|\psi_{ij}\rangle}[Q_{ij} + \kappa\langle\psi_{ij}|\psi_{ij}\rangle] = 0, \tag{1.57}$$

where κ is a Lagrange multiplier. In this way we arrive at the equation for the vector $|\psi_{ij}\rangle$:

$$|\psi_{ij}\rangle = \frac{1}{\kappa}\left(\frac{|a_i\rangle}{\langle\psi_{ij}|a_i\rangle} + \frac{|b_j\rangle}{\langle\psi_{ij}|b_j\rangle}\right). \tag{1.58}$$

Taking the scalar product of this equation with $\langle\psi_{ij}|$, we obtain from the normalization condition that $\kappa = 2$. To find the solution of (1.58) we multiply this equation by the adjoined vectors $\langle a_i|$ and $\langle b_j|$. In this way we obtain the following algebraic equations for two complex variables $a = \langle\psi_{ij}|a_i\rangle$ and $b = \langle\psi_{ij}|b_j\rangle$ with one fixed complex parameter $\langle a_i|b_j\rangle$.

$$|a|^2 = \frac{1}{2} + \frac{a\langle a_i|b_j\rangle}{2b}, \qquad |b|^2 = \frac{1}{2} + \frac{b\langle b_j|a_i\rangle}{2a}. \tag{1.59}$$

The last term in each equation must be real since the rest is real. Next, we divide both sides of the first equation by those of the second. After some simple rearrangements, we arrive at:

$$\frac{|a|^2}{|b|^2} - 1 = \frac{a}{b}\langle a_i|b_j\rangle - \frac{a^*}{b^*}\langle a_i|b_j\rangle^*. \tag{1.60}$$

Since both terms on the right hand side are real, they cancel out and we obtain $|a| = |b|$. Inserting this into any of the two equations (1.59), we obtain

$$|a|^2 = \frac{1}{2}\left[1 \pm |\langle a_i|b_j\rangle|\right]. \tag{1.61}$$

The choice of the minus sign gives the maximal value of Q_{ij} but only in the subspace spanned by the vectors $|a_i\rangle$ and $|b_j\rangle$. Choosing the plus sign, we obtain

$$|\psi_{ij}\rangle = \frac{\exp(i\phi)}{\sqrt{2(1+|\langle a_i|b_j\rangle|)}}[|a_i\rangle + \exp(-i\arg\langle a_i|b_j\rangle)|b_j\rangle], \tag{1.62}$$

where $\exp(i\phi)$ is an arbitrary phase factor. We can insert this solution into (1.55), and using the fact that this solution is minimizer, we obtain the inequality:

$$H^{(a)} + H^{(b)} \geq -2\sum_{i=1}^{N}\sum_{j=1}^{N} q_i p_j \ln\left[\frac{1}{2}(1+|\langle a_i|b_j\rangle|)\right]. \tag{1.63}$$

This inequality will still hold if we replace each modulus $|\langle a_i|b_j\rangle|$ by the maximal value C_B,

$$C_B = \sup_{(i,j)} |\langle a_i|b_j\rangle|. \tag{1.64}$$

After this replacement, we can do summations over i and j and arrive at the Deutsch result [48]:

$$H^{(a)} + H^{(b)} \geq -2\ln\left[\frac{1}{2}(1+C_B)\right]. \tag{1.65}$$

1.4.1.2 Maassen-Uffink Inequalities

Even though the bound in the Deutsch uncertainty relation is saturated when the two bases have common vectors, there is room for improvement. Of course, the improved inequality must coincide with (1.65) when it is saturated. An improved relation was conjectured by Kraus [49] and it has the form:

$$H^{(a)} + H^{(b)} \geq -2\ln C_B. \tag{1.66}$$

This inequality was later proved by Maassen and Uffink [50]. We shall present a different version of the proof based on the Rényi entropy. The mathematical tools needed to prove the inequality (1.66) are the same as those used for the entropies with continuous variables. At the bottom of every uncertainty relation there is some mathematical theorem. In the case of the Maassen-Uffink relation this role is played by the Riesz theorem [51, 52] which says that for every N-dimensional complex vector X and a unitary transformation matrix $\hat{T}$ with coefficients t_{ji}, the following inequality between the norms is valid:

$$c^{1/\mu}\|X\|^{\mu} \leq c^{1/\nu}\|\hat{T}X\|^{\nu}, \tag{1.67}$$

where the constant $c = \sup_{(i,j)} |t_{ji}|$ and the coefficients μ and ν obey the relation

$$\frac{1}{\mu} + \frac{1}{\nu} = 1, \quad 1 \leq \nu \leq 2. \tag{1.68}$$

The norms are defined as usual,

$$\|X\|^{\mu} = \left[\sum_k |x_k|^{\mu}\right]^{1/\mu}, \tag{1.69}$$

but we do not use the traditional symbols (p,q) for the norms since this would interfere with our usage of p and q to denote the probabilities.

To prove the inequality (1.66) we shall take $x_i = \langle a_i|\psi\rangle$ and

$$t_{ji} = \langle b_j|a_i\rangle. \tag{1.70}$$

From the resolution of the identity ($\sum_i |a_i\rangle\langle a_i| = 1$), we obtain:

$$\sum_{i=1}^{N} t_{ji}x_i = \sum_{i=1}^{N} \langle b_j|a_i\rangle\langle a_i|\psi\rangle = \langle b_j|\psi\rangle. \tag{1.71}$$

Using the definitions (1.53), we can rewrite the inequality (1.67) in the form:

$$c^{1/\mu}\left[\sum_{j=1}^{N} q_j^{\mu/2}\right]^{1/\mu} \leq c^{1/\nu}\left[\sum_{i=1}^{N} p_i^{\nu/2}\right]^{1/\nu}, \tag{1.72}$$

where

$$c = \sup_{(i,j)} |t_{ij}| = \sup_{(i,j)} |\langle a_i|b_j\rangle| = C_B. \tag{1.73}$$

The parameters μ and ν differ by a factor of 2 from the parameters α and β appearing in (1.37),

$$\mu = 2\alpha, \qquad \nu = 2\beta. \tag{1.74}$$

Next, we take the logarithm of both sides of the inequality (1.72) and with the use of (1.74), we arrive at the uncertainty relation for the Rényi entropies:

$$H_\alpha^{(a)} + H_\beta^{(b)} \geq -2\ln C_B. \tag{1.75}$$

When two observables associated with the two bases have a common vector then $C_B = 1$ and the right hand side in the last inequality vanishes. Thus, this uncertainty relation becomes empty. In the limit $\alpha \to 1$, $\beta \to 1$ this inequality reduces to the Maassen-Uffink result (1.66).

1.4.1.3 Discrete Fourier Transform

In previous subsections we dealt with two orthonormal vector bases $\{|a_k\rangle\}$ and $\{|b_m\rangle\}$. We can expand every vector of the first basis in the second basis,

$$|a_i\rangle = \sum_{j=1}^{N} t_{ji}|b_j\rangle, \tag{1.76}$$

where t_{ji} are the coefficients of the unitary transformation (1.70), and for each $|a_i\rangle$

$$1 = \langle a_i|a_i\rangle = \sum_{j=1}^{N} |t_{ji}|^2 \leq N \sup_{(j)} |t_{ji}|^2 \leq N C_B^2. \tag{1.77}$$

It follows from this inequality that $C_B \geq 1/\sqrt{N}$. There are bases, called mutually unbiased bases, for which the sum (1.75) not only has the lowest possible value but the moduli of the matrix elements are *all* the same,

$$|\langle a_i|b_j\rangle| = \frac{1}{\sqrt{N}}. \tag{1.78}$$

One can check by a direct calculation that for the observables associated with two mutually unbiased bases, for every state represented by a basis vector, either $|a_i\rangle$ or $|b_j\rangle$, the sum of the Rényi entropies saturates the lower bound,

$$H_\alpha^{(a)} + H_\beta^{(b)} = \ln N. \tag{1.79}$$

Mutually unbiased bases were introduced by Schwinger [53]. More recently they were studied in more detail [54] and were shown to be useful in the description of entanglement and other fundamental properties of quantum systems. In the fourth chapter we shall say more about the mutually unbiased bases. The most important example of mutually unbiased bases is present in the discrete Fourier transformation. In this case, the two bases are related by the unitary transformation with the coefficients:

$$f_{kl} = \frac{1}{\sqrt{N}} \exp\left[\frac{2\pi i\, kl}{N}\right], \tag{1.80}$$

which satisfy the conditions (1.78).

The physical interpretation of the inequality (1.79) is similar to the Heisenberg uncertainty relation. Two observables A and B, characterized by the bases $|a_i\rangle$ and $|b_j\rangle$ cannot have simultaneously sharp values since they do not have common basis vectors. Moreover, every state described by a basis vector has the sum of the uncertainties equal to the same value $1/\sqrt{N}$. The observables A and B are finite-dimensional analogs of canonically conjugate physical quantities.

1.4.2 Shannon and Rényi Entropies for Continuous Distributions

In a rather crude manner we can say that the entropies for continuous distributions of probability are obtained by replacing the sums in the expressions (1.14) and (1.36) by the integrals. We have already encountered such expressions ((1.28) and (1.29)), in our discussion of the Shannon entropies for position and momentum. Naively, we would like to define the Shannon entropy for a continuous probability distribution $\rho(X)$ in the form:

$$H^{(X)} \stackrel{?}{=} -\int dX \rho(X) \ln \rho(X), \tag{1.81}$$

where X is some random variable. However, in most cases we encounter a problem since X is a dimensional quantity and the result will depend on the unit that is used to measure X. In order to obtain a correct formula, let us start with the original Shannon definition for a discrete set of probabilities (1.14). Next, we insert in this formula the discrete probabilities p_k derived from some continuous distribution. For example, let us choose the distribution function for position $\rho(x)$. If $\rho(x)$ does not change appreciably over the distance δx, the Shannon entropy defined by (1.15) and (1.16) can be approximated by,

$$H^{(x)} \approx - \sum_{k=-\infty}^{\infty} \delta x \rho(x_k) \ln[\rho(x_k)\delta x], \tag{1.82}$$

where $x_k = k\delta x$.

We can write the right hand side as a sum of two terms

$$- \sum_{k=-\infty}^{\infty} \delta x \rho(x_k) \ln[\rho(x_k)L] - \ln \frac{\delta x}{L}, \tag{1.83}$$

where L is some fixed unit of length. The first term is a Riemann sum and in the limit, when $\delta x \to 0$, it tends to the following integral:

$$S^{(x)} = - \int_{\mathbb{R}} dx\, \rho(x) \ln[\rho(x)L]. \tag{1.84}$$

This integral may be called the entropy of the continuous distribution $\rho(x)$ or the continuous entropy.

The second term $-\ln(\delta x/L)$ measures the difference between $H^{(x)}$ and $S^{(x)}$. This difference must tend to ∞ because the information measured by $H^{(x)}$ grows indefinitely when $\delta x \to 0$, while $S^{(x)}$ remains finite.

The same reasoning leads to the following definition of the Rényi entropy for a continuous distribution

$$S_\alpha^{(x)} = \frac{1}{1-\alpha} \ln \left(\int_{\mathbb{R}} dx [\rho(x)]^\alpha L^{\alpha-1} \right). \tag{1.85}$$

The difference between the Rényi entropies $H_\alpha^{(x)}$ and $S_\alpha^{(x)}$ does not depend on α and is, therefore, the same as for the Shannon entropies.

1.4.3 Uncertainty Relations for Continuous Entropies

We can formulate now uncertainty relations in terms of the Rényi and Shannon entropies for continuous variables, although the operational meaning of these relations is not as clear as for discrete bins. In the proof we shall make use of the Babenko-Beckner inequality (1.46). This inequality may, at first, seem to violate the requirement that it should be invariant under a change of the physical units used to measure the probability distributions $\rho(x)$ and $\tilde{\rho}(k)$. However, it turns out that this inequality is dimensionally correct. To see this, let us assume that we measure position using the unit L. It follows from the definition of the Fourier transform (1.18) that to keep this relation invariant, we have to choose $1/L$ as the unit of k. Now, we multiply the left hand side of (1.46) by 1 written in the form $L^{2-1/\alpha-1/\beta}$. After distributing the powers of L, we arrive at the inequality

$$\left(\int_{\mathbb{R}} dx[\rho(x)L]^{\alpha}/L\right)^{1/\alpha} \leq n(\alpha,\beta)\left(\int_{\mathbb{R}} dk[\tilde{\rho}(k)/L]^{\beta}L\right)^{1/\beta}. \tag{1.86}$$

This inequality is not only explicitly dimensionally correct but it also shows its invariance under a change of the unit. Taking the logarithm of both sides of this inequality and using the relation (1.37), we arrive at the uncertainty relation for the continuous Rényi entropies,

$$\begin{aligned} S_{\alpha}^{(x)} + S_{\beta}^{(k)} &= \frac{1}{1-\alpha}\ln\left(\int_{\mathbb{R}} dx[\rho(x)]^{\alpha}L^{\alpha-1}\right) + \frac{1}{1-\beta}\ln\left(\int_{\mathbb{R}} dk[\tilde{\rho}(k)]^{\beta}L^{1-\beta}\right) \\ &\geq -\frac{1}{2}\left(\frac{1}{1-\alpha}\ln\frac{\alpha}{\pi} + \frac{1}{1-\beta}\ln\frac{\beta}{\pi}\right). \end{aligned} \tag{1.87}$$

Due to the relation

$$\frac{\alpha}{1-\alpha} + \frac{\beta}{1-\beta} = 0, \tag{1.88}$$

the uncertainty relation is invariant under a rescaling of the wave function. In the limit, when both α and β tend to 1, we obtain the uncertainty relation for the Shannon entropies proved in [38] and in [39],

$$S^{(x)} + S^{(k)} = -\int_{\mathbb{R}} dx\,\rho(x)\ln\rho(x) - \int_{\mathbb{R}} dk\,\tilde{\rho}(k)\ln\tilde{\rho}(k) \geq 1 + \ln\pi. \tag{1.89}$$

In evaluating this limit by the L'Hôspital rule we assumed that the wave function is normalized,

$$\int_{\mathbb{R}} dx\,|\psi(x)|^2 = 1. \tag{1.90}$$

After an arbitrary rescaling of the wave function, the uncertainty relation takes on the form:

$$S^{(x)} + S^{(k)} = -\int_{\mathbb{R}} dx\,\rho(x)\ln\rho(x) - \int_{\mathbb{R}} dk\,\tilde{\rho}(k)\ln\tilde{\rho}(k) \geq N^2(1 + \ln\pi - 4\ln N), \tag{1.91}$$

where N is the norm of the wave function. We dropped in (1.89) the scale factor L because the sum of the two integrals is invariant under a change of units.

1.4.3.1 Gaussian Wave Functions

Gaussian wave functions play a distinguished role in uncertainty relations. They saturate the standard uncertainty relations expressed in terms of standard deviations.

They also saturate the uncertainty relations (1.87) expressed in terms of the Rényi entropies for continuous variables. Thus, unlike their discrete counterparts (1.24) or (1.50), the continuous uncertainty relations are sharp. Gaussians provide also a direct link between the Shannon entropies for continuous variables and the uncertainties measured by the standard deviation. Namely, the Gaussian distribution function gives the maximal value of the Shannon entropy subjected to the conditions of normalization and a given value σ_x of the standard deviation. To prove this property we shall search for the maximum of the functional:

$$\int_{\mathbb{R}} dx[-\rho(x)\ln\rho(x) + \lambda x^2\rho(x) + \mu\rho(x)], \tag{1.92}$$

where λ and μ are two Lagrange multipliers introduced to satisfy the constraints:

$$\int_{\mathbb{R}} dx\, \rho(x) = 1, \tag{1.93a}$$

$$\int_{\mathbb{R}} dx\, x^2\rho(x) = \sigma_x^2. \tag{1.93b}$$

Varying the functional (1.92) with respect to $\rho(x)$ we obtain the expression:

$$-1 - \ln\rho(x) + \lambda x^2 + \mu. \tag{1.94}$$

From the requirement that this expression vanishes, we obtain a Gaussian distribution. Finally, we determine the Lagrange multipliers λ and μ by imposing the constraints (1.93a), (1.93b), and we obtain:

$$\rho(x) = \frac{1}{\sqrt{2\pi}\sigma_x} e^{-(x^2/2\sigma_x^2)}. \tag{1.95}$$

This extremum is a true maximum since the entropy is a strictly concave functional of $\rho(x)$ [38].

1.5 Generalizations and Extensions

In this chapter we present a generalization of the entropic uncertainty relations to cover mixed states and an extension of these relations to angular variables. We shall also introduce uncertainty relations in phase space. Next, we return to the problem of mutually unbiased bases, mentioned already in Sect. 1.4.1.3. Finally, we show a connection between the entropic uncertainty relations and the logarithmic Sobolev inequalities. Other extensions of mathematical nature of the entropic uncertainty relations were recently published by Zozor et al. [55].

1.5.1 Uncertainty Relations for Mixed States

Let us take a set of normalized wave functions $\psi_i(x)$ that determine the probability distributions $\rho_i(x)$. Using these distributions we may define the probability distribution for a mixed state:

$$\rho_{\text{mix}}(x) = \sum_i \lambda_i \rho_i(x). \tag{1.96}$$

The sum may be finite or infinite. Alternatively, we may start from the density operator $\hat{\rho}$ and define the probability density as the diagonal matrix element $\rho_{\text{mix}}(x) = \langle x|\hat{\rho}|x\rangle$. The positive coefficient λ_i determines the probability with which the state described by $\psi_i(x)$ enters the mixture. The normalization condition $\text{Tr}\{\hat{\rho}\} = 1$, or the basic property of probabilities, require that

$$\sum_i \lambda_i = 1. \tag{1.97}$$

For each $\psi_i(x)$ we may introduce its Fourier transform $\tilde{\psi}_i(k)$ and then define the probability distribution $\tilde{\rho}_i(k)$ in momentum space.

$$\tilde{\rho}_{\text{mix}}(k) = \sum_i \lambda_i \tilde{\rho}_i(k). \tag{1.98}$$

This function can also be viewed as the momentum representation of the density operator, $\tilde{\rho}_{\text{mix}}(k) = \langle k|\hat{\rho}|k\rangle$.

We begin with the following sum of Babenko-Beckner inequalities (1.46) with wave functions $\psi_i(x)$ and $\tilde{\psi}_i(k)$ and with some weights λ_i:

$$\sum_i \lambda_i \left(\int_{\mathbb{R}} dx|\rho_i(x)|^\alpha\right)^{1/\alpha} \leq n(\alpha,\beta) \sum_i \lambda_i \left(\int_{\mathbb{R}} dk|\tilde{\rho}_i(k)|^\beta\right)^{1/\beta}. \tag{1.99}$$

To find the uncertainty relation for the mixed state we shall use now the Minkowski inequality [33] as was done in [44]. For $\alpha \geq 1$ and for an arbitrary set of nonnegative functions $f_i(x)$ the Minkowski inequality reads:

$$\left(\int_{\mathbb{R}} dx\left|\sum_i f_i(x)\right|^\alpha\right)^{1/\alpha} \leq \sum_i \left(\int_{\mathbb{R}} dx|f_i(x)|^\alpha\right)^{1/\alpha}. \tag{1.100}$$

Since for $\alpha \geq 1$ we have $\beta \leq 1$, the Minkowski inequality gets inverted. Choosing another set of nonnegative functions $g_i(k)$ we obtain:

$$\sum_i \left(\int_{\mathbb{R}} dk\,|g_i(k)|^\beta\right)^{1/\beta} \leq \left(\int_{\mathbb{R}} dk\left|\sum_i g_i(k)\right|^\beta\right)^{1/\beta}. \tag{1.101}$$

Substituting now $f_i(x) = \lambda_i \rho_i(x)$ in (1.100) and $g_i(k) = \lambda_i \tilde{\rho}_i(k)$ in (1.101) we obtain two inequalities involving the densities for mixed states:

$$\left(\int_{\mathbb{R}} dx |\rho_{\text{mix}}(x)|^\alpha\right)^{1/\alpha} \leq \sum_i \lambda_i \left(\int_{\mathbb{R}} dx |\rho_i(x)|^\alpha\right)^{1/\alpha}, \tag{1.102}$$

and

$$\sum_i \lambda_i \left(\int_{\mathbb{R}} dk |\tilde{\rho}_i(k)|^\beta\right)^{1/\beta} \leq \left(\int_{\mathbb{R}} dk |\tilde{\rho}_{\text{mix}}(k)|^\beta\right)^{1/\beta}. \tag{1.103}$$

Putting together the inequalities (1.99), (1.102), and (1.103) we obtain the generalization of the Babenko-Beckner inequality for mixed states:

$$\left(\int_{\mathbb{R}} dx [n\rho_{\text{mix}}(x)]^\alpha\right)^{1/\alpha} \leq n(\alpha,\beta) \left(\int_{\mathbb{R}} dk [\tilde{\rho}_{\text{mix}}(k)]^\beta\right)^{1/\beta}. \tag{1.104}$$

From this inequality we can prove the uncertainty relations (1.50) and (1.51) for mixed states in the same manner as we have done it for pure states.

1.5.2 Uncertainty Relations for Angles and Angular Momenta

In Sect. 1.2.5 we gave the uncertainty relation for angle φ and angular momentum M in terms of the Shannon entropy. Now we are going to prove this relation in a more general case of the Rényi entropy. To this end we shall repeat and extend the previous definitions.

1.5.2.1 The Uncertainty Relation for a Particle on a Circle

We have discussed this problem already in Sect. 1.2.5 and now we shall provide the proofs of the uncertainty relations. The proof of the uncertainty relation (1.34) in the general case of the Rényi entropy will be given in two steps.

First, we quote the Young-Hausdorff inequality for the Fourier series [44]:

$$\left(\int_0^{2\pi} d\varphi |\psi(\varphi)|^{2\alpha}\right)^{1/\alpha} \leq (2\pi)^{1/4\alpha - 1/4\beta} \left(\sum_{m=-\infty}^{\infty} |c_m|^{2\beta}\right)^{1/\beta}. \tag{1.105}$$

The coefficients α and β fulfill the standard relations (1.37). Upon comparing the left hand side in (1.105) with the definition (1.32), we see that we can use again the Jensen inequality for convex functions to obtain:

$$\delta\varphi^{1-\alpha} q_n^\alpha = \delta\varphi^{1-\alpha} \left(\int_{n\delta\varphi}^{(n+1)\delta\varphi} d\varphi |\psi(\varphi)|^2\right)^\alpha \leq \int_{n\delta\varphi}^{(n+1)\delta\varphi} d\varphi |\psi(\varphi)|^{2\alpha}. \tag{1.106}$$

Next, we do summation over n:

$$\delta\varphi^{1-\alpha}\sum_{n=0}^{N-1} q_n^\alpha \le \int_0^{2\pi} d\varphi|\psi(\varphi)|^{2\alpha}. \tag{1.107}$$

Putting the inequalities (1.105) and (1.107) together, we find:

$$(\delta\varphi)^{1/\alpha-1}\left(\sum_{n=0}^{N-1} q_n^\alpha\right)^{1/\alpha} \le (2\pi)^{1/4\alpha-1/4\beta}\left(\sum_{m=-\infty}^{\infty}|c_m|^{2\beta}\right)^{1/\beta}. \tag{1.108}$$

This inequality will lead us directly to the final form of the uncertainty relation for the Rényi entropy. To this end let us define the Rényi entropy for angle and angular momentum in a following form:

$$H_\alpha^{(\varphi)} = \frac{1}{1-\alpha}\ln\left[\sum_{n=0}^{N-1} q_n^\alpha\right], \tag{1.109}$$

$$H_\beta^{(M)} = \frac{1}{1-\beta}\ln\left[\sum_{m=-\infty}^{\infty} p_m^\beta\right], \tag{1.110}$$

where $p_m = |c_m|^2$. Taking the logarithm of both sides of (1.108), multiplying by $\beta/(1-\beta)$, and identifying the Rényi entropies (1.109) and (1.110), we obtain final inequality:

$$H_\alpha^{(\varphi)} + H_\beta^{(M)} \ge -\ln\frac{\delta\varphi}{2\pi} = \ln N. \tag{1.111}$$

As in the case of Maassen-Uffink result (1.79), the right hand side of the inequality (1.111) is independent on the parameters α and β.

This uncertainty relation can be also applied to a different physical situation. Let us consider the state vectors of a harmonic oscillator expanded in the Fock basis,

$$|\psi\rangle = \sum_{n=0}^{\infty} c_n|n\rangle. \tag{1.112}$$

Every such state can be described by a function of φ defined by [56]:

$$\psi(\varphi) = \sum_{n=0}^{\infty} c_n e^{im\varphi}. \tag{1.113}$$

The only difference between this wave function and the wave function for a particle on a circle is the restriction to nonnegative values of n. Therefore, the inequalities (1.35) and (1.111) still hold but they have now a different physical interpretation. The value of n determines the amplitude and the angle φ is the phase of the harmonic oscillation. This interpretation is useful in quantum electrodynamics where n

becomes the number of photons in a given mode and φ is the phase of the electromagnetic field.

1.5.2.2 The Uncertainty Relation for a Particle on the Surface of a Sphere

The uncertainty relation (1.111) may be called "the uncertainty relation on a circle", because the variable φ varies from 0 to 2π. We may ask, whether there is an uncertainty relation for a particle on the surface of a sphere. The answer is positive [57] and to find the uncertainty relation we shall parameterize this surface by the azimuthal angle $\varphi \in [0, 2\pi]$ and the polar angle $\theta \in [0, \pi]$. The wave functions on a sphere $\psi(\theta, \varphi)$ can be expanded into spherical harmonics:

$$\psi(\theta, \varphi) = \frac{1}{\sqrt{2\pi}} \sum_{l=0}^{\infty} \sum_{m=-l}^{l} c_{lm} Y_l^m(\theta, \varphi). \tag{1.114}$$

This expansion gives us a set of probabilities $|c_{lm}|^2$ which can be used to construct the Rényi entropy for the square of the angular momentum, determined by l, and for the projection of the angular momentum on the z axis, determined by m. As in the case of a particle moving on a circle the two characteristics of the state—angular position and angular momentum—are complementary.

There is one mathematical property of spherical harmonics that will help us to derive the uncertainty relation. Namely, when $m = l$ then for large l the function Y_l^m is to a very good approximation localized in the neighborhood of the equator where $\theta = \pi/2$. In other words, if we divide the θ range into equal parts with length $\delta\theta$ then for every ε we can find sufficiently large l such that the following relation holds (the modulus of the spherical harmonic does not depend on φ):

$$\int_{\pi/2-\delta\theta/2}^{\pi/2+\delta\theta/2} d\theta |Y_l^l(\theta, \varphi)|^2 = 1 - \varepsilon. \tag{1.115}$$

This property follows from the fact that $|Y_l^l(\theta, \varphi)|^2 = N|\sin\theta|^{2l}$. Of course, the smaller is the value of $\delta\theta$, the larger l is required to localize the wave function.

To define the Rényi entropies in the standard way we need the probability distributions in angular momentum $p_{lm} = |c_{lm}|^2$ and in the position on the sphere

$$q_{ij} = \int_{i\delta\theta}^{(i+1)\delta\theta} d\theta \sin\theta \int_{j\delta\varphi}^{(j+1)\delta\varphi} d\varphi\, |\psi(\theta, \varphi)|^2. \tag{1.116}$$

From the argument about localizability of the wave function near the equatorial plane we deduce that for a *fixed* (sufficiently large) value of l we can have the localization in one bin in the variable θ. Therefore, l and θ can be eliminated and we are left with the uncertainty relation in the projection of the angular momentum on the z axis and the azimuthal angle φ. The uncertainty relation in these variables has already been established and it will have the same form (1.111) for the sphere.

A complete proof of this fact is a bit more subtle and can be found in [57]. The introduction of a preferred direction associated with the projection of the angular momentum is rather artificial and it would be interesting to find a rotationally invariant form of the uncertainty relation. It would also be of interest to find generalizations of the uncertainty relation (1.111) to more dimensions and to functions defined on manifolds different from the surface of a sphere.

1.5.3 Uncertainty Relations in Phase Space

The next topic in this review is a "phase-space approach" to the uncertainty relations. We shall rewrite the sum of the Shannon entropies for position and momentum in a more symmetric way and we shall extend this symmetric description to the mixed states.

To this end we rewrite the sum of the expressions (1.16) and (1.19) for the Shannon entropies in the following, compact form:

$$H^{(x)} + H^{(k)} = -\sum_{i=-\infty}^{\infty} \sum_{j=-\infty}^{\infty} f_{ij} \ln(f_{ij}), \tag{1.117}$$

where:

$$f_{ij} = q_i p_j = \int_{(i-1/2)\delta x}^{(i+1/2)\delta x} dx \int_{(j-1/2)\delta k}^{(j+1/2)\delta k} dk |\psi(x)\tilde{\psi}(k)|^2. \tag{1.118}$$

In this way, the uncertainty relation (1.24) becomes an inequality involving just one function $f_1(x,k) = \psi(x)\tilde{\psi}(k)^*$ defined on the phase space. This rearrangement is not so trivial when we extend it to the mixed states. For a density operator $\hat{\rho}$ we define the function:

$$f(x,k) = \langle x|\hat{\rho}|k\rangle. \tag{1.119}$$

If the density operator represents a pure state, so that $\hat{\rho} = |\psi\rangle\langle\psi|$ we obtain the previous case $f(x,k) = f_1(x,k)$ but for mixed states the function (1.119) is not a simple product. Next, we define the two dimensional Fourier transform of $f(x,k)$:

$$\begin{aligned} \tilde{f}(\lambda,\mu) &= \frac{1}{2\pi}\int_{\mathbb{R}} dx\, e^{-i\lambda x} \int_{\mathbb{R}} dk\, e^{ik\mu} f(x,k) \\ &= \langle\lambda| \left(\int_{\mathbb{R}} dx\, |x\rangle\langle x|\right) \hat{\rho} \left(\int_{\mathbb{R}} dk\, |k\rangle\langle k|\right) |\mu\rangle \\ &= \langle\lambda|\hat{\rho}|\mu\rangle = \langle\mu|\hat{\rho}|\lambda\rangle^* = f(\mu,\lambda)^*, \end{aligned} \tag{1.120}$$

where we used the following representation of the Fourier transform kernels:

$$\frac{1}{\sqrt{2\pi}} e^{-i\lambda x} = \langle\lambda|x\rangle, \tag{1.121a}$$

$$\frac{1}{\sqrt{2\pi}}e^{ik\mu} = \langle k|\mu\rangle. \tag{1.121b}$$

The quantities in parentheses in this formula are the resolutions of the identity and were replaced by 1. Thus, the function f and its Fourier transform $\tilde{f}$ differ only in phase,

$$|f| = |\tilde{f}|, \tag{1.122}$$

and we obtain:

$$H^{(x,k)} = H^{(\mu,\lambda)}. \tag{1.123}$$

For the function f_1 this is, of course, a trivial conclusion. Because of (1.123) the uncertainty relation for the Shannon entropies becomes a relation that involves only one entropy, defined in terms of a single function f. The price paid for this simpler form of the uncertainty relation is the additional condition (1.122) that must be obeyed by the function $f(x,k)$.

1.5.4 Mutually Unbiased Bases

In Sect. 1.4.1.3 we have introduced the notion of mutually unbiased bases (MUB) and we discussed the Maassen-Uffink lower bound for these bases. In this subsection we are going to present several other uncertainty relations of this type. To this end, we introduce in a D-dimensional Hilbert space $\mathcal{H}_D$ a system of M orthonormal bases $B^m = \{|b_i^m\rangle, i \in 1,\ldots,D\}, m \in 1,\ldots,M$. Two bases B^m and B^n are mutually unbiased bases if and only if [58] for each pair of vectors:

$$|\langle b_i^m|b_j^n\rangle|^2 = \frac{1}{D}. \tag{1.124}$$

One may ask how many mutually unbiased bases can be found in $\mathcal{H}_D$? In other words how does the number $M_{\max}$ depends on D. There is no general answer to this question. It was found in [59] that if D is a prime number then $M_{\max} = D+1$.

We have already given the uncertainty relations (1.79) for two MUB's related by the discrete Fourier transformation. A straightforward generalization of these relations involves a sum over several MUB's. Let us denote by $H^{(m)}$ the Shannon entropy for the m-th base:

$$H^{(m)} = -\sum_{i=1}^{D} p_i^m \ln p_i^m, \tag{1.125}$$

$$p_i^m = |\langle b_i^m|\psi\rangle|^2. \tag{1.126}$$

Adding the uncertainty relations (1.79) for every pair of MUB's, we obtain [60]:

$$\sum_{m=1}^{M} H^{(m)} \geq \frac{M}{2} \ln D. \tag{1.127}$$

When D is a prime number and we put $M = D + 1$ then the relation (1.127) reads:

$$\sum_{m=1}^{D+1} H^{(m)} \geq \frac{(D+1)}{2} \ln D, \tag{1.128}$$

but this bound is rather poor. Sánchez [61] found a better lower bound,

$$\sum_{m=1}^{D+1} H^{(m)} \geq (D+1) \ln \frac{D+1}{2}. \tag{1.129}$$

The reader can find extensions of the uncertainty relations of this type in [60] and [58]. Recently in [62] another refinement of the inequality (1.127) has been obtained,

$$\sum_{m=1}^{M} H^{(m)} \geq M \ln \frac{MD}{M+D-1}. \tag{1.130}$$

For $M \geq 1 + \sqrt{D}$ this inequality is stronger than (1.127).

1.5.5 Logarithmic Sobolev Inequality

In this subsection we find a direct relation between the standard Heisenberg uncertainty relation and the entropic uncertainty relation. This will be done with the use of a different mathematical tool—the logarithmic Sobolev inequality and its inverse. We start with the inequalities for the continuous entropy (1.84). The logarithmic Sobolev inequality for the Shannon entropy proved in [63] reads:

$$S^{(x)} \geq \frac{1}{2}(1 + \ln 2\pi) - \frac{1}{2} \ln \left[L^2 \int_{\mathbb{R}} dx \frac{1}{\rho(x)} \left| \frac{d\rho(x)}{dx} \right|^2 \right]. \tag{1.131}$$

The important feature of this inequality is that we have only one function $\rho(x)$. Therefore, we do not need conjugate variables (momentum) to obtain a lower bound of the Shannon entropy (1.84). On the other hand, the right hand side of (1.131) depends on $\rho(x)$, so it is a functional relation rather than an uncertainty relation. In order to obtain an uncertainty relation, we shall exploit the inverse logarithmic Sobolev inequality proved in [64]:

$$S^{(x)} \leq \frac{1}{2}(1 + \ln 2\pi) + \ln L\sigma_x, \tag{1.132}$$

where σ_x is the standard deviation (1.4). Since this inequality involves only one distribution function, we can write it also for the probability density in momentum space $\tilde{\rho}(k)$:

$$S^{(k)} \leq \frac{1}{2}(1 + \ln 2\pi) + \ln(\sigma_k/L). \tag{1.133}$$

Evaluating the exponential function of the sum of the inequalities (1.132) and (1.133), we obtain a refined version of the Heisenberg uncertainty relation [38, 65]:

$$\sigma_x \sigma_k \geq \frac{1}{2} \exp(S^{(x)} + S^{(k)} - 1 - \ln \pi) \geq \frac{1}{2}. \tag{1.134}$$

This uncertainty relation is stronger than the standard Heisenberg uncertainty relation. Whenever the sum of the Shannon entropies exceeds its lower bound $1 + \ln \pi$, we obtain a stronger bound for $\sigma_x \sigma_k$ than the standard $1/2$. Note that the uncertainty relation (1.134) holds for any pair of distributions. They do not have to be related by the Fourier transformation of the wave functions.

References

1. Adesso G, Serafini A, Illuminati F (2004) Extremal entanglement and mixedness in continuous variable systems. Phys Rev A 70:022318
2. Bengtsson I, Życzkowski K (2006) Geometry of quantum states. Cambridge University Press, Cambridge
3. Bovino FA, Castagnoli G, Ekert A, Horodecki P, Alves CM, Sergienko AV (2005) Direct measurement of nonlinear properties of bipartite quantum states. Phys Rev Lett 95:240407
4. Gühne O, Lewenstein M (2004) Entropic uncertainty relations and entanglement. Phys Rev A 70:022316
5. Terhal BM (2002) Detecting quantum entanglement. J Theor Comput Sci 287:313
6. Giovannetti V, Lloyd S (2004) Additivity properties of a Gaussian channel. Phys Rev A 69:062307
7. Renner R, Gisin N, Kraus B (2005) Information-theoretic security proof for quantum-key-distribution protocols. Phys Rev A 72:012332
8. Lévay P, Nagy S, Pipek J (2005) Elementary formula for entanglement entropies of fermionic systems. Phys Rev A 72:022302
9. Beck C, Graudenz D (1992) Symbolic dynamics of successive quantum-mechanical measurements. Phys Rev A 46:6265
10. Kohler S, Hänggi P (2002) In: Leuchs G, Beth T (eds) Quantum information processing. Wiley-VCH, Berlin. arXiv:quant-ph/0206189
11. Białas A, Czyż W (2000) Event by event analysis and entropy of multiparticle systems. Phys Rev D 61:074021
12. Białas A, Czyż W, Zalewski K (2005) Moments of the particle phase-space density at freeze-out and coincidence probabilities. Acta Phys Pol B 36:3109
13. Białas A, Czyż W, Zalewski K (2006) Moments of the Wigner function and Rényi entropies at freeze-out. Phys Rev C 73:034912
14. Majka A, Wiślicki W (2003) Uniformity of the phase space and fluctuations in thermal equilibrium. Physica A 322C:313
15. Cybulski O, Matysiak D, Babin V, Hołyst R (2004) Pattern formation in nonextensive thermodynamics: selection criteria based on the Rényi entropy production. Phys Rev E 69:016110

16. Cybulski O, Babin V, Hołyst R (2005) Minimization of the Rényi entropy production in the stationary states of the Brownian process with matched death and birth rates. J Chem Phys 122:174105
17. Arbó DG, Reinhold CO, Burgdörfer J, Pattanayak AK, Stokely CL, Zhao W, Lancaster JC, Dunning FB (2003) Pulse-induced focusing of Rydberg wave packets. Phys Rev A 67:063401
18. Gnutzmann S, Życzkowski K (2001) Rényi-Wehrl entropies as measures of localization in phase space. J Phys A, Math Gen 34:10123
19. Verstraete F, Cirac JI (2006) Matrix product states represent ground states faithfully. Phys Rev B 73:094423
20. Dehesa JS, Martínez-Finkelshtein A, Sorokin VN (2002) Quantum-information entropies for highly excited states of single-particle systems with power-type potentials. Phys Rev A 66:062109
21. De Nicola S, Fedele R, Man'ko MA, Man'ko VI (2009) Entropic uncertainty relations for electromagnetic beams. Phys Scr 135:014053
22. Salcedo LL (2009) Phase space localization of antisymmetric functions. J Math Phys 50:012106
23. Varga I, Pipek J (2003) Rényi entropies characterizing the shape and the extension of the phase space representation of quantum wave functions in disordered systems. Phys Rev E 68:026202
24. Brukner C, Zeilinger A (2001) Conceptual inadequacy of the Shannon information in quantum measurements. Phys Rev A 63:022113
25. Timpson CG (2003) On a supposed Conceptual inadequacy of the Shannon information in quantum mechanics. Stud Hist Philos Mod Phys 33:441. arXiv:quant-ph/0112178
26. Shannon CE, Weaver W (1949) The mathematical theory of communication. University of Illinois Press, Urbana
27. Heisenberg W (1927) Über den Anschaulichen Inhalt der quanten-theoretischen Kinematik und Mechanik. Z Phys 43:172
28. Bialynicki-Birula I (2007) Rényi entropy and the uncertainty relations. In: Adenier G, Fuchs CA, Khrennikov AYu (eds) Foundations of probability and physics. AIP Conf Proc, vol 889. AIP, New York
29. Peres A (1995) Quantum theory: concepts and methods. Kluwer, Dordrecht
30. Partovi MH (1983) Entropic formulation of uncertainty for quantum measurements. Phys Rev Lett 50:1883
31. Heisenberg W (1930) The physical properties of the quantum theory. Dover, New York
32. Bialynicki-Birula I (1984) Entropic uncertainty relations. Phys Lett 103 A:253
33. Hardy G, Littlewood JL, Pólya G (1934) Inequalities. Cambridge University Press, Cambridge
34. Jensen JLWV (1906) Sur les fonctions convexes et les inégalités entre les valeurs moyennes. Acta Math 30:175
35. Hirschman II (1957) A note on entropy. Am J Math 79:152
36. Everett H III (1957) "Relative State" formulation of quantum mechanics. Rev Mod Phys 29:454
37. Everett H III (1973) The theory of the universal wave function. In: DeWitt BS, Graham N (eds) The many-world interpretation of quantum mechanics. Princeton University Press, Princeton. PhD thesis
38. Bialynicki-Birula I, Mycielski J (1975) Uncertainty relations for information entropy in wave mechanics. Commun Math Phys 44:129
39. Beckner W (1975) Inequalities in Fourier analysis. Ann Math 102:159
40. Řehaček J, Bouchal Z, Čelechovský R, Hradil Z, Sánchez-Soto LL (2008) Experimental test of uncertainty relations for quantum mechanics on a circle. Phys Rev A 77:032110
41. Rényi A (1960) Some fundamental questions of information theory. MTA III Oszt Közl 251
42. Rényi A (1960) On measures of information and entropy. In: Proceedings of the 4th Berkeley symposium on mathematics, statistics and probability, p 547
43. Rényi A (1970) Probability theory. North-Holland, Amsterdam

44. Bialynicki-Birula I (2006) Formulation of the uncertainty relations in terms of the Rényi entropies. Phys Rev A 74:052101
45. Babenko KI (1961) An inequality in the theory of Fourier integrals. Izv Akad Nauk SSSR, Ser Mat 25:531 (in Russian)
46. Wilk G, Włodarczyk Z (2009) Uncertainty relations in terms of the Tsallis entropy. Phys Rev A 79:062108
47. Bialynicki-Birula I, Rudnicki Ł (2010) Comment on "Uncertainty relations in terms of the Tsallis entropy". Phys Rev A 81:026101
48. Deutsch D (1983) Uncertainty in quantum measurements. Phys Rev Lett 50:631
49. Kraus K (1987) Complementary observables and uncertainty relations. Phys Rev D 35:3070
50. Maassen H, Uffink JBM (1988) Generalized entropic uncertainty relations. Phys Rev Lett 60:1103
51. Reed M, Simon B (1975) Methods of modern mathematical physics, vol II. Academic Press, New York
52. Riesz M (1927) Sur les maxima des formes bilinéaires et sur les fonctionnelles linéaires. Acta Math 49:465
53. Schwinger J (1960) Unitary operator bases. Proc Natl Acad Sci USA 46:570
54. Bengtsson I (2007) Three ways to look at mutually unbiased bases. In: Adenier G, Fuchs CA, Khrennikov AYu (eds) Foundations of probability and physics. AIP Conf Proc, vol 889. AIP, New York. arXiv:quant-ph/0610216
55. Zozor S, Portesi M, Vignat C (2008) Some extensions of the uncertainty principle. Physica A 387:4800–4808
56. Bialynicki-Birula I, Bialynicka-Birula Z (1976) Quantum electrodynamics of intense photon beams. New approximation method. Phys Rev A 14:1101
57. Bialynicki-Birula I, Madajczyk J (1985) Entropic uncertainty relations for angular distributions. Phys Lett 108 A:384
58. Wehner S, Winter A (2009) Entropic uncertainty relations—a survey. arXiv:0907.3704v1 [quant-ph]
59. Ivonovic ID (1981) Geometrical description of quantal state determination. J Phys A 14:3241–3245
60. Azarchs A (2004) Entropic uncertainty relations for incomplete sets of mutually unbiased observables. arXiv:quant-ph/0412083v1
61. Sánchez J (1993) Entropic uncertainty and certainty relations for complementary observables. Phys Lett A 173:233–239
62. Wu S, Yu S, Mølmer K (2009) Entropic uncertainty relation for mutually unbiased bases. Phys Rev A 79:022104
63. Gross L (1975) Logarithmic Sobolev inequalities. Am J Math 97:1061–1083
64. Chafai D (2002) Gaussian maximum of entropy and reversed log-Sobolev inequality Séminaire de probabilitiés. Strasbourg 36:194–200
65. Dodonov VV, Man'ko VI (1989) Generalized uncertainty relations in quantum mechanics. In: Markov MA (ed) Invariants and evolution of nonstationary quantum systems. Proceedings of the Lebedev Physics Institute, vol 183. Nova Science, Commack

Chapter 2
Derivation of Generalized von Weizsäcker Kinetic Energies from Quasiprobability Distribution Functions

Debajit Chakraborty and Paul W. Ayers

Abstract The Fisher Information of the electronic distribution functions is closely related to the von Weizsäcker kinetic energy functional. We show how generalizations of the Weizsäcker kinetic energy density functional can be derived from the canonical momentum-space expression for the kinetic energy and extend this result to higher-order electron distribution functions.

2.1 Introduction

One of the biggest challenges in density-functional theory (DFT) and its many-electron generalizations is formulating an approximate kinetic energy functional. The Weizsäcker family of functionals [1–3] is particularly useful, at least at a theoretical level, because the form of the functional,

$$T_w[p] \propto \int \frac{\nabla p(\tau) \cdot \nabla p(\tau)}{8p(\tau)} d\tau \tag{2.1}$$

is preserved, whether $p(\boldsymbol{\tau})$ is a one-electron, two-electron, or many-electron distribution function. Moreover, for the N-electron distribution function, $p(\boldsymbol{\tau}) = |\Psi(\boldsymbol{\tau})|^2$, the Weizsäcker functional form is exact [4, 5]. In this book chapter we will elucidate some properties of the Weizsäcker kinetic energy form, elucidating its link to information theory (Fisher information) and also to the momentum-space equation for the kinetic energy.

2.2 Fisher Information

The Fisher information, $I[p]$, of a probability distribution function, $p(\mathbf{x})$, measures the local inhomogeneity of the system [6]. For a unimodal distribution, it is a mea-

P.W. Ayers (✉)
McMaster University, Hamilton, Ontario, Canada, L8S 4M1
e-mail: ayers@mcmaster.ca

K.D. Sen (ed.), *Statistical Complexity*,
DOI 10.1007/978-90-481-3890-6_2,

sure of the compactness of $p(\mathbf{x})$. For a multimodal distribution, $I[p]$ is a measure of "peakiness." As stressed by Frieden, nature seems to favor extreme values of the Fisher information, and many of the laws and governing equations of physics can be obtained by minimizing/maximizing the Fisher information subject to appropriate physical constraints [7]. An early derivation of the Schrödinger equation by minimizing the Fisher information was given by Sears, Parr, and Dinur [5]. A recent review by Nalewajski features the many ways Fisher information is used in quantum chemistry and, more generally, molecular electronic structure theory [8].

Suppose that a probability distribution function depends parametrically on parameters $\boldsymbol{\theta}$. Denote the probability of observing data value $\mathbf{x}$ given that the parameters have values $\boldsymbol{\theta}$ as $p(\mathbf{x}|\boldsymbol{\theta})$. In general, $\boldsymbol{\theta}$ is a vector containing multiple parameters. The Fisher information indicates how much information we gain about the value of the parameters by measuring $\mathbf{x}$. For one data value and one parameter, the Fisher information is simply,

$$\begin{aligned} I[p] &= \int p(\mathbf{x}|\theta)\left(\frac{\partial \ln(p(\mathbf{x}|\theta))}{d\theta}\right)^2 d\mathbf{x} \\ &= \int \frac{1}{p(\mathbf{x}|\theta)}\left(\frac{\partial p(\mathbf{x}|\theta)}{d\theta}\right)^2 d\mathbf{x}. \end{aligned} \tag{2.2}$$

For multiple parameters, this expression generalizes to

$$I[p] = \int \frac{\nabla_\theta p(\mathbf{x}|\theta) \cdot \nabla_\theta p(\mathbf{x}|\theta)}{p(\mathbf{x}|\theta)} d\mathbf{x}. \tag{2.3}$$

Owing to the Heisenberg momentum-position uncertainty principle, in quantum mechanics we cannot measure the position of particles exactly. What can we say about the true position of a particle if we observe it at the point $\mathbf{x}$? Let $\boldsymbol{\theta}$ be the position of the particle; this is the quantity we are trying to estimate. The fluctuation of the observed position of the particle from its true position must be translationally invariant. This means that $p(\mathbf{x}|\boldsymbol{\theta}) = p(\mathbf{x}-\boldsymbol{\theta})$ [7]. Inserting this expression into (2.3) and making the substitution $\mathbf{y} = \mathbf{x} - \boldsymbol{\theta}$ gives

$$\begin{aligned} I[p] &= \int \frac{\nabla p(\mathbf{y}) \cdot \nabla p(\mathbf{y})}{p(\mathbf{y})} d\mathbf{y} \\ &= \int p(\mathbf{y})|\nabla(\ln p(\mathbf{y}))|^2 d\mathbf{y}. \end{aligned} \tag{2.4}$$

This particular manifestation of the Fisher information is sometimes called the Fisher information of locality, because it captures the inherent delocalization of quantum mechanical particles [4].

The preceding derivation may be criticized in the context of molecular electronic structure theory because when one makes the Born-Oppenheimer approximation, the electronic coordinates $\mathbf{x}$ are no longer translationally invariant because the nuclei are at fixed positions in space. We will not worry about the epistemological issue in this paper, because (2.4) may be interpreted even if one questions the appropriateness of this derivation in electronic structure theory. The Fisher information measure from (2.4) measures the amount of local inhomogeneity in the system [6]. For a unimodal distribution, it is a measure of the compactness of $p(\mathbf{x})$. For a multimodal distribution, $I[p]$ is a measure of "peakiness."

As stressed by Frieden, nature seems to favor extreme values of the Fisher information, and many of the laws and governing equations of physics can be obtained by minimizing/maximizing the Fisher information subject to appropriate physical constraints [7]. An early derivation of the Schrödinger equation from minimizing the Fisher information was given by Sears, Parr, and Dinur [5].

In his prescient work, Fisher introduced the probability amplitude, $|\psi(\mathbf{x})|^2 = p(\mathbf{x})$. In terms of the probability amplitude, (2.4) takes a form reminiscent of the quantum mechanical kinetic energy,

$$I[\psi] = 4\int \nabla\psi^*(\mathbf{x})\cdot\nabla\psi(\mathbf{x})d\mathbf{x} \sim 8T[\psi]. \tag{2.5}$$

This form is the starting point for the derivation of the Schrödinger equation from the principle of extreme physical information [4, 5, 7]. Equation (2.5) is more general, however, because $\psi(\mathbf{x})$ does not have to be a wavefunction.

Now let us consider the Fourier transform of the probability amplitude,

$$\begin{aligned} \widehat{\psi}(\mathbf{p}) &= \left(\frac{1}{2\pi}\right)^{d/2} \int e^{i\mathbf{p}\cdot\mathbf{x}}\psi(\mathbf{x})d\mathbf{x}, \\ \psi(\mathbf{x}) &= \left(\frac{1}{2\pi}\right)^{d/2} \int e^{-i\mathbf{p}\cdot\mathbf{x}}\widehat{\psi}(\mathbf{p})d\mathbf{p}. \end{aligned} \tag{2.6}$$

Using identities from Fourier analysis and defining the momentum-space probability distribution function in the obvious way, $p(\mathbf{p}) = |\widehat{\psi}(\mathbf{p})|^2$, we can write

$$I[\psi] = 4\int \mathbf{p}\cdot\mathbf{p}|\widehat{\psi}(\mathbf{p})|^2 d\mathbf{p} \sim 8T[\psi]. \tag{2.7}$$

The derivation of (2.7) will be expounded upon later. For now it suffices to note that a very similar derivation may be found in chapter three of Frieden's book [7]. Equation (2.7) is intuitive: the amount of information that can be obtained about position is proportional to the variance of the momentum. It is appealing that the most common of all measures of uncertainty—the variance—Fourier transforms into the Fisher information.

2.3 Kinetic Energy

Even though the electronic kinetic energy is readily computed from the N-electron wavefunction,

$$
\begin{aligned}
T[\Psi] &= \iint \cdots \int \Psi^*(\mathbf{r}_1, \mathbf{r}_2, \ldots, \mathbf{r}_N)\left(\sum_{i=1}^{N} -\frac{1}{2}\nabla_i^2\right) \\
&\quad \times \Psi(\mathbf{r}_1, \mathbf{r}_2, \ldots, \mathbf{r}_N) d\mathbf{r}_1 d\mathbf{r}_2 \ldots d\mathbf{r}_N \\
&= \iint \cdots \int \sum_{i=1}^{N} \frac{1}{2}\nabla_i \Psi^*(\mathbf{r}_1, \mathbf{r}_2, \ldots, \mathbf{r}_N) \\
&\quad \times \nabla_i \Psi(\mathbf{r}_1, \mathbf{r}_2, \ldots, \mathbf{r}_N) d\mathbf{r}_1 d\mathbf{r}_2 \ldots d\mathbf{r}_N,
\end{aligned}
\tag{2.8}
$$

the problem of approximating the kinetic energy directly from the electron density persists. In (2.8), and throughout the remainder of this article, atomic units (where $\hbar = m_e = 1$) are used.

It is impossible to review even a small fraction of the literature on kinetic energy density functionals. The discipline started with the work of Thomas and Fermi [9, 10] followed soon after by von Weizsäcker [1]. In the late 1970's, the gradient expansion approximation for kinetic energy of nearly uniform electron densities was performed to high order, and shown to diverge for atomic and molecular electron densities [11, 12]. Most recent work has focused on approaches that incorporate information about the exact linear response function of the uniform electron gas [13–16]. We refer the interested reader to recent articles with a review-like character [14, 17–22].

We wish to focus on two aspects of kinetic energy functionals in this paper. First, we will focus on the Weizsäcker functional. Many authors have suggested that the Weizsäcker functional is a good starting point for kinetic energy functionals [13, 17, 23–35] partly because it ensures the correct behavior at the nuclear-electron cusps and also in the asymptotic decaying tails of the electron density. The Weizsäcker functional is, in its spin-resolved form,

$$
T_w^{(1)}[\rho_\sigma] = \sum_{\sigma=\alpha,\beta} \int \frac{\nabla\rho_\sigma(\mathbf{r}) \cdot \nabla\rho_\sigma(\mathbf{r})}{8\rho_\sigma(\mathbf{r})} d\mathbf{r}
\tag{2.9}
$$

where the spin density is given by the expression

$$
\rho_\sigma(\mathbf{r}) = \langle\Psi|\sum_{i=1}^{N} |\sigma(i)\rangle\delta(\mathbf{r}_i - \mathbf{r})\langle\sigma(i)||\Psi\rangle
\tag{2.10}
$$

and is normalized to the number of electrons with the specified spin,

$$
N_\sigma = \int \rho_\sigma(\mathbf{r}) d\mathbf{r}.
\tag{2.11}
$$

The electron density is nonnegative but because it is not normalized to one, it is not a probability distribution function. For this reason, it is sometimes more convenient to work with the so-called shape functions [36–39]

$$
p_\sigma(\mathbf{r}) = \frac{\rho_\sigma(\mathbf{r})}{N_\sigma}.
\tag{2.12}
$$

The Weizsäcker functional is then

$$T_w^{(1)}[p_\sigma] = \sum_{\sigma=\alpha,\beta} N_\sigma \int \frac{\nabla p_\sigma(\mathbf{r}) \cdot \nabla p_\sigma(\mathbf{r})}{8 p_\sigma(\mathbf{r})} d\mathbf{r}. \tag{2.13}$$

The shape function is usually denoted $\sigma(\mathbf{r})$, but we will use the nonstandard notation in (2.12) to avoid confusion with the spin index.

The second aspect of kinetic energy functionals we wish to focus on is the momentum-space representation. In momentum space, the kinetic energy is a simple and explicit functional of the momentum density,

$$T = \sum_{\sigma=\alpha,\beta} \int \frac{1}{2}(\mathbf{p}\cdot\mathbf{p})\rho_\sigma(\mathbf{p})d\mathbf{p}. \tag{2.14}$$

The momentum density is defined by an expression just like (2.10), but now the Fourier-transformed wavefunctions (denoted $\widehat{\Psi}$; cf. (2.6)) are used,

$$\Pi_\sigma(\mathbf{p}) = \langle \widehat{\Psi} | \sum_{i=1}^{N} |\sigma(i)\rangle \delta(\mathbf{p}_i - \mathbf{p}) \langle \sigma(i)| | \widehat{\Psi} \rangle. \tag{2.15}$$

Notice that the momentum density is not the Fourier transform of the position density; this will be important later in our analysis when we try to approximate the kinetic energy functional.

There is another perspective that is intermediate between the position-space and momentum-space approach; this perspective is based on quasiprobability distribution functions [40–42]. In classical mechanics, one can generate a phase-space distribution function, $f(\mathbf{r},\mathbf{p})$ that represents the probability of observing a particle at the point $\mathbf{r}$ with momentum $\mathbf{p}$. In quantum mechanics, the Heisenberg uncertainty principle forbids measuring the position and momentum of a particle simultaneously, and there are innumerably many choices for $f(\mathbf{r},\mathbf{p})$. Given a quasiprobability distribution function, however, the local kinetic energy [43–46],

$$t_\sigma(\mathbf{r}) = \int \frac{1}{2}(\mathbf{p}\cdot\mathbf{p}) f_\sigma(\mathbf{r},\mathbf{p}) d\mathbf{p}, \tag{2.16}$$

and the total kinetic energy,

$$T = \sum_{\sigma=\alpha,\beta} \int t_\sigma(\mathbf{r}) d\mathbf{r} \tag{2.17}$$

are readily evaluated.

The one-electron quasiprobability distribution function can be computed from the one-electron reduced density matrix, [42, 46–48]

$$\gamma_{\sigma\sigma}(\mathbf{r},\mathbf{r}') = N_\sigma \iiint \cdots \int \begin{bmatrix} \Psi^*(\mathbf{r}_1', \mathbf{r}_2, \ldots, \mathbf{r}_N) \\ \times (|\sigma(1)\rangle \delta(\mathbf{r}_1' - \mathbf{r}') \delta(\mathbf{r}_1 - \mathbf{r}) \langle \sigma(1)|) \\ \times \Psi(\mathbf{r}_1', \mathbf{r}_2, \ldots, \mathbf{r}_N) d\mathbf{r}_1 d\mathbf{r}_1' d\mathbf{r}_2 \ldots d\mathbf{r}_N \end{bmatrix}, \tag{2.18}$$

by the equation,

$$f_\sigma(\mathbf{r},\mathbf{p}) = \left(\frac{1}{2\pi}\right)^6 \iiint e^{-i\tau\cdot\mathbf{p}} e^{-i\theta\cdot(\mathbf{r}-\mathbf{u})} g(\theta,\tau)\gamma_{\sigma\sigma}\left(\mathbf{u}+\frac{1}{2}\tau,\mathbf{u}-\frac{1}{2}\tau\right) d\mathbf{u}\,d\theta\,d\tau, \tag{2.19}$$

where the function $g(\boldsymbol{\theta},\boldsymbol{\tau})$ is any function that is well-behaved enough for the integral to exist that satisfies the constraints

$$\begin{aligned} &g(\theta,\tau=0) = g(\theta=0,\tau) = 1, \\ &(g(\theta,\tau))^* = g(-\theta,-\tau). \end{aligned} \tag{2.20}$$

The most popular choice, $g = 1$, corresponds to the Wigner distribution [40]. Quasiprobability distribution functions for many-electron reduced density matrices are computed from very similar formulas.

Every kinetic energy density functional is based on some choice, whether implicit or explicit, for the momentum density and/or the quasiprobability distribution function.

2.4 Generalized Weizsäcker Forms of the Kinetic Energy

Notice that the Weizsäcker kinetic energy functional, (2.13), strongly resembles the form of the Fisher information, (2.4). Similarly, the *exact* kinetic energy functional, (2.14) recalls the momentum-space formula for the Fisher information, (2.7). Is there some way to, using the link to momentum space, improve the Weizsäcker functional so that it is more accurate? Can formulating a momentum-density or quasiprobability distribution version of the Weizsäcker functional give some insight into the functional?

Since the Weizsäcker functional depends only on the electron density in position space, the corresponding momentum density must also be a density functional. Consider the momentum density of a "piece" of the electron density,

$$\pi_\sigma^{(a)}(\mathbf{p}) = \left(\frac{1}{2\pi}\right)^{3/2} \int e^{i\mathbf{p}\cdot\mathbf{r}} (\rho_\sigma(\mathbf{r}))^a d\mathbf{r}. \tag{2.21}$$

A reasonable, if highly approximate, formula for the momentum density is then

$$\tilde{\Pi}_\sigma^{(a)}(\mathbf{p}) = \frac{1}{2}\left(\pi_\sigma^{(a)}(\mathbf{p})[\pi_\sigma^{(1-a)}(\mathbf{p})]^* + [\pi_\sigma^{(a)}(\mathbf{p})]^*\pi_\sigma^{(1-a)}(\mathbf{p})\right). \tag{2.22}$$

This form is motivated by the idea that the square root of the density has the units of the wavefunction. So using the square root of the electron density instead of $\psi(\mathbf{r})$ in (2.6) seems analogous to the usual procedure for deriving the momentum density and, moreover, is exact for one-electron systems; we will see that this is equivalent to the Weizsäcker approximation. Equation (2.22) is just the generalization of this idea, and we hoped that by optimizing the value of a we could obtain better results.

Now we derive the kinetic energy density functional that is built from the approximate momentum-space density in (2.22),

$$\tilde{T}_\sigma^{(a)} = \int \frac{1}{2}(\mathbf{p} \cdot \mathbf{p})\tilde{\Pi}_\sigma^a(\mathbf{p})d\mathbf{p}. \tag{2.23}$$

Substituting in the definition of the approximate momentum density, this simplifies to

$$\begin{aligned}\tilde{T}_\sigma^{(a)} &= \frac{1}{4}\left(\int (\mathbf{p} \cdot \mathbf{p})\pi_\sigma^{(a)}(\mathbf{p})(\pi_\sigma^{(1-a)}(\mathbf{p}))^* d\mathbf{p} + \text{c.c.}\right) \\ &= \frac{1}{4}\left(\frac{1}{2\pi}\right)^3 \int (\mathbf{p} \cdot \mathbf{p})\left[\iint e^{i(\mathbf{p}\cdot\mathbf{r})}(\rho_\sigma(\mathbf{r}))^a e^{-i(\mathbf{p}\cdot\mathbf{r}')}(\rho_\sigma(\mathbf{r}'))^{1-a} d\mathbf{r}d\mathbf{r}'\right]d\mathbf{p} + \text{c.c.}\end{aligned} \tag{2.24}$$

Here c.c. denotes the addition of the complex conjugate of the preceding term. Interchanging the order of integration and using the Fourier transform form of the derivative,

$$(-1)^n \delta^{(n)}(x - x') = \frac{1}{2\pi}\int e^{ip(x-x')}(ip)^n dp \tag{2.25}$$

gives

$$\begin{aligned}\tilde{T}_\sigma^{(a)} &= -\frac{1}{4}\iint \delta^{(2)}(\mathbf{r} - \mathbf{r}')(\rho_\sigma(\mathbf{r}))^a(\rho_\sigma(\mathbf{r}'))^{1-a} d\mathbf{r}d\mathbf{r}' + \text{c.c.} \\ &= -\frac{1}{4}\int (\rho_\sigma(\mathbf{r}'))^{1-a}\nabla^2(\rho_\sigma(\mathbf{r}'))^a d\mathbf{r}' + \text{c.c.} \\ &= -\frac{1}{2}\int (\rho_\sigma(\mathbf{r}'))^{1-a}\nabla^2(\rho_\sigma(\mathbf{r}'))^a d\mathbf{r}'.\end{aligned} \tag{2.26}$$

This formula simplifies to

$$\begin{aligned}\tilde{T}_\sigma^{(a)} &= -\frac{a(a-1)}{2}\int (\rho_\sigma(\mathbf{r}'))^{1-a}(\rho_\sigma(\mathbf{r}'))^{a-2}(\nabla\rho_\sigma(\mathbf{r}') \cdot \nabla\rho_\sigma(\mathbf{r}'))d\mathbf{r}' \\ &\quad - \frac{a}{2}\int \nabla^2\rho(\mathbf{r})d\mathbf{r} \\ &= -\frac{a(a-1)}{2}\int \frac{\nabla\rho_\sigma(\mathbf{r}') \cdot \nabla\rho_\sigma(\mathbf{r}')}{\rho_\sigma(\mathbf{r}')}d\mathbf{r}'.\end{aligned} \tag{2.27}$$

The Laplacian term in the second line vanishes because of the rapidly decaying nature of the electron density (which is, in turn, forced by the boundary conditions on the electronic wavefunction). If we wish to interpret (2.23) and (2.27) as manifestations of the Fisher information, it is better to write instead

$$\tilde{T}_\sigma^{(a)} = -\frac{a(a-1)}{2}N_\sigma \int \frac{\nabla p_\sigma(\mathbf{r}') \cdot \nabla p_\sigma(\mathbf{r}')}{p_\sigma(\mathbf{r}')}d\mathbf{r}'. \tag{2.28}$$

The approximate kinetic energy functional $\tilde{T}_\sigma^{(a)}$ is a parabola in a with maximum value at $a = 1/2$. The maximum value is precisely the Weizsäcker functional [3, 49, 50]. Since the Weizsäcker functional is a lower bound to the true kinetic energy, the most accurate member of this family of generalized Weizsäcker functionals is the conventional Weizsäcker functional itself.

Notice also that $\tilde{T}_\sigma^{(0)} = \tilde{T}_\sigma^{(1)} = 0$. This follows from (2.28) and the fact that the following momentum density has zero kinetic energy,

$$\tilde{\Pi}_\sigma^{(0)}(\mathbf{p}) = \tilde{\Pi}_\sigma^{(1)}(\mathbf{p}) = \frac{1}{2}\left(\delta(\mathbf{p})\int e^{i\mathbf{p}\cdot\mathbf{r}}\rho_\sigma(\mathbf{r})d\mathbf{r} + \text{c.c.}\right). \tag{2.29}$$

The reader may wonder why we did not consider the straightforward Fourier transform of the electron density,

$$\widehat{\rho}(\mathbf{p}) = \frac{1}{2}\left(\int e^{i\mathbf{p}\cdot\mathbf{r}}\rho_\sigma(\mathbf{r})d\mathbf{r} + \text{c.c.}\right). \tag{2.30}$$

(Notice: this is *not* the ansatz in (2.21) and (2.22).) This momentum density gives an entirely different, and seemingly absurd, value for the kinetic energy. Namely,

$$\begin{aligned}
\tilde{T}_\sigma[\widehat{\rho}_\sigma] &= \int \frac{1}{2}(\mathbf{p}\cdot\mathbf{p})\widehat{\rho}_\sigma(\mathbf{p})d\mathbf{p} \\
&= \int \frac{1}{2}(\mathbf{p}\cdot\mathbf{p})\left(\frac{1}{2}\right)\left[\left(\frac{1}{2\pi}\right)^{3/2}\int e^{i\mathbf{p}\cdot\mathbf{r}}\rho_\sigma(\mathbf{r})d\mathbf{r} + \text{c.c.}\right]d\mathbf{p} \\
&= -\frac{1}{2}(2\pi)^{3/2}\int \delta^{(2)}(\mathbf{r}-\mathbf{0})\rho_\sigma(\mathbf{r})d\mathbf{r} \\
&= -\sqrt{2\pi^3}\nabla^2\rho_\sigma(\mathbf{0}).
\end{aligned} \tag{2.31}$$

For an atom centered at the origin, this kinetic energy actually diverges. Distressingly, this formula for the kinetic energy appears not to be translationally invariant, but this reveals an oversight in the derivation—the integrand is too ill-conditioned to permit interchange of the order of differentiation in the second step of (2.31)—rather than a physical inconsistency. (In general, the Laplacian of the density is evaluated at a point in space that is determined by the electron density itself.)

Returning to the Weizsäcker form, the density matrix corresponding to the Weizsäcker functional is,

$$\tilde{\gamma}_{\sigma\sigma}(\mathbf{r},\mathbf{r}') = \sqrt{\rho_\sigma(\mathbf{r})\rho_\sigma(\mathbf{r}')}. \tag{2.32}$$

This density matrix usually violates the Pauli principle (it is not N-representable) because there are N_σ electrons in the first natural orbital [51]. This is why the correction to the Weizsäcker functional is often made using the Pauli potential [23, 24, 52, 53].

It is interesting, and disappointing, that even though the Weizsäcker model for the density matrix, (2.32), is very simple, the quasiprobability distribution cannot generally be expressed in closed form, even for the simplest $g(\boldsymbol{\theta}, \boldsymbol{\tau}) = 1$ case (corresponding to the Wigner distribution). To understand this, consider that even for the ground state of the hydrogen atom; the Wigner distribution has a very complicated analytic form [54–56]. Thus, while the naïveté of the Weizsäcker form is obvious from the mathematical form of the reduced density matrix, the momentum distribution function, $\tilde{\Pi}_\sigma^{(a)}(\mathbf{p})$, and quasiprobability distribution function have complicated forms that seem to obscure the inappropriateness of this choice and for which, in general, there is no explicit expression.

2.5 Extension to Many-Electron Distribution Functions

As attempts to find accurate, variational stable, explicit kinetic energy functionals of the electron density have so far been unsuccessful, it is reasonable to consider descriptors that contain more information that the electron density. In *ab initio* quantum chemistry, there is a hierarchy of methods, starting with single-particle methods, then electron pair methods, etc. There is also a hierarchy of k-density functional theories based on the electron distribution functions, [3, 57, 58] starting with the electron density ($k = 1$) and moving to the pair density ($k = 2$) and even higher-order electron distribution functions. The most common of these "extended" density-functional theories is based on electron pair density, [59–61]

$$\rho_{\sigma\sigma'}^{(2)}(\mathbf{r}, \mathbf{r}') = \langle \Psi | \sum_{i=1}^{N} \sum_{\substack{j=1 \\ j \neq i}}^{N} |\sigma(i)\sigma'(j)\rangle \delta(\mathbf{r}_i - \mathbf{r}) \delta(\mathbf{r}_j - \mathbf{r}') \langle \sigma'(j)\sigma(i)| |\Psi\rangle. \tag{2.33}$$

The theoretical properties of the kinetic energy functional of the pair density have been thoroughly explored [3, 59, 62–64]; some practical formulas have also been presented [2, 3, 65–73]. The most popular functional seems to be the two-electron Weizsäcker function, [2, 3, 66, 74, 75]

$$\begin{aligned}
T_w^{(2)}[\rho_{\sigma\sigma'}^{(2)}] &= \sum_{\sigma=\alpha,\beta} \frac{1}{N_\sigma - 1} \iint \frac{\nabla_1 \rho_{\sigma\sigma}^{(2)}(\mathbf{r}_1, \mathbf{r}_2) \cdot \nabla_1 \rho_{\sigma\sigma}^{(2)}(\mathbf{r}_1, \mathbf{r}_2)}{8\rho_{\sigma\sigma}^{(2)}(\mathbf{r}_1, \mathbf{r}_2)} d\mathbf{r} d\mathbf{r}' \\
&= \sum_{\sigma=\alpha,\beta} N_\sigma \iint \frac{\nabla_1 p_{\sigma\sigma}^{(2)}(\mathbf{r}_1, \mathbf{r}_2) \cdot \nabla_1 p_{\sigma\sigma}^{(2)}(\mathbf{r}_1, \mathbf{r}_2)}{8 p_{\sigma\sigma}^{(2)}(\mathbf{r}_1, \mathbf{r}_2)} d\mathbf{r} d\mathbf{r}' \\
&= \sum_{\sigma=\alpha,\beta} \frac{N_\sigma}{2} \iint \frac{\nabla_\tau p_{\sigma\sigma}^{(2)}(\tau) \cdot \nabla_\tau p_{\sigma\sigma}^{(2)}(\tau)}{8 p_{\sigma\sigma}^{(2)}(\tau)} d\tau.
\end{aligned} \tag{2.34}$$

The second equality in (2.34) uses the unit-normalized many-electron shape functions,

$$p_{\sigma\sigma'}(\mathbf{r},\mathbf{r}') = \begin{cases} \frac{\rho_{\sigma\sigma'}(\mathbf{r},\mathbf{r}')}{N_\sigma N_{\sigma'}}, & \sigma \neq \sigma', \\ \frac{\rho_{\sigma\sigma'}(\mathbf{r},\mathbf{r}')}{N_\sigma (N_\sigma - 1)}, & \sigma = \sigma'. \end{cases} \tag{2.35}$$

The third equality in (2.34) uses the 6-dimensional gradient. This form is especially useful for extending the Weizsäcker functional to the higher-order electron distribution functions needed in general k-density functional theories,

$$T_w^{(k)}[\rho^{(k)}_{\sigma_1\sigma_2\ldots\sigma_k}] = \sum_{\sigma=\alpha,\beta} \frac{N_\sigma}{k} \iint \frac{\nabla_\tau p^{(k)}_{\sigma\sigma\cdots\sigma}(\tau) \cdot \nabla_\tau p^{(k)}_{\sigma\sigma\cdots\sigma}(\tau)}{8 p^{(k)}_{\sigma\sigma\cdots\sigma}(\tau)} d\tau. \tag{2.36}$$

The extended Weizsäcker functionals from (2.36) form an increasing sequence of lower bounds to the exact kinetic energy,

$$T_w^{(1)} \leq T_w^{(2)} \leq T_w^{(3)} \leq \cdots \leq T_w^{(N)} = T_{\text{exact}} \tag{2.37}$$

with the $k = N$ formula being exact for any N-electron distribution function that arises from a real-valued wavefunction [3]. Unfortunately, this series of bounds converges slowly [76].

Comparing (2.28) and (2.36) it is clear that the entire analysis from the previous section can be extended to many-electron distribution functions. In particular, we can define a k-particle spin-momentum probability distribution function by

$$\begin{aligned} \pi^{(a,k)}_{\sigma_1\sigma_2\ldots\sigma_k}(\mathbf{P}) &= \left(\frac{1}{2\pi}\right)^{3k/2} \int e^{i\mathbf{P}\cdot\tau} (p^{(k)}_{\sigma_1\sigma_2\cdots\sigma_k}(\tau))^a d\tau, \\ \tilde{\Pi}^{(a,k)}_{\sigma_1\sigma_2\cdots\sigma_k}(\mathbf{P}) &= \frac{1}{2}\big(\pi^{(a,k)}_{\sigma_1\sigma_2\ldots\sigma_k}(\mathbf{P})[\pi^{(1-a,k)}_{\sigma_1\sigma_2\ldots\sigma_k}(\mathbf{P})]^* \\ &\quad + [\pi^{(a,k)}_{\sigma_1\sigma_2\ldots\sigma_k}(\mathbf{P})]^* \pi^{(1-a,k)}_{\sigma_1\sigma_2\ldots\sigma_k}(\mathbf{P})\big). \end{aligned} \tag{2.38}$$

One of the many possible expressions for the kinetic energy that can be written using these approximate k-particle momentum distributions is,

$$\tilde{T}^{(k)}[\tilde{\Pi}^{(a,k)}_{\sigma_1\sigma_2\cdots\sigma_k}] = \sum_{\sigma=\alpha,\beta} \frac{N_\sigma}{k} \int \frac{1}{2}(\mathbf{P}\cdot\mathbf{P}) \tilde{\Pi}^{(a,k)}_{\sigma\sigma\cdots\sigma}(\mathbf{P}) d\mathbf{P}. \tag{2.39}$$

As before, the most accurate functional is obtained for $a = 1/2$, which is the Fourier-transformed form of the extended Weizsäcker functional in (2.36).

The other aspects of the analysis of one-particle Weizsäcker kinetic energy functional also generalize. For example, the k-electron reduced density matrix that is implicit in (2.36) is

$$\tilde{\Gamma}^{(k)}_{\sigma\cdots\sigma;\sigma\cdots\sigma}(\tau,\tau') = \binom{N_\sigma}{k} \sqrt{p^{(k)}_{\sigma\sigma\cdots\sigma}(\tau)\, p^{(k)}_{\sigma\sigma\cdots\sigma}(\tau')}. \tag{2.40}$$

This density matrix is usually not N-representable because the maximum occupation number of a k-particle state is $\binom{N_\sigma}{k}\frac{1}{N_\sigma-k+1}$ [77–79]. The extended Weizsäcker functionals can be exact only when the number of electrons of a given spin is equal to the number of electrons in the functional, k.

2.6 Summary

The expression for the Fisher information in coordinate space is similar to the form of the Weizsäcker kinetic energy functional; compare (2.4) and (2.13). The expression for the Fisher information in reciprocal (momentum) space is reminiscent of the quantum mechanical kinetic energy; compare (2.7) and (2.14). These similarities motivated us to find a momentum representation for the Weizsäcker functional and to, moreover, attempt to generalize the Weizsäcker functional. The form of momentum distribution in (2.21) and (2.22) recovers the Weizsäcker functional for $a = 1/2$. Unfortunately, even though this family of momentum densities gives a generalized Weizsäcker function (2.28), all of the other functionals in this family are less accurate than the Weizsäcker functional.

Is it possible to generalize (2.22) still further, so that we can perhaps obtain an improved kinetic energy density functional? We tried to use the more general form,

$$\tilde{\tilde{\Pi}}_\sigma^{(a,b,c,d)}(\mathbf{p}) = \frac{1}{2}\left((\pi_\sigma^{(a)}(\mathbf{p}))^c[(\pi_\sigma^{(b)}(\mathbf{p}))^d]^* + [(\pi_\sigma^{(a)}(\mathbf{p}))^c]^*(\pi_\sigma^{(b)}(\mathbf{p}))^d\right),$$
$$ac + bd = 1. \tag{2.41}$$

We were unable to find the position-space representation of the kinetic energy functional for this form. Perhaps some functionals in the extended family derived from (2.41) are more accurate than the usual Weizsäcker form. We note that approximating the momentum density by simply Fourier transforming the spatial electron density gives seemingly absurd, and certainly inaccurate, results.

The same reasoning applies to the extended Weizsäcker functionals that are used in what is often called k-density functional theory, where the fundamental descriptor of an electronic system is the k-electron distribution function. The extended Weizsäcker functionals are also proportional to the Fisher information of the k-electron distribution function, and they also have a compact momentum-space representation that follows directly from the momentum-space representation of the Fisher information density via the probability amplitude. To our knowledge, this is the first time that the momentum-space representation for the many-electron Weizsäcker family of functionals has been presented. We find it intuitively appealing that the second moment (ergo, the variance, and also the Weizsäcker kinetic energy) of the momentum is closely related to the Fisher information of a many-electron distribution function.

Acknowledgements Financial and computational support from NSERC, the Canada Research Chairs, the Sloan Foundation, and Sharcnet are acknowledged.

References

1. von Weizsäcker CF (1935) Zur Theorie der Kernmassen. Z Phys 96:431
2. Furche F (2004) Towards a practical pair-density functional theory for many-electron systems. Phys Rev A 70:022514
3. Ayers PW (2005) Generalized density functional theories using the k-electron densities: Development of kinetic energy functionals. J Math Phys 46:062107
4. Sears SB (1980) Applications of information theory in chemical physics. PhD thesis, University of North Carolina at Chapel Hill
5. Sears SB, Parr RG, Dinur U (1980) On the quantum-mechanic kinetic-energy as a measure of the information in a distribution. Isr J Chem 19(1–4):165
6. Fisher RA (1918) The correlation between relatives on the supposition of Mendelian inheritance. Trans R Soc Edinb 52:399
7. Frieden BR (1998) Physics from Fisher information: a unification. Cambridge University Press, Cambridge
8. Nalewajski RF (2008) Use of Fisher information in quantum chemistry. Int J Quant Chem 108:2230
9. Thomas LH (1927) The calculation of atomic fields. Proc Camb Philol Soc 23:542
10. Fermi E (1928) Eine statistische Methode zur Bestimmung einiger Eigenschaften des Atoms und ihre Anwendung auf die Theorie des periodischen Systems der Elemente. Z Phys 48:73
11. Wang WP, Parr RG, Murphy DR, Henderson GA (1976) Chem Phys Lett 43(3):409
12. Murphy DR (1981) Sixth-order term of the gradient expansion of the kinetic-energy density functional. Phys Rev A 24(4):1682
13. Wang YA, Govind N, Carter EA (1998) Orbital-free kinetic-energy functionals for the nearly free electron gas. Phys Rev B 58:13465
14. Wang YA, Carter EA, Schwartz SD (2000) Theoretical methods in condensed phase chemistry. Kluwer, Dordrecht, p 117
15. Garcia-Gonzalez PJ, Alvarellos E, Chacon E (1998) Kinetic-energy density functionals based on the homogeneous response function applied to one-dimensional fermion systems. Phys Rev A 57(6):4192
16. Garcia-Aldea D, Alvarellos JE (2007) Kinetic-energy density functionals with nonlocal terms with the structure of the Thomas-Fermi functional. Phys Rev A 76:052504
17. Ho GS, Huang C, Carter EA (2007) Describing metal surfaces and nanostructures with orbital-free density functional theory. Curr Opin Solid State Mater Sci 11:57
18. Chen HJ, Zhou AH (2008) Orbital-free density functional theory for molecular structure calculations. Numer Math-Theory Method Appl 1:1
19. March NH (2010) Some pointers towards a future orbital-free density functional theory of inhomogeneous electron liquids. Phys Chem Liq 48:141
20. Wesolowski TA (2004) Quantum chemistry 'Without orbitals'—an old idea and recent developments. Chimia 58(5):311
21. Garcia-Aldea D, Alvarellos JE (2007) Kinetic energy density study of some representative semilocal kinetic energy functionals. J Chem Phys 127:144109
22. Garcia-Aldea D, Alvarellos JE (2008) Fully nonlocal kinetic energy density functionals: A proposal and a general assessment for atomic systems. J Chem Phys 129:074103
23. Kozlowski PM, March NH (1989) Approximate density-external potential relation and the Pauli potential for systems with Coulombic interaction. Int J Quant Chem 36(6):741
24. Holas A, March NH (1991) Construction of the Pauli potential, Pauli energy, and effective potential from the electron density. Phys Rev A 44:5521
25. Gazquez JL, Robles JJ (1982) On the atomic kinetic energy functionals with full Weizsacker correction. Chem Phys 76:1467
26. Acharya PK, Bartolotti LJ, Sears SB, Parr RG (1980) An atomic kinetic energy functional with full Weizsacker correction. Proc Natl Acad Sci USA 77:6978
27. Garcia-Aldea D, Alvarellos JE (2008) Approach to kinetic energy density functionals: Nonlocal terms with the structure of the von Weizsäcker functional. Phys Rev A 77:022502

28. Zhou BJ, Carter EA (2005) First principles local pseudopotential for silver: Towards orbital-free density-functional theory for transition metals. J Chem Phys 122:184108
29. Zhou BJ, Ligneres VL, Carter EA (2005) Improving the orbital-free density functional theory description of covalent materials. J Chem Phys 122:044103
30. Zhou BJ, Wang YA (2006) Orbital-corrected orbital-free density functional theory. J Chem Phys 124:081107
31. King RA, Handy NC (2001) Kinetic energy functionals for molecular calculations. Mol Phys 99:1005
32. Chacon E, Alvarellos JE, Tarazona P (1985) Nonlocal kinetic energy functional for nonhomogeneous electron systems. Phys Rev B 32:7868
33. Chai JD, Weeks JD (2007) Orbital-free density functional theory: Kinetic potentials and ab initio local pseudopotentials. Phys Rev B 75:205122
34. Karasiev VV, Trickey SB, Harris FE (2006) Born-Oppenheimer interatomic forces from simple, local kinetic energy density functionals. J Comput-Aided Mater Des 13:111
35. Karasiev VV, Jones RS, Trickey SB, Harris FE (2009) Properties of constraint-based single-point approximate kinetic energy functionals. Phys Rev B 80:245120
36. Parr RG, Bartolotti LJ (1983) Some remarks on the density functional theory of few-electron systems. J Phys Chem 87:2810
37. Cedillo A (1994) A new representation for ground states and its Legendre transforms. Int J Quant Chem 52:231
38. Ayers PW (2000) Density per particle as a descriptor of Coulomb systems. Proc Natl Acad Sci USA 97:1959
39. Ayers PW, Cedillo A (2009) Chemical reactivity theory: A density functional view. Taylor and Francis, Boca Raton, p 269 (edited by PK Chattaraj)
40. Wigner E (1932) On the quantum correction for thermodynamic equilibrium. Phys Rev 40:749
41. Wigner EP, Yourgrau W, van der Merwe A (1971) Perspectives in quantum theory. MIT, Cambridge, p 25
42. Cohen L (1966) Can quantum mechanics be formulated as a classical probability theory? Philos Sci 33:317
43. Cohen L (1979) Local kinetic energy in quantum mechanics. J Chem Phys 70:788
44. Cohen L (1984) Representable local kinetic energy. J Chem Phys 80:4277
45. Ghosh SK, Berkowitz M, Parr RG (1984) Transcription of ground-state density-functional theory into a local thermodynamics. Proc Natl Acad Sci USA 81:8028
46. Ayers PW, Parr RG, Nagy A (2002) Local kinetic energy and local temperature in the density-functional theory of electronic structure. Int J Quant Chem 90:309
47. Cohen L (1966) Generalized phase-space distribution functions. J Math Phys 7:781
48. Anderson JSM, Rodriguez Hernandez JI, Ayers PW (2010) How ambiguous is the local kinetic energy? J Phys Chem A 114:8884–8895
49. Hoffmann-Ostenhof M, Hoffmann-Ostenhof T (1977) "Schrödinger inequalities" and asymptotic behavior of the electron density of atoms and molecules. Phys Rev A 16:1782
50. Sagvolden E, Perdew JP (2008) Discontinuity of the exchange-correlation potential: Support for assumptions used to find it. Phys Rev A 77:012517
51. Coleman J (1963) Structure of Fermion density matrices. Rev Mod Phys 35:668
52. March NH (1986) The local potential determining the square root of the ground-state electron density of atoms and molecules from the Schrödinger equation. Phys Lett A 113(9):476
53. Levy M, Ouyang H (1988) Exact properties of the Pauli potential for the square root of the electron density and the kinetic energy functional. Phys Rev A 38(2):625
54. Dahl JP, Springborg M (1982) Wigner's phase space function and atomic structure. Mol Phys 47:1001
55. Dahl JP, Springborg M (1999) Comment on "Wigner phase-space distribution function for the hydrogen atom". Phys Rev A 59:4099
56. Mostowski LJ, Wodkiewicz K (2006) Hydrogen atom in phase space: the Wigner representation. J Phys A, Math Gen 39:14143

57. Ayers PW (2006) Using the classical many-body structure to determine electronic structure: an approach using k-electron distribution functions. Phys Rev A 74:042502
58. Ayers PW (2008) Constraints for hierarchies of many-electron distribution functions. J Math Chem 44:311
59. Ziesche P (1994) Pair density functional theory—a generalized density functional theory. Phys Lett A 195:213
60. Ziesche P (1996) Attempts toward a pair density functional theory. Int J Quant Chem 60:1361
61. Ayers PW, Davidson ER (2007) Linear inequalities for diagonal elements of density matrices. Adv Chem Phys 134:443
62. Levy M, Ziesche P (2001) The pair density functional of the kinetic energy and its simple scaling property. J Chem Phys 115:9110
63. Ayers PW, Levy M (2005) Generalized density-functional theory: conquering the N-representability problem with exact functionals for the electron pair density and the second-order reduced density matrix. J Chem Sci 117:507
64. Ayers PW, Golden S, Levy M (2006) Generalizations of the Hohenberg-Kohn theorem. J Chem Phys 124:054101
65. March NH, Santamaria R (1991) Non-local relation between kinetic and exchange energy densities in Hartree-Fock theory. Int J Quant Chem 39(4):585
66. Nagy A (2002) Density-matrix functional theory. Phys Rev A 66:022505
67. Ayers PW, Levy M (2005) Using the Kohn-Sham formalism in pair density-functional theories. Chem Phys Lett 416:211
68. Higuchi M, Higuchi K (2007) A proposal of the approximate scheme for calculating the pair density. Physica B, Condens Matter 387:117
69. Higuchi M, Miyasita M, Kodera M, Higuchi K (2007) Simultaneous equations for calculating the pair density. J Phys, Condens Matter 19:365219
70. Higuchi M, Miyasita M, Kodera M, Higuchi K (2007) Density functional scheme for calculating the ground-state pair density. J Magn Magn Mater 310:990
71. Higuchi M, Higuchi K (2007) Pair density-functional theory by means of the correlated wave function. Phys Rev A 75:042510
72. Higuchi M, Higuchi K (2008) Pair density functional theory utilizing the non-interacting reference system: An effective initial theory. Phys Rev B 78:125101
73. Higuchi K, Higuchi M (2009) J Phys, Condens Matter 21:064206
74. Nagy A, Amovilli C (2004) Effective potential in density matrix functional theory. J Chem Phys 121:6640
75. Nagy A (2006) Spherically and system-averaged pair density functional theory. J Chem Phys 125:184104
76. Chakraborty D, Ayers PW (2011) Failure of the Weizsäcker kinetic energy functionals for higher-order electron distribution functions. J Math Chem (accepted). doi:10.1007/s10910-011-9860-1
77. Sasaki F (1965) Eigenvalues of Fermion density matrices. Phys Rev 138:B1338
78. Davidson ER (1976) Reduced density matrices in quantum chemistry. Academic Press, New York
79. Coleman AJ, Yukalov VI (2000) Reduced density matrices: Coulson's challenge. Springer, Berlin

Chapter 3
Atomic Statistical Complexity

C.P. Panos, K.C. Chatzisavvas, C.C. Moustakidis, N. Nikolaidis, S.E. Massen, and K.D. Sen

Abstract Applications of the Shannon information entropy, Fisher measure, Onicescu energy, and two of the simple statistical complexity measures on atoms are presented using the Hartree-Fock density in position and momentum space. Interesting correlations with the experimental properties such as the ionization potential and static dipole polarizability are reported using the information measures. The statistical complexity analysis reveals interesting insights into the shell structure of electron density. The net Shannon information entropy is found to obey an approximate linear dependence on $\ln N$, where N gives the number of particles in the quantum system.

3.1 Introduction

Information-theoretical measures have been extensively used in recent years to study a variety of quantum mechanical systems [1]. A large number of studies have remained focused on atomic systems [2–20]. In this chapter we shall present a review of the results obtained in this area by the *Thessaloniki group*. We shall begin our presentation with the definition of the Shannon [21, 22], Fisher [23] and Onicescu [24] information measures followed by two of the complexity measures defined by Shiner, Davison, Landsberg (SDL) [25], and López-Ruiz, Mancini, Calbet (LMC) [26–28], respectively. Following this, in each case, the applications of these measures for atoms and other quantum systems will be discussed. Throughout our calculations we have used the electron probability density derived from the wave functions of Hartree-Fock (HF) quality [29].

C.P. Panos (✉)
Department of Theoretical Physics, Aristotle University of Thessaloniki, 54124 Thessaloniki, Greece
e-mail: chpanos@auth.gr

K.D. Sen (ed.), *Statistical Complexity*,
DOI 10.1007/978-90-481-3890-6_3, © Springer Science+Business Media B.V. 2011

3.2 Shannon Information Entropy

The Shannon information entropy S_r of the electron density $\rho(\mathbf{r})$ in coordinate space is defined as

$$S_r = -\int \rho(\mathbf{r}) \ln \rho(\mathbf{r})\, d\mathbf{r}, \tag{3.1}$$

and the corresponding momentum space entropy S_k is given by

$$S_k = -\int n(\mathbf{k}) \ln n(\mathbf{k})\, d\mathbf{k}, \tag{3.2}$$

where $n(\mathbf{k})$ denotes the momentum density. The densities $\rho(\mathbf{r})$ and $n(\mathbf{k})$ are respectively normalized to unity and all quantities are given in atomic units. The Shannon entropy sum $S_T = S_r + S_k$ contains the net information and obeys the well known lower bound due to Bialynicki-Birula and Mycielski [3] defining the entropic uncertainty relation (EUR). It represents a stronger version of the Heisenberg uncertainty principle of quantum mechanics, in the sense that the EUR leads to Heisenberg relation, while the inverse is not true. Additionally, the right-hand side of Heisenberg relation depends on the quantum state of the system, while EUR does not. Accordingly, the entropy sum in D-dimensions satisfies the inequality [3, 30, 31]

$$S_T = S_r + S_k \geq D(1 + \ln \pi). \tag{3.3}$$

Individual entropies S_r and S_k depend on the units used to measure r and k respectively, but their sum S_T does not i.e. it is invariant to uniform scaling of coordinates. Further, we note that S_T, S_r and S_k obey the following rigorous inequalities [9]

$$S_{r\,\min} \leq S_r \leq S_{r\,\max}, \tag{3.4}$$

$$S_{k\,\min} \leq S_k \leq S_{k\,\max}, \tag{3.5}$$

$$S_{T\,\min} \leq S \leq S_{T\,\max}. \tag{3.6}$$

The lower and the upper limits can be written, for density distributions normalized to one

$$\begin{aligned} S_{r\,\min} &= \frac{3}{2}(1 + \ln \pi) - \frac{3}{2} \ln\left(\frac{4}{3} K\right), \\ S_{r\,\max} &= \frac{3}{2}(1 + \ln \pi) + \frac{3}{2} \ln\left(\frac{2}{3} \langle r^2 \rangle\right), \end{aligned} \tag{3.7}$$

$$\begin{aligned} S_{k\,\min} &= \frac{3}{2}(1 + \ln \pi) - \frac{3}{2} \ln\left(\frac{2}{3} \langle r^2 \rangle\right), \\ S_{k\,\max} &= \frac{3}{2}(1 + \ln \pi) + \frac{3}{2} \ln\left(\frac{4}{3} K\right), \end{aligned} \tag{3.8}$$

$$\begin{aligned} S_{T\,\min} &= 3(1 + \ln \pi), \\ S_{T\,\max} &= 3(1 + \ln \pi) + \frac{3}{2} \ln\left(\frac{8}{9} \langle r^2 \rangle K\right), \end{aligned} \tag{3.9}$$

where $\langle r^2 \rangle$ is the mean square radius and K is the kinetic energy.

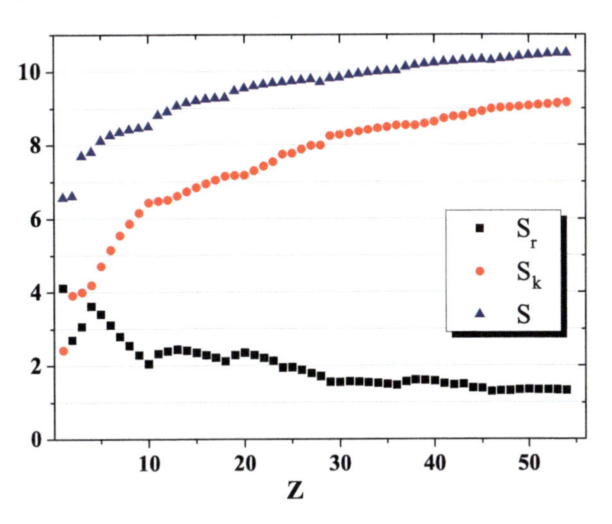

Fig. 3.1 Shannon information entropy in coordinate-space S_r, momentum space S_k and their sum S, as functions of Z

3.2.1 Approximate Linearity of S_T with $\ln N$

For various quantum systems such as atoms, nuclei, atomic clusters and bosons in a trap, it has been shown empirically [5, 6, 8] that an approximate relationship given by $S_T = a + b \ln N$ holds good. These are systems with widely ranging numbers of particles N, from a few to millions, and have various sizes, with constituent particles obeying different interactions and different statistics such as fermions and bosons. The parameters a, b depend on the system under consideration. S_T is connected with the kinetic energy [7]. We illustrate the N-dependence of the Shannon entropy sum using the numerical HF data on neutral atoms. In Fig. 3.1 we plot the Shannon information entropy for both the coordinate-space S_r, and momentum-space S_k as functions of the electron number Z. For the S_r values, we observe an average decreasing behavior carrying the shell structure. For atoms with the completely filled shells, such as He, Ne, Ar, Kr there exist minima in the curve $S_r(Z)$. This is due to the fact that compared to the neighboring atoms, $\rho(r)$ for these atoms is most compact. The values of S_k show a monotonic increase with Z. Interestingly, the behavior of S_k also reveals the local shell effect.

In the same figure we have displayed the total Shannon information entropy $S = S_T$ which is a monotonically increasing function of Z with just two exceptions corresponding to Ni and Pd. These exceptions are due to the fact that S_r and S_k depend on the arrangement of the electrons in shells. There is a delicate balance between S_r and S_k resulting in the general rule that $S_T = S_r + S_k$ is a monotonic increasing function of Z except in Ni and Pd where the electron arrangement in shells is such that the decrease of the value of S_r cannot be balanced by a corresponding increase of S_k. Thus the strict monotonicity of S_T is broken in such cases. A shell effect is also obvious in the behavior of S_T i.e. minima at closed shells. Figure 3.2 illustrates the trend of $S = S_T$ as a function of $\ln Z$. The corresponding linear fit is also plotted in the same figure where $S = 6.33 + 1.046 \ln Z$. It is noted that this result is not new but has been already obtained using other wave functions in [8, 9].

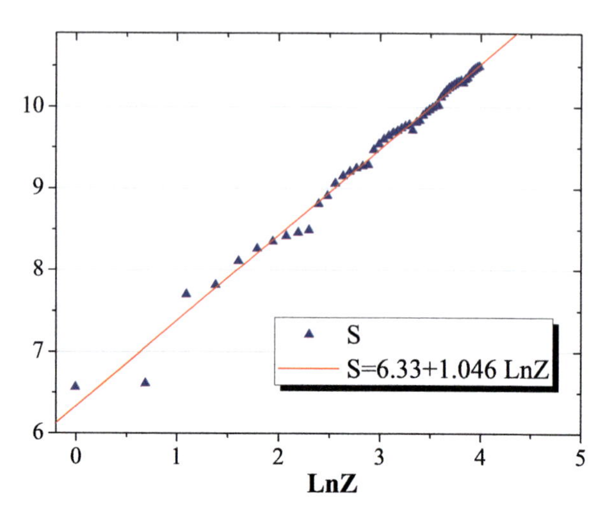

Fig. 3.2 The universal relation $S = a + b \ln N$ for atoms

In the present work we verify this result with RHF electron wave functions [29, 32] and we employ this framework for new calculations.

For a variety of other applications of Shannon information entropy on electronic structure we refer the reader to the literature [33–50].

3.3 The Fisher Information Measure

The Fisher information measure or intrinsic accuracy in position space is defined as

$$I_r = \int \frac{|\nabla \rho(\mathbf{r})|^2}{\rho(\mathbf{r})} d\mathbf{r}, \tag{3.10}$$

and the corresponding momentum space measure is given by

$$I_k = \int \frac{|\nabla n(\mathbf{k})|^2}{n(\mathbf{k})} d\mathbf{k}. \tag{3.11}$$

The individual Fisher measures are bounded through the Cramer-Rao inequality [51, 52] according to $I_r \geq \frac{1}{V_r}$ and $I_k \geq \frac{1}{V_k}$, where V's denote the corresponding spatial and momentum variances respectively. The Fisher information, in position space, measures the sharpness of probability density and for a Gaussian distribution is exactly equal to the inverse of the variance. The Fisher measure in this sense is complementary to the Shannon entropy and their reciprocal proportionality is, in fact, utilized in obtaining a correlation with the experimental ionization potentials in the next section. The Fisher measure has the desirable properties that it is always positive and reflects the localization characteristics of the density more sensitively than the Shannon information entropy [53, 54]. The lower bounds of Shannon sum $S_r + S_k$ and Fisher product $I_r I_k$ get saturated for the Gaussian distributions. For a variety of applications of the Fisher information measure we refer to the recent book

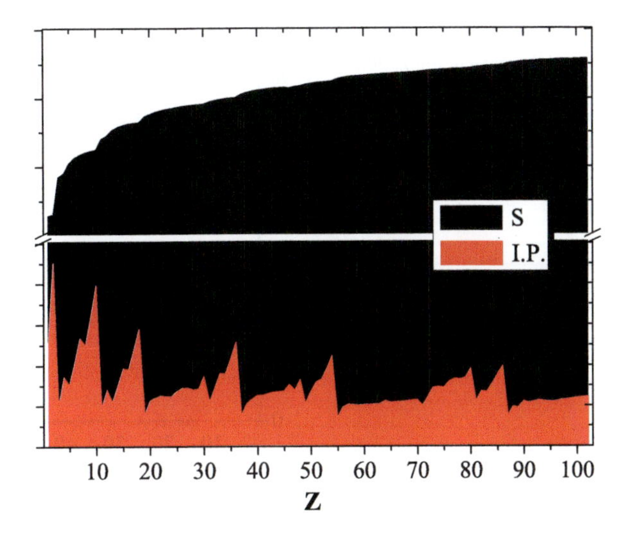

Fig. 3.3 Net Shannon information entropy S_T and the ionization potential as a function of nuclear charge Z of neutral atoms

[55] and for applications to the electronic structure of atoms, to the pioneering work of Dehesa and coworkers [56–61].

In the context of density functional theory (DFT), Sears, Parr and Dinur [62] were the first to underline the importance of Fisher information. These authors showed explicitly that the quantum mechanical kinetic energy is a measure of the information content of a distribution. The electron localization function [63] which has been widely successful in revealing the localization properties of electron density in molecules has been interpreted in terms of Fisher information [64, 65]. Recently, the Euler equation of density functional theory has been derived [66] from the principle of the minimum Fisher information within the time dependent versions. For the real wave functions in the position and momentum space, it has been shown that [67] the net or composite Fisher information measure, in D-dimensions, obeys the following lower bound

$$I_T = I_r I_k \geq 4D^2. \tag{3.12}$$

3.3.1 Ionization Potential, Dipole Polarizability and I_T

The important electronic properties of ionization potential and the static electric dipole polarizability of atoms correlate well with the composite Fisher measure, I_T. In Fig. 3.3 we first plot S_T and I.P. as functions of Z. It is observed that S_T does not perform as a sensitive information measure reproducing the details of the trends in I.P. It does show the gross atomic periodicity in terms of the shell structure as the humps.

In Fig. 3.4 the values of I_T (instead of S_T as in Fig. 3.3) and the *inverse* of I.P. are plotted as functions of Z. Compared to Fig. 3.3 the two curves in Fig. 3.4, resemble each other in far more details. It has been shown earlier that I_k behaves similarly to

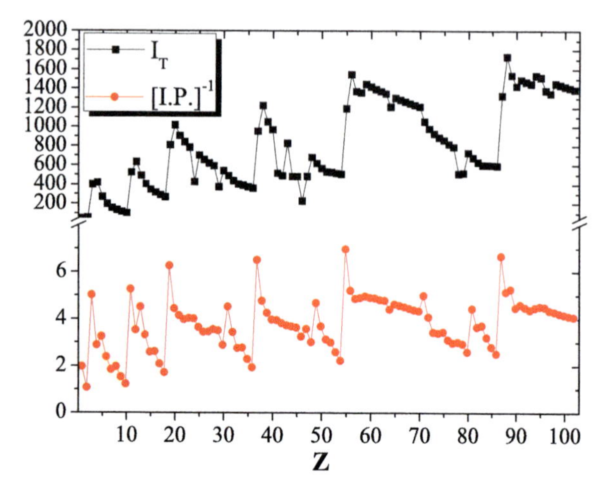

Fig. 3.4 Net Fisher measure I_T and the *inverse* ionization potential as a function of nuclear charge Z of neutral atoms

I.P. for atoms [68]. The net Fisher information amplifies the details of correlation by approximately two orders of magnitude. Our aim of plotting I_T versus inverse of I.P. is to show the similarities in the two curves in the *upward* direction and also lay emphasis on the significance of the net Fisher information, I_T. Very recently, the idea of taking the relative Shannon entropy of an element within a group of the periodic table with respect to the inert gas atom, located at the end of the group has been proposed [69] in order to get a more sensitive quantum similarity measure of density distributions. It would be interesting to investigate whether the correlations found in Fig. 3.4 can be further improved by using such a similarity measure for each group using the Fisher information according to

$$\Omega(Z) = 1 - \left[\frac{I_{T(\mathrm{ref})}}{I_{T(Z)}}\right], \tag{3.13}$$

where $\Omega(Z)$ measures the distance in compactness of the element Z from the most compact ideal gas atom in the same group, used as reference. A larger value of $\Omega(Z)$ would correspond to smaller I.P.

In Fig. 3.5 we present Ω and $[\mathrm{I.P.}]^{-1}$ as functions of Z. It is found that the correlation is more direct than that obtained in Fig. 3.4. This observation suggests that $\Omega(Z)$ can be used as a measure of quantum similarity of atoms and opens up a new application of the net Fisher information measure I_T.

We shall now consider the correlation of the static dipole polarizability of atoms with the information measures. The variation of polarizability α_d of atoms with S_T is found to be essentially similar to that of S_T versus I.P., as already given in Fig. 3.3. In the background of this rather mild sensitivity of S_T, we shall instead consider the correlation of I_T with the experimental estimates of polarizability α_d. The experimental values have been taken from the compilation of Miller and Bederson [70] for atoms with $Z = 1$–88. The variation of I_T and α_d with Z for atoms with $Z = 1$–88 is shown in Fig. 3.6. The overall correlation is found to be excellent with the maximum polarizability elements of the alkali atoms immediately following the sharply

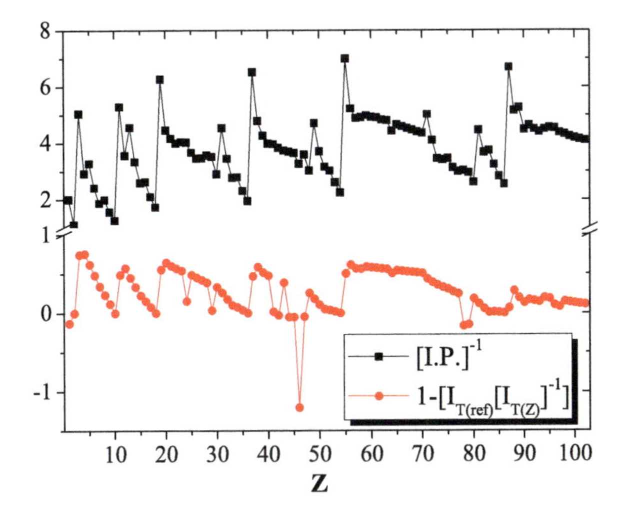

Fig. 3.5 A plot of Ω and $[\mathrm{I.P.}]^{-1}$ as functions of Z for neutral atoms

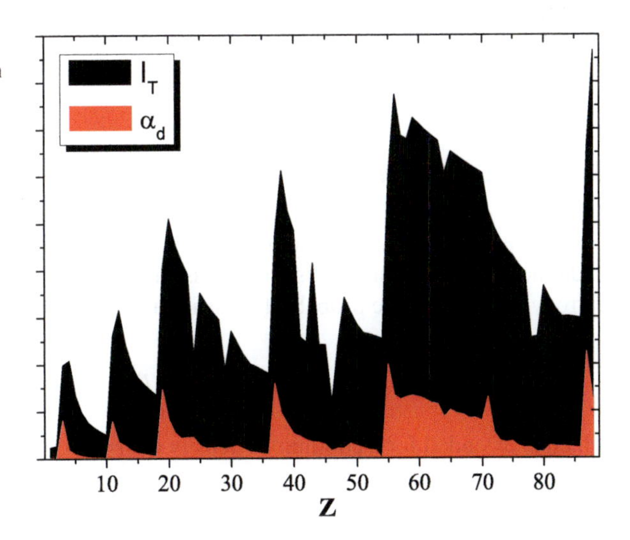

Fig. 3.6 The variation of I_T and α_d with Z for atoms with $Z = 1$–88

increasing values of I_T just after the inert gas atoms. The polarizability predicted by I_T for the alkaline atoms present themselves as the only examples which are not sufficiently well discriminated against the neighbouring atoms, in this case, the alkali metal atoms. It appears that the compactness described by I_T in going from the valence electron configuration of $(ns)^1$ to $(ns)^2$ does not increase sharply enough to quantitatively reflect the changes in polarizability from alkali to alkaline earth atoms. For these examples it is advisable to carry out further computations of the Fisher information using wave functions which include electron correlation effects.

3.4 Statistical Complexity of Atoms

The concept of complexity in chemistry in the graph theoretical context is described [71] in the literature. In this section we shall specifically discuss two of the statistical complexity measures and apply them to atoms in the periodic table. There are various measures of complexity in the literature. A quantitative measure of complexity is useful to estimate the ability of a variety of physical or biological systems for organization. According to [72] a complex world is interesting because it is highly structured. Some of the proposed measures of complexity are difficult to compute, although they are intuitively attractive, e.g. the algorithmic complexity [73, 74] defined as the length of the shortest possible program necessary to reproduce a given object. The fact that a given program is indeed the shortest one, is hard to prove. In contrast, there is a class of definitions of complexity, which can be calculated easily i.e. the *simple* measure of complexity $\Gamma_{\alpha,\beta}$ according to Shiner, Davison, Landsberg (SDL) [25], and the *statistical* measure of complexity C, defined by López-Ruiz, Mancini, Calbet (LMC) [26–28]. We shall be specially interested to know whether the features of periodicity in terms of the shell structure of the radial density is revealed by two of the simplest complexity measures, SDL and LMC when studied as functions of the atomic number Z. In order to define these measures of complexity we need to introduce the Onicescu information measure.

3.4.1 *Onicescu Information Measure*

The concept of information energy E was introduced by Onicescu [24] in an attempt to define a finer measure of dispersion distribution than that of Shannon information entropy . For a discrete probability distribution $(p_1, p_2, \ldots, p_k)$, E is defined as

$$E = \sum_{i=1}^{k} p_i^2, \tag{3.14}$$

which is extended for a continuous density distribution $\rho(x)$ as

$$E = \int \rho^2(x)\, dx. \tag{3.15}$$

So far, only the mathematical aspects of the concept have been developed, while the physical aspects have been ignored. A recent study of E for atomic nuclei has been carried out in [75].

The meaning of (3.15) can be seen by the following simple argument: For a Gaussian distribution of mean value μ, standard deviation σ and normalized density

$$\rho(x) = \frac{1}{\sqrt{2\pi}\sigma} \exp\left[-\frac{(x-\mu)^2}{2\sigma^2}\right], \tag{3.16}$$

relation (3.15) gives

$$E = \frac{1}{2\pi\sigma^2} \int_{-\infty}^{\infty} \exp\left[-\frac{(x-\mu)^2}{\sigma^2}\right] dx = \frac{1}{2\sigma\sqrt{\pi}}. \tag{3.17}$$

E is maximum if one of the $p_i's$ equals 1 and all the others are equal to zero i.e. $E_{\max} = 1$, while E is minimum when $p_1 = p_2 = \cdots = p_k = \frac{1}{k}$, hence $E_{\min} = \frac{1}{k}$ (total disorder). The fact that E becomes minimum for equal probabilities (total disorder), by analogy with thermodynamics, is the reason it has been called information energy, although it does not have the dimension of energy [76].

It is seen from (3.17) that the greater the information energy, the more concentrated is the probability distribution, while the information content decreases. Thus, one can define a measure of information content analogous to Shannon's S, by the relation

$$O = \frac{1}{E}. \tag{3.18}$$

Relation (3.15) is extended for a 3-dimensional spherically symmetric density distribution $\rho(r)$:

$$\begin{aligned} E_r &= \int_0^\infty \rho^2(r) 4\pi r^2\, dr, \\ E_k &= \int_0^\infty n^2(k) 4\pi k^2\, dk, \end{aligned} \tag{3.19}$$

in position and momentum space respectively, where $n(k)$ is the corresponding density distribution in momentum space.

E_r has dimension of inverse volume, while E_k of volume. Thus the product $E_r E_k$ is dimensionless and can serve as a measure of concentration (or information content) of a quantum system. It is also seen from (3.17) that E increases as σ decreases (or concentration increases) and Shannon's information entropy (or uncertainty) S decreases. Thus S and E are reciprocal. In order to be able to compare them, we redefine the quantity O by

$$O = \frac{1}{E_r E_k}, \tag{3.20}$$

as a measure of the information content of a quantum system in both position and momentum spaces.

3.4.2 SDL Measure of Complexity

Landsberg [77] defined the order parameter Ω (or disorder Δ) as

$$\Omega = 1 - \Delta = 1 - \frac{S}{S_{\max}}, \tag{3.21}$$

where S is the information entropy (actual) of the system and $S_{\max}$ the maximum entropy accessible to the system. Thus, the concepts of entropy and disorder are decoupled and it is possible for the entropy and order to increase simultaneously. It is noted that $\Omega = 1$ corresponds to perfect order and predictability, while $\Omega = 0$ means complete disorder and randomness.

In [25] a measure of complexity $\Gamma_{\alpha,\beta}$ was defined of the form

$$\Gamma_{\alpha,\beta} = \Delta^{\alpha}\Omega^{\beta} = \Delta^{\alpha}(1-\Delta)^{\beta} = \Omega^{\beta}(1-\Omega)^{\alpha}, \tag{3.22}$$

which is called the "simple complexity of disorder strength α and order strength β". When $\beta = 0$ and $\alpha > 0$ "complexity" is an increasing function of "disorder", and we have a measure of category I (Fig. 1 of [25]). When $\alpha = 0$ and $\beta > 0$, "complexity" is an increasing function of "order" and we have a measure of category III. When both α and β are non-vanishing and positive ($\alpha > 0$, $\beta > 0$), "complexity" vanishes at zero "disorder" and zero "order" and has a maximum of

$$(\Gamma_{\alpha,\beta})_{\max} = \alpha^{\alpha}\beta^{\beta}/(\alpha+\beta)^{(\alpha+\beta)}, \tag{3.23}$$

at $\Delta = \alpha/(\alpha+\beta)$ and $\Omega = \beta/(\alpha+\beta)$. This is complexity of category II according to [25].

Several cases for both α and β non-negative are shown in Fig. 2 of [25], where $\Gamma_{\alpha,\beta}$ is plotted as function of Δ. In the present work we can find $\Delta = S/S_{\max}$ or $\Omega = 1-\Delta$ as function of Z. Thus we are able to plot the dependence of $\Gamma_{\alpha,\beta}$ on the atomic number Z.

3.4.3 LMC Measure of Complexity

The complexity C is measured accordingly to the prescription due to Lopez-Ruiz, Manchini and Calbet (LMC) [26, 78] as

$$C = S_T E_T, \tag{3.24}$$

where $E_T = E_r E_k$.

S_T denotes the information content stored in the system and E_T corresponds to the disequilibrium of the system, i.e. the distance from its actual state to equilibrium, according to [26]. Shiner, Davison and Landsberg (SDL) [25] and LMC measures were criticised in [79–81]. A related discussion can be found in [18, 82].

In the light of its sensitivity to describe the localization property it is useful to consider in the above equation I_T^{-1} instead of S_T, to define LMC complexity measure based on the net Fisher information. Thus a new definition of complexity measure (LMC-like) is the following

$$C = E_T I_T^{-1}. \tag{3.25}$$

We note here that a slightly different measure of shape complexity, e.g., in position space [27, 83–85] can be defined as $C = \mathrm{e}^{S_r} E_r$, and has been used in the more recent literature.

In Fig. 3.7(a), (b) we plot the complexity measure $\Gamma_{\alpha,\beta}$ in atoms for various values of parameters α and β. It is seen that for all sets of α and β, $\Gamma_{\alpha,\beta}$ shows qualitatively the same trend as function of Z, i.e. it fluctuates around an average value and shows local minima for atoms with closed shells. These results compare favourably with intuition, i.e. complexity is less at closed shells which can be considered more compact than neighbouring nuclei and consequently less complex. It

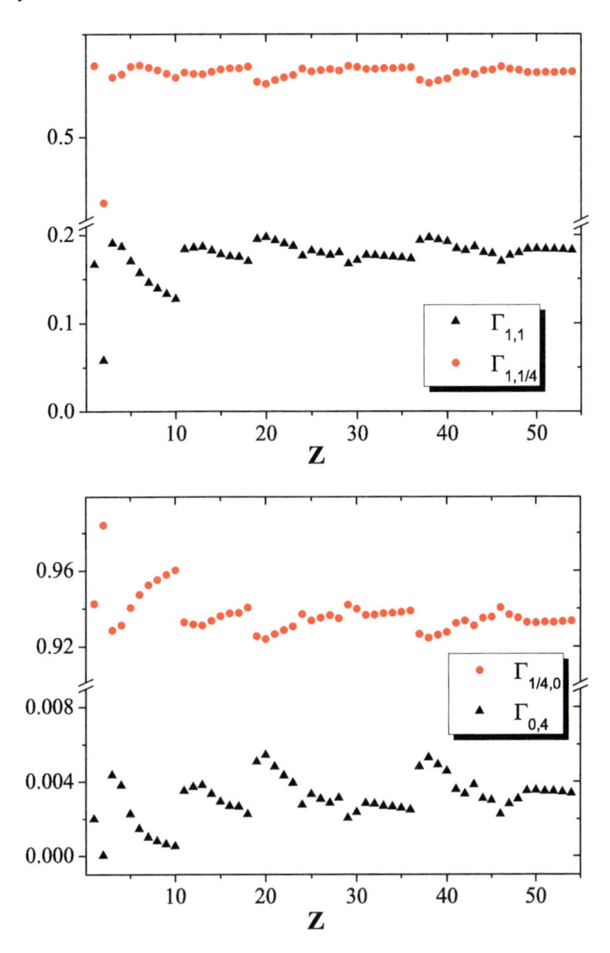

Fig. 3.7 SDL measure of complexity $\Gamma_{\alpha,\beta}$ for atoms, for various values of α and β, as functions of Z

is noted that this result comes from a procedure which is not trivial, i.e. first we calculate $\rho(\mathbf{r})$ and $n(\mathbf{k})$, second we find $S = S_r + S_k$ from the Shannon definition and S_{max} employing rigorous inequalities and third we obtain the complexity measure introduced in [25].

In Fig. 3.8 we plot in the same footing the Onicescu information content O and the ionization potential I.P. The first three maxima of O correspond to the fully closed shells (He, Ne and Ar) where in the case of the next closed shell (Kr) a local maximum exists. It is indicated that O and I.P. are correlated in the sense that there is a similarity in the trend of values of O and I.P. as functions of Z. This similarity is more obvious in regions of small Z where linear relations $O = a + b \ln Z$ can be extracted for regions Li-Ne and Na-Ar. However, it seems that, there is no universal relation between them. There are many entropic measures of spread of probability densities e.g. S, E, etc., but researchers prefer S because of its unique properties, while E was introduced by Onicescu as a sensitive measure of information. However, S and E are different functionals of the density and their relation is difficult to find.

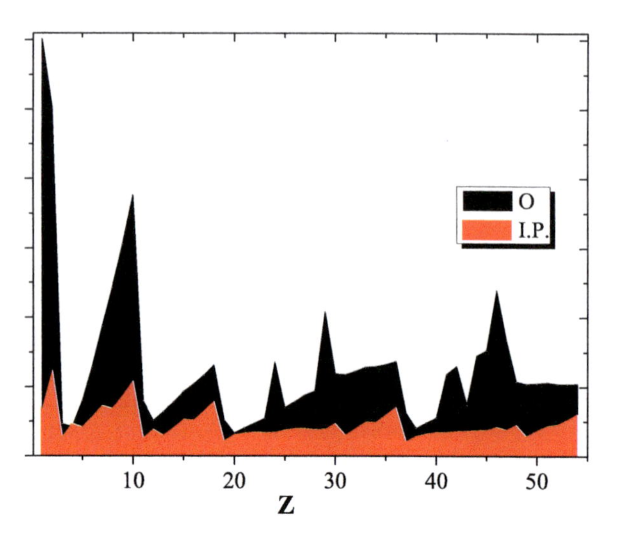

Fig. 3.8 Onisescu information O and first ionization potential I.P. (in Hartree units) as functions of Z

The dependence of $\Gamma_{\alpha,\beta}$ on Z for atoms has been calculated for the first time in [78]. In the present section we calculate $C(Z)$ employing the same RHF wave functions in the same region $1 \leq Z \leq 54$, for the sake of comparison. Our results are shown in Figs. 3.9 and 3.10. We compare them with $\Gamma_{\alpha,\beta}(Z)$ shown in Fig. 3 of [78] for $(\alpha,\beta) = (1,1)$, $(1,1/4)$, $(1/4,0)$, $(0,4)$. In all (six) cases we observe that the measures of complexity show local minima at closed shells atoms, namely for $Z = 10$ (Ne), 18 (Ar), 36 (Kr). The physical meaning of that behaviour is that the electron density for those atoms is the most compact one compared to neighbouring atoms. The local maxima can be interpreted as being far from the most compact distribution (low ionization potential systems). This does not contradict common sense and satisfies our intuition. There are also local minima for $Z = 24$ (Cr), 29 (Cu), 42 (Mo). Those minima are due to a specific change of the arrangement of electrons in shells. For example, going from $Z = 24$ (Cr) with electron configuration [Ar]$4s^1 3d^5$ to the next atom $Z = 25$ (Mn), with configuration [Ar]$4s^2 3d^5$, it is seen that one electron is added in an s-orbital (highest). The situation is similar for $Z = 29$ (Cu) and $Z = 42$ (Mo). The local minimum for $Z = 46$ (Pd) is due to the fact that Pd has a $4d^{10}$ electron configuration with extra stability of electron density. It has no 5s electron, unlike the neighbouring atoms. There are also fluctuations of the complexity measures within particular subshells. This behaviour can be understood in terms of screening effects within the subshell. The question naturally arises if the values of complexity correlate with properties of atoms in the periodic table. An example is the correlation of Onicescu information content O with the ionization potential (Fig. 4 of [78]). A more detailed/systematic study is needed, which is beyond the scope of the present report.

Our calculations in [78] show a dependence of complexity on the indices of disorder α and order β. In [78], we made a general comment that there are fluctuations of complexity around an average value and atoms cannot grow in complexity as Z increases. The second part of our comment needs to be modified. Various values

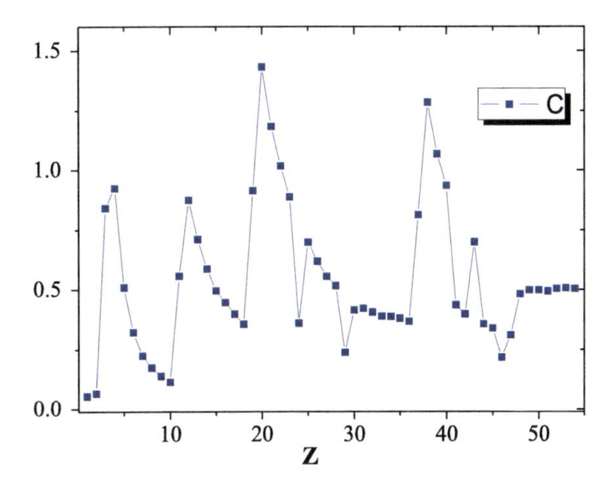

Fig. 3.9 LMC measure of complexity C as function of Z

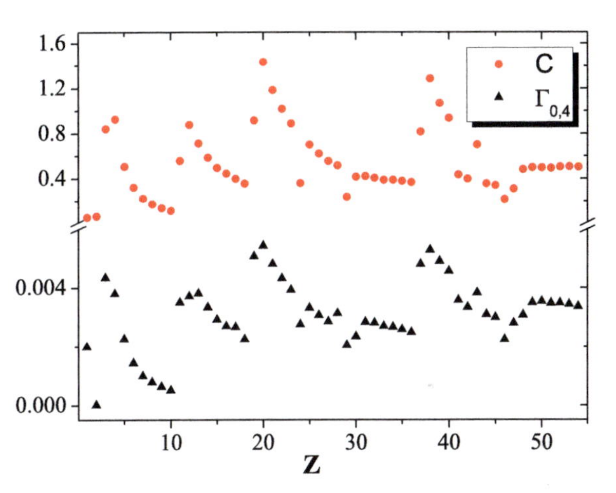

Fig. 3.10 LMC measure of complexity C and SDL measure $\Gamma_{0,4}$ as functions of Z

of (α, β) lead to different trends of $\Gamma_{\alpha,\beta}(Z)$ i.e. increasing, decreasing or approximately constant. In addition, in [82] we compare $C(Z)$ with $\Gamma_{\alpha,\beta}(Z)$ and we find a significant overall similarity between the curves $\Gamma_{0,4}(Z)$ and $C(Z)$ by plotting $C(Z)$ and $\Gamma_{0,4}(Z)$ in the same Fig. 3.10. The numerical values are different but a high degree of similarity is obvious by simple inspection. There is also the same succession of local maxima and minima at the same values of Z. Less striking similarities are observed for other values of (α, β) as well, e.g. $\Gamma_{1,1}(Z)$ and $C(Z)$.

Concluding, the behavior of SDL complexity depends on the values of the parameters α and β. The statistical measure LMC displays an increasing trend as Z increases. An effort to connect the aforementioned measures, implies that LMC measure corresponds to SDL when the magnitude of disorder $\alpha \simeq 0$ and of order $\beta \simeq 4$. In other words, if one insists that SDL and LMC behave similarly as functions of Z, then we can conclude that complexity shows an overall increasing behaviour with Z. Their correlation gives for atoms the strength of disorder $a \simeq 0$ and order $\beta \simeq 4$.

A final comment seems appropriate: An analytical comparison of the similarity of $\Gamma_{\alpha,\beta}(Z)$ and $C(Z)$ is not trivial. Combining (3.24) and (3.25) we find for the SDL measure

$$\Gamma_{\alpha,\beta}(Z) = \left(\frac{S}{S_{\max}}\right)^{\alpha} \left(1 - \frac{S}{S_{\max}}\right)^{\beta}, \tag{3.26}$$

while for the LMC one has

$$C = S \cdot (E_r E_k). \tag{3.27}$$

S and $S_{\max}$ depend on Z as follows

$$S(Z) = S_r(Z) + S_k(Z) = 6.33 + 1.046 \ln Z \tag{3.28}$$

(almost exact fitted expression) [8, 9, 78], while

$$S_{\max}(Z) = S_{r\max}(Z) + S_{k\max}(Z) = 7.335 + 1.658 \ln Z \tag{3.29}$$

(a rough approximation). We mention that S_r, S_k, E_r, E_k are known but different functionals of $\rho(\mathbf{r})$ and $n(\mathbf{k})$ according to relations (3.1), (3.2) and (3.20) respectively. It is noted that our numerical calculations were carried out with exact values of S_r, S_k, E_r, E_k, while our fitted expressions for $S(Z)$, $S_{\max}(Z)$ are presented in order to help the reader to appreciate approximately the trend of $\Gamma_{\alpha,\beta}(Z)$ and $C(Z)$.

Acknowledgements K.D.S. is very grateful to Professor Ioannis Antoniou for constant encouragement and support during his brief stay in the Thessaloniki group. K.Ch.Ch. has been supported by a Post-Doctoral Research Fellowship of the Hellenic State Institute of Scholarships (IKY).

References

1. Gadre SR (2002) Reviews of modern quantum chemistry. A celebration of the contributions of Robert G Parr, vol 1. World Scientific, Singapore, pp 108–147. Chapter: Information theoretical approaches to quantum chemistry
2. Ohya M, Petz D (1993) Quantum entropy and its use. Springer, Berlin, New York
3. Bialynicki-Birula I, Mycielski J (1975) Commun Math Phys 44:129
4. Panos CP, Massen SE (1997) Int J Mod Phys E 6:497
5. Massen SE, Moustakidis ChC, Panos CP (2002) Phys Lett A 64:131
6. Massen SE, Panos CP (1998) Phys Lett A 246:530
7. Massen SE, Panos CP (2001) Phys Lett A 280:65
8. Gadre SR, Sears SB, Chakravorty SJ, Bendale RD (1985) Phys Rev A 32:2602
9. Gadre SR, Bendale RD (1987) Phys Rev A 36:1932
10. Ghosh SK, Berkowitz M, Parr RG (1984) Proc Natl Acad Sci USA 81:8028
11. Lalazissis GA, Massen SE, Panos CP, Dimitrova SS (1998) Int J Mod Phys E 7:485
12. Moustakidis ChC, Massen SE, Panos CP, Grypeos ME, Antonov AN (2001) Phys Rev C 64:014314
13. Panos CP, Massen SE, Koutroulos CG (2001) Phys Rev C 63:064307
14. Panos CP (2001) Phys Lett A 289:287
15. Massen SE (2003) Phys Rev C 67:014314
16. Moustakidis ChC, Massen SE (2003) Phys Rev B 71:045102
17. Massen SE, Moustakidis ChC, Panos CP (2005) In: Ling AV (ed) Focus on Boson research. Nova Publishers, New York, p 115

18. Chatzisavvas KCh, Panos CP (2005) Int J Mod Phys E 14:653
19. Massen SE, Psonis VP, Antonov AN, e-print nucl-th/0502047
20. Sen KD, Panos CP, Chatzisavvas KCh, Moustakidis ChC (2007) Net Fisher information measure versus ionization potential and dipole polarizability in atoms. Phys Lett A 364:286
21. Shannon CE (1948) A mathematical theory of communication. Bell Syst Tech J 27:379–423
22. Shannon CE (1948) Bell Syst Tech J 27:623
23. Fisher RA (1925) Theory of statistical estimation. Proc Camb Philos Soc 22:700–725
24. Onicescu O (1966) C R Acad Sci Paris A 263:25
25. Shiner JS, Davison M, Landsberg PT (1999) Phys Rev E 59:1459
26. López-Ruiz R, Mancini HL, Calbet X (1995) Phys Lett A 209:321
27. Catalan RG, Garay J, Lopez-Ruiz R (2002) Phys Rev E 66:011102
28. Sánchez JR, López-Ruiz R (2005) Physica A 355:633
29. Koga T, Kanayama K, Watanabe S, Imai T, Thakkar AJ (2002) Theor Chem Acc 104:411
30. Sears SB (1980) Applications of information theory in chemical physics. PhD thesis, University of North Carolina at Chapel Hill
31. Yanez RJ, Van Assche W, Dehesa JS (1994) Phys Rev A 50:3065
32. Bunge CF, Barrientos JA, Bunge AV (1993) At Data Nucl Data Tables 53:113
33. Gadre SR (1984) Phys Rev A 30:620
34. Gadre SR, Bendale RD (1985) Int J Quant Chem 28:311
35. Gadre SR, Kulkarni SA, Shrivastava IH (1990) Chem Phys Lett 16:445
36. Gadre SR, Bendale RD, Gejji SP (1985) Chem Phys Lett 117:138
37. Gadre SR, Bendale RD (1985) Curr Sci (India) 54:970
38. Hô M, Sagar RP, Pérez-Jordá JM, Smith VH Jr, Esquivel RO (1994) Chem Phys Lett 219:15
39. Hô M, Schmider H, Sagar RP, Weaver DE, Smith VH Jr (1995) Int J Quant Chem 53:627
40. Tripathi AN, Smith VH Jr, Sagar RP, Esquivel RO (1996) Phys Rev A 54:1877
41. Ho M, Weaver DF, Smith VH Jr, Sagar RP, Esquivel RO (1998) Phys Rev A 57:4512
42. Ho M, Smith VH Jr, Weaver DF, Gatti C, Sagar RP, Esquivel RO (1998) J Chem Phys 108:5469
43. Ramirez JC, Perez JMH, Sagar RP, Esquivel RO, Ho M, Smith VH Jr (1998) Phys Rev A 58:3507
44. Ho M, Weaver DF, Smith VH Jr, Sagar RP, Esquivel RO, Yamamoto S (1998) J Chem Phys 109:10620
45. Hô M, Clark BJ, Smith VH, Weaver DF, Gatti C, Sagar RP, Esquivel RO (2000) J Chem Phys 112:7572
46. Sagar RP, Ramirez JC, Esquivel RO, Ho M, Smith VH Jr (2001) Phys Rev A 63:022509
47. Guevara NL, Sagar RP, Esquivel RO (2003) J Chem Phys 119:7030
48. Guevara NL, Sagar RP, Esquivel RO (2005) J Chem Phys 122:084101
49. Shi Q, Kais S (2004) J Chem Phys 121:5611
50. Shi Q, Kais S (2005) J Chem Phys 309:127
51. Rao CR (1965) Linear statistical interference and its applications. Wiley, New York
52. Stam A (1959) Inf Control 2:101
53. Carroll RW (2006) Fluctuations, information, gravity and quantum potential. Springer, Dordrecht, p 240
54. Garbacjewski P (2006) J Stat Phys 123:315, 317, 328
55. Frieden BR (2004) Science from Fisher information. Cambridge University Press, Cambridge
56. Dehesa JS, Martinez-Finkelstein A, Sanchez-Ruiz J (2001) J Comput Appl Math 133:23
57. Romera E, Sanchez-Moreno P, Dehesa JS (2005) Chem Phys Lett 414:468
58. Romera E, Dehesa JS (2004) J Chem Phys 120:8906
59. Dehesa JS, Martinez-Finkelstein A, Sorokin VN (2006) Mol Phys 104:613
60. Dehesa JS, Lopez-Rosa S, Olmos B, Yanez RJ (2006) J Math Phys 47:052104
61. Dehesa JS, Olmos B, Yanez RJ (2006) arXiv:math.CA/0606133v1
62. Sears SB, Parr RG, Dinur U (1980) Isr J Chem 19:165
63. Becke AD, Edgecombe KE (1990) J Chem Phys 92:5397
64. Nalewajski RF, Koester AM, Escalante S (2005) J Phys Chem A 109:10038

65. Nalewajski RF (2006) Information theory of molecular systems. Elsevier, Amsterdam
66. Nagy A (2003) J Chem Phys 119:9401
67. Sanchez-Moreno P, Plastino AR, Dehesa JS (2011) J Phys A 44:065301
68. Romera E (2002) Mol Phys 100:3325
69. Borgoo A, Godfroid M, Sen KD, De Proft F, Greelings P (2004) Chem Phys Lett 399:363
70. Miller TH, Bederson B (1977) Adv At Mol Opt Phys 13:1
71. Bonchev D (2003) Shannon information and complexity. In: Bonchev D, Rouvray DH (eds) Complexity in chemistry. Mathematical chemistry series, vol 7. Taylor and Francis, London, p 55. See also other chapters therein
72. Goldenfeld N, Kadanoff L (1999) Science 284:87
73. Kolmogorov AN (1965) Probl Inf Transm 1:3
74. Chaitin G (1966) J ACM 13:547
75. Moustakidis ChC, Chatzisavvas KCh, Panos CP (2005) Int J Mod Phys E 14:1087
76. Lepadatu C, Nitulescu E (2004) Acta Chim Slov 50:539
77. Landsberg PT (1984) Phys Lett A 102:171
78. Chatzisavvas KCh, Moustakidis ChC, Panos CP (2005) J Chem Phys 123:174111
79. Crutchfield J, Feldman DP, Shalizi CR (2000) Phys Rev E 62:2996
80. Feldman DP, Crutchfield J (1998) Phys Lett A 238:244
81. Stoop R, Stoop N, Kern A, Steeb W-H (2005) J Stat Mech: Theory Exp 11009
82. Panos CP, Chatzisavvas KC, Moustakidis CC, Kyrkou EG (2006) Phys Lett A 363:78
83. Pipek J, Varga I, Nagy T (1990) Int J Quant Chem 37:529
84. Pipek J, Varga I (1992) Phys Rev A 46:3148
85. Pipek J, Varga I (2002) Phys Rev E 68:026202. The structural entropy S_{str} introduced here as a localization quantity characteristic of the decay of the distribution function is related to the shape complexity as $\ln C$

Chapter 4
Statistical Complexity and Fisher-Shannon Information: Applications

Ricardo López-Ruiz, Jaime Sañudo, Elvira Romera, and Xavier Calbet

Abstract In this chapter, a statistical measure of complexity and the Fisher-Shannon information product are introduced and their properties are discussed. These measures are based on the interplay between the Shannon information, or a function of it, and the separation of the set of accessible states to a system from the equiprobability distribution, i.e. the disequilibrium or the Fisher information, respectively. Different applications in discrete and continuous systems are shown. Some of them are concerned with quantum systems, from prototypical systems such as the H-atom, the harmonic oscillator and the square well to other ones such as He-like ions, Hooke's atoms or just the periodic table. In all of them, these statistical indicators show an interesting behavior able to discern and highlight some conformational properties of those systems.

4.1 A Statistical Measure of Complexity. Some Applications

This century has been told to be the century of *complexity* [1]. Nowadays the question *"what is complexity?"* is circulating over the scientific crossroads of physics, biology, mathematics and computer science, although under the present understanding of the world could be no urgent to answer this question. However, many different points of view have been developed to this respect and hence a lot of different answers can be found in the literature. Here we explain in detail one of these options.

On the most basic grounds, an object, a procedure, or system is said to be "complex" when it does not match patterns regarded as simple. This sounds rather like an oxymoron but common knowledge tells us what is simple and complex: simplified systems or idealizations are always a starting point to solve scientific problems. The notion of "complexity" in physics [2, 3] starts by considering the perfect crystal

R. López-Ruiz (✉)
Department of Computer Science, Faculty of Science and BIFI, University of Zaragoza, Zaragoza 50009, Spain
e-mail: rilopez@unizar.es

K.D. Sen (ed.), *Statistical Complexity*,
DOI 10.1007/978-90-481-3890-6_4, © Springer Science+Business Media B.V. 2011

and the isolated ideal gas as examples of simple models and therefore as systems with zero "complexity". Let us briefly recall their main characteristics with "order", "information" and "equilibrium".

A perfect crystal is completely ordered and the atoms are arranged following stringent rules of symmetry. The probability distribution for the states accessible to the perfect crystal is centered around a prevailing state of perfect symmetry. A small piece of "information" is enough to describe the perfect crystal: the distances and the symmetries that define the elementary cell. The "information" stored in this system can be considered minimal. On the other hand, the isolated ideal gas is completely disordered. The system can be found in any of its accessible states with the same probability. All of them contribute in equal measure to the "information" stored in the ideal gas. It has therefore a maximum "information". These two simple systems are extrema in the scale of "order" and "information". It follows that the definition of "complexity" must not be made in terms of just "order" or "information".

It might seem reasonable to propose a measure of "complexity" by adopting some kind of distance from the equiprobable distribution of the accessible states of the system. Defined in this way, "disequilibrium" would give an idea of the probabilistic hierarchy of the system. "Disequilibrium" would be different from zero if there are privileged, or more probable, states among those accessible. But this would not work. Going back to the two examples we began with, it is readily seen that a perfect crystal is far from an equidistribution among the accessible states because one of them is totally prevailing, and so "disequilibrium" would be maximum. For the ideal gas, "disequilibrium" would be zero by construction. Therefore such a distance or "disequilibrium" (a measure of a probabilistic hierarchy) cannot be directly associated with "complexity".

In Fig. 4.1 we sketch an intuitive qualitative behavior for "information" H and "disequilibrium" D for systems ranging from the perfect crystal to the ideal gas. This graph suggests that the product of these two quantities could be used as a measure of "complexity": $C = H \cdot D$. The function C has indeed the features and asymtotical properties that one would expect intuitively: it vanishes for the perfect crystal and for the isolated ideal gas, and it is different from zero for the rest of the systems of particles. We will follow these guidelines to establish a quantitative measure of "complexity".

Before attempting any further progress, however, we must recall that "complexity" cannot be measured univocally, because it depends on the nature of the description (which always involves a reductionist process) and on the scale of observation. Let us take an example to illustrate this point. A computer chip can look very different at different scales. It is an entangled array of electronic elements at microscopic scale but only an ordered set of pins attached to a black box at a macroscopic scale.

We shall now discuss a measure of "complexity" based on the statistical description of systems. Let us assume that the system has N accessible states $\{x_1, x_2, \ldots, x_N\}$ when observed at a given scale. We will call this an N-system.

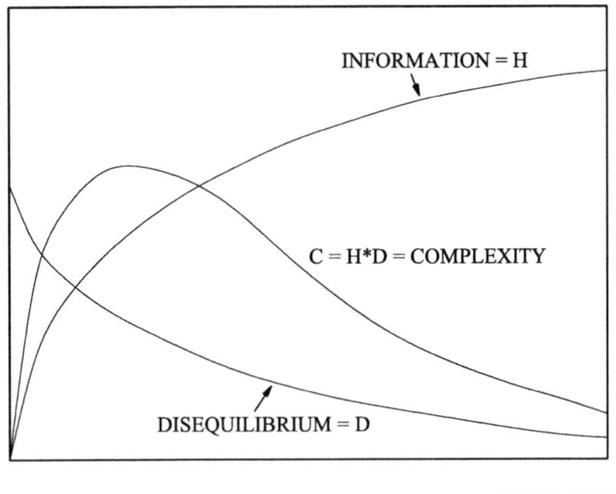

Fig. 4.1 Sketch of the intuitive notion of the magnitudes of "information" (H) and "disequilibrium" (D) for the physical systems and the behavior intuitively required for the magnitude "complexity". The quantity $C = H \cdot D$ is proposed to measure such a magnitude

Our understanding of the behavior of this system determines the corresponding probabilities $\{p_1, p_2, \ldots, p_N\}$ (with the condition $\sum_{i=1}^{N} p_i = 1$) of each state ($p_i > 0$ for all i). Then the knowledge of the underlying physical laws at this scale is incorporated into a probability distribution for the accessible states. It is possible to find a quantity measuring the amount of "information". Under to the most elementary conditions of consistency, Shannon [4] determined the unique function $H(p_1, p_2, \ldots, p_N)$ that accounts for the "information" stored in a system:

$$H = -K \sum_{i=1}^{N} p_i \log p_i, \tag{4.1}$$

where K is a positive constant. The quantity H is called *information*. The redefinition of information H as some type of monotone function of the Shannon entropy can be also useful in many contexts as we shall show in the next sections. In the case of a crystal, a state x_c would be the most probable $p_c \sim 1$, and all others x_i would be very improbable, $p_i \sim 0$, $i \neq c$. Then $H_c \sim 0$. On the other side, equiprobability characterizes an isolated ideal gas, $p_i \sim 1/N$ so $H_g \sim K \log N$, i.e., the maximum of information for a N-system. (Notice that if one assumes equiprobability and $K = \kappa \equiv$ *Boltzmann constant*, H is identified with the thermodynamic entropy, $S = \kappa \log N$). Any other N-system will have an amount of information between those two extrema.

Let us propose a definition of *disequilibrium* D in a N-system [5]. The intuitive notion suggests that some kind of distance from an equiprobable distribution should be adopted. Two requirements are imposed on the magnitude of D: $D > 0$ in order to have a positive measure of "complexity" and $D = 0$ on the limit of equiprobability.

The straightforward solution is to add the quadratic distances of each state to the equiprobability as follows:

$$D = \sum_{i=1}^{N} \left(p_i - \frac{1}{N} \right)^2 . \tag{4.2}$$

According to this definition, a crystal has maximum disequilibrium (for the dominant state, $p_c \sim 1$, and $D_c \to 1$ for $N \to \infty$) while the disequilibrium for an ideal gas vanishes ($D_g \sim 0$) by construction. For any other system D will have a value between these two extrema.

We now introduce the definition of *complexity* C of a N-system [6, 7]. This is simply the interplay between the information stored in the system and its disequilibrium:

$$C = H \cdot D = -\left(K \sum_{i=1}^{N} p_i \log p_i \right) \cdot \left(\sum_{i=1}^{N} \left(p_i - \frac{1}{N} \right)^2 \right) . \tag{4.3}$$

This definition fits the intuitive arguments. For a crystal, disequilibrium is large but the information stored is vanishingly small, so $C \sim 0$. On the other hand, H is large for an ideal gas, but D is small, so $C \sim 0$ as well. Any other system will have an intermediate behavior and therefore $C > 0$.

As was intuitively suggested, the definition of complexity (4.3) also depends on the *scale*. At each scale of observation a new set of accessible states appears with its corresponding probability distribution so that complexity changes. Physical laws at each level of observation allow us to infer the probability distribution of the new set of accessible states, and therefore different values for H, D and C will be obtained. The straightforward passage to the case of a continuum number of states, x, can be easily inferred. Thus we must treat with probability distributions with a continuum support, $p(x)$, and normalization condition $\int_{-\infty}^{+\infty} p(x)dx = 1$. Disequilibrium has the limit $D = \int_{-\infty}^{+\infty} p^2(x)dx$ and the complexity could be defined by:

$$C = H \cdot D = -\left(K \int_{-\infty}^{+\infty} p(x) \log p(x) dx \right) \cdot \left(\int_{-\infty}^{+\infty} p^2(x) dx \right) . \tag{4.4}$$

As we shall see, other possibilities for the continuous extension of C are also possible.

Direct simulations of the definition give the values of C for general N-systems. The set of all the possible distributions $\{p_1, p_2, \ldots, p_N\}$ where an N-system could be found is sampled. For the sake of simplicity H is normalized to the interval $[0, 1]$. Thus $H = \sum_{i=1}^{N} p_i \log p_i / \log N$. For each distribution $\{p_i\}$ the normalized information $H(\{p_i\})$, and the disequilibrium $D(\{p_i\})$ (4.2) are calculated. In each case the normalized complexity $C = H \cdot D$ is obtained and the pair (H, C) stored. These two magnitudes are plotted on a diagram $(H, C(H))$ in order to verify the qualitative behavior predicted in Fig. 4.1. For $N = 2$ an analytical expression for the curve $C(H)$ is obtained. If the probability of one state is $p_1 = x$, that of the

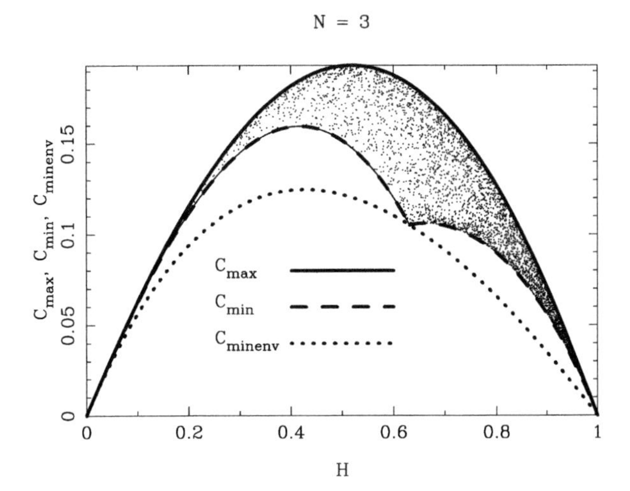

Fig. 4.2 In general, dependence of complexity (C) on normalized information (H) is not univocal: many distributions $\{p_i\}$ can present the same value of H but different C. This is shown in the case $N = 3$

second one is simply $p_2 = 1 - x$. The complexity of the system will be:

$$\begin{aligned} C(x) &= H(x) \cdot D(x) \\ &= -\frac{1}{\log 2}\left[x \log\left(\frac{x}{1-x}\right) + \log(1-x)\right] \cdot 2\left(x - \frac{1}{2}\right)^2. \end{aligned} \tag{4.5}$$

Complexity vanishes for the two simplest 2-systems: the crystal ($H = 0$; $p_1 = 1$, $p_2 = 0$) and the ideal gas ($H = 1$; $p_1 = 1/2$, $p_2 = 1/2$). Let us notice that this curve is the simplest one that fulfills all the conditions discussed in the introduction. The largest complexity is reached for $H \sim 1/2$ and its value is: $C(x \sim 0.11) \sim 0.151$. For $N > 2$ the relationship between H and C is not univocal anymore. Many different distributions $\{p_i\}$ store the same information H but have different complexity C. Figure 4.2 displays such a behavior for $N = 3$. If we take the maximum complexity $C_{max}(H)$ associated with each H a curve similar to the one for a 2-system is recovered. Every 3-system will have a complexity below this line and upper the line of $C_{min}(H)$ and also upper the minimum envelope complexity C_{minenv}. These lines will be analytically found in a next section. In Fig. 4.3 curves $C_{max}(H)$ for the cases $N = 3, \ldots, 10$ are also shown. Let us observe the shift of the complexity-curve peak to smaller values of entropy for rising N. This fact agrees with the intuition telling us that the biggest complexity (number of possibilities of 'complexification') be reached for lesser entropies for the systems with bigger number of states.

Let us return to the point at which we started this discussion. Any notion of complexity in physics [2, 3] should only be made on the basis of a well defined or operational magnitude [6, 7]. But two additional requirements are needed in order to obtain a good definition of complexity in physics: (1) the new magnitude must be measurable in many different physical systems and (2) a comparative relationship and a physical interpretation between any two measurements should be possible.

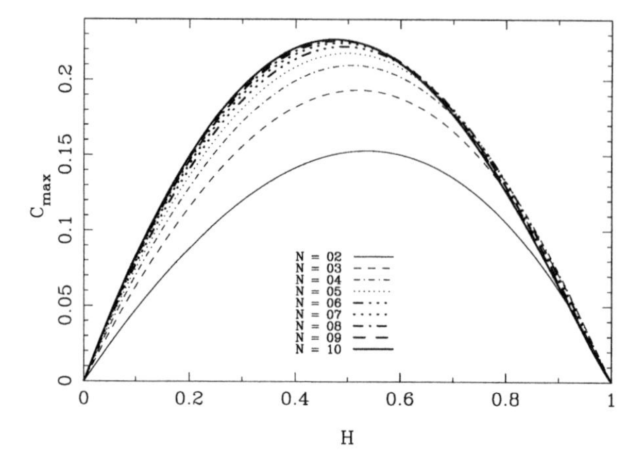

Fig. 4.3 Complexity ($C = H \cdot D$) as a function of the normalized information (H) for a system with two accessible states ($N = 2$). Also curves of maximum complexity (C_{max}) are shown for the cases: $N = 3, \ldots, 10$

Many different definitions of complexity have been proposed to date, mainly in the realm of physical and computational sciences. Among these, several can be cited: algorithmic complexity (Kolmogorov-Chaitin) [8–10], the Lempel-Ziv complexity [11], the logical depth of Bennett [12], the effective measure complexity of Grassberger [13], the complexity of a system based in its diversity [14], the thermodynamical depth [15], the ε-machine complexity [16], the physical complexity of genomes [17], complexities of formal grammars, etc. The definition of complexity (4.3) proposed in this section offers a new point of view, based on a statistical description of systems at a given *scale*. In this scheme, the knowledge of the physical laws governing the dynamic evolution in that scale is used to find its accessible states and its probability distribution. This process would immediately indicate the value of complexity. In essence this is nothing but an interplay between the information stored by the system and the *distance from equipartition* (measure of a probabilistic hierarchy between the observed parts) of the probability distribution of its accessible states. Besides giving the main features of a "intuitive" notion of complexity, we will show in this chapter that we can go one step further and to compute this quantity in other relevant physical situations and in continuum systems. The most important point is that the new definition successfully enables us to discern situations regarded as complex. For example, we show here two of these applications in complex systems with some type of discretization: one of them is the study of this magnitude in a phase transition in a coupled map lattice [18] and the other one is its calculation for the time evolution of a discrete gas out of equilibrium [19]. Other applications to more realistic systems can also be found in the literature [20].

4.1.1 Complexity in a Phase Transition: Coupled Map Lattices

If by complexity it is to be understood that property present in all systems attached under the epigraph of 'complex systems', this property should be reasonably quantified by the measures proposed in the different branches of knowledge. As discussed above, this kind of indicators is found in those fields where the concept of information is crucial, from physics [13, 15] to computational sciences [8–11, 16].

In particular, taking into account the statistical properties of a system, the indicator called the *LMC (LópezRuiz-Mancini-Calbet) complexity* has been introduced [6, 7] in the former section. This magnitude identifies the entropy or information H stored in a system and its disequilibrium D, i.e. the distance from its actual state to the probability distribution of equilibrium, as the two basic ingredients for calculating its complexity. Hence, the LMC complexity C is given by the formula (4.3), $C(\bar{p}) = H(\bar{p}) \cdot D(\bar{p})$, where $\bar{p} = \{p_i\}$, with $p_i > 0$ and $i = 1, \ldots, N$, represents the distribution of the N accessible states to the system, and k is a constant taken as $1/\log N$.

As well as the Euclidean distance D is present in the original LMC complexity, other kinds of disequilibrium measures have been proposed in order to remedy some statistical characteristics considered troublesome for some authors [21]. In particular, some attention has been focused [22, 23] on the Jensen-Shannon divergence D_{JS} as a measure for evaluating the distance between two different distributions $(\bar{p}_1, \bar{p}_2)$. This distance reads:

$$D_{JS}(\bar{p}_1, \bar{p}_2) = H(\pi_1 \bar{p}_1 + \pi_2 \bar{p}_2) - \pi_1 H(\bar{p}_1) - \pi_2 H(\bar{p}_2), \tag{4.6}$$

with π_1, π_2 the weights of the two probability distributions $(\bar{p}_1, \bar{p}_2)$ verifying $\pi_1, \pi_2 \geq 0$ and $\pi_1 + \pi_2 = 1$. The ensuing statistical complexity

$$C_{JS} = H \cdot D_{JS} \tag{4.7}$$

becomes intensive and also keeps the property of distinguishing among distinct degrees of periodicity [24]. In this section, we consider $\bar{p}_2$ the equiprobability distribution and $\pi_1 = \pi_2 = 0.5$.

As it can be straightforwardly seen, all these LMC-like complexities vanish both for completely ordered and for completely random systems as it is required for the correct asymptotic properties of a such well-behaved measure. Recently, they have been successfully used to discern situations regarded as complex in discrete systems out of equilibrium [19, 25–31].

Here, the local transition to chaos via intermittency [32] in the logistic map, $x_{n+1} = \lambda x_n(1 - x_n)$ presents a sharp transition when C is plotted versus the parameter λ in the region around the instability for $\lambda \sim \lambda_t = 3.8284$. When $\lambda < \lambda_t$ the system approaches the laminar regime and the bursts become more unpredictable. The complexity increases. When the point $\lambda = \lambda_t$ is reached a drop to zero occurs for the magnitude C. The system is now periodic and it has lost its complexity. The dynamical behavior of the system is finally well reflected in the magnitude C as it has been studied in [7].

When a one-dimensional array of such maps is put together a more complex behavior can be obtained depending on the coupling among the units. Ergo the phenomenon called *spatio-temporal intermittency* can emerge [33–35]. This dynamical regime corresponds with a situation where each unit is weakly oscillating around a laminar state that is aperiodically and strongly perturbed for a traveling burst. In this case, the plot of the one-dimensional lattice evolving in time gives rise to complex patterns on the plane. If the coupling among units is modified the system can settle down in an absorbing phase where its dynamics is trivial [36, 37] and then homogeneous patterns are obtained. Therefore an abrupt transition to spatio-temporal intermittency can be depicted by the system [38, 39] when modifying the coupling parameter.

Now we are concerned with measuring C and C_{JS} in a such transition for a coupled map lattice of logistic type. Our system will be a line of sites, $i = 1, \dots, L$, with periodic boundary conditions. In each site i a local variable x_i^n evolves in time (n) according to a discrete logistic equation. The interaction with the nearest neighbors takes place via a multiplicative coupling:

$$x_i^{n+1} = (4 - 3pX_i^n)x_i^n(1 - x_i^n), \tag{4.8}$$

where p is the parameter of the system measuring the strength of the coupling ($0 < p < 1$). The variable X_i^n is the digitalized local mean field,

$$X_i^n = \text{nint}\left[\frac{1}{2}(x_{i+1}^n + x_{i-1}^n)\right], \tag{4.9}$$

with nint(.) the integer function rounding its argument to the nearest integer. Hence $X_i^n = 0$ or 1.

There is a biological motivation behind this kind of systems [40, 41]. It could represent a *colony of interacting competitive individuals*. They evolve randomly when they are independent ($p = 0$). If some competitive interaction ($p > 0$) among them takes place the local dynamics loses its erratic component and becomes chaotic or periodic in time depending on how populated the vicinity is. Hence, for bigger X_i^n more populated is the neighborhood of the individual i and more constrained is its free action. At a first sight, it would seem that some particular values of p could stabilize the system. In fact, this is the case. Let us choose a number of individuals for the colony ($L = 500$ for instance), let us initialize it randomly in the range $0 < x_i < 1$ and let it evolve until the asymptotic regime is attained. Then the *black/white* statistics of the system is performed. That is, the state of the variable x_i is compared with the critical level 0.5 for $i = 1, \dots, L$: if $x_i > 0.5$ the site i is considered *white* (high density cell) and a counter N_w is increased by one, or if $x_i < 0.5$ the site i is considered *black* (low density cell) and a counter N_b is increased by one. This process is executed in the stationary regime for a set of iterations. The *black/white* statistics is then the rate $\beta = N_b/N_w$. If β is plotted versus the coupling parameter p Fig. 4.4 is obtained.

The region $0.258 < p < 0.335$ where β vanishes is remarkable. As stated above, β represents the rate between the number of black cells and the number of white

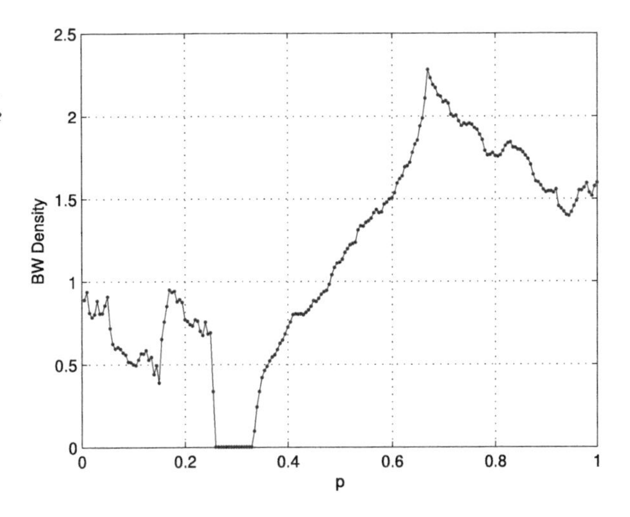

Fig. 4.4 β versus p. The β-statistics (or BW density) for each p is the rate between the number of *black* and *white* cells depicted by the system in the two-dimensional representation of its after-transient time evolution. (Computations have been performed with $\Delta p = 0.005$ for a lattice of 10000 sites after a transient of 5000 iterations and a running of other 2000 iterations)

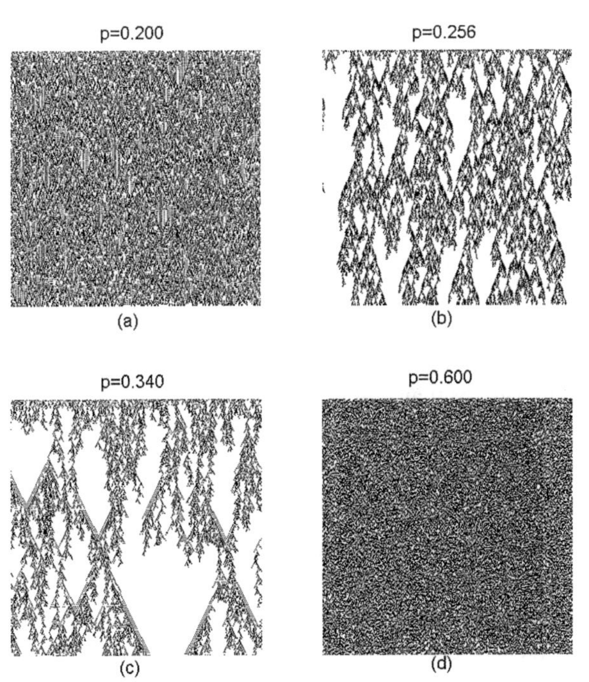

Fig. 4.5 Digitalized plot of the one-dimensional coupled map lattice (axe OX) evolving in time (axe OY) according to (4.8): if $x_i^n > 0.5$ the (i,n)-cell is put in *white color* and if $x_i^n < 0.5$ the (i,n)-cell is put in *black color*. The discrete time n is reset to zero after the transitory. (Lattices of 300×300 sites, i.e., $0 < i < 300$ and $0 < n < 300$)

cells appearing in the two-dimensional digitalized representation of the colony evolution. A whole white pattern is obtained for this range of p. The phenomenon of spatio-temporal intermittency is displayed by the system in the two borders of this parameter region (Fig. 4.5). Bursts of low density (black color) travel in an irregular way through the high density regions (white color). In this case two-dimensional complex patterns are shown by the time evolution of the system (Fig. 4.5b–c). If the coupling p is far enough from this region, i.e., $p < 0.25$ or $p > 0.4$, the absorbent

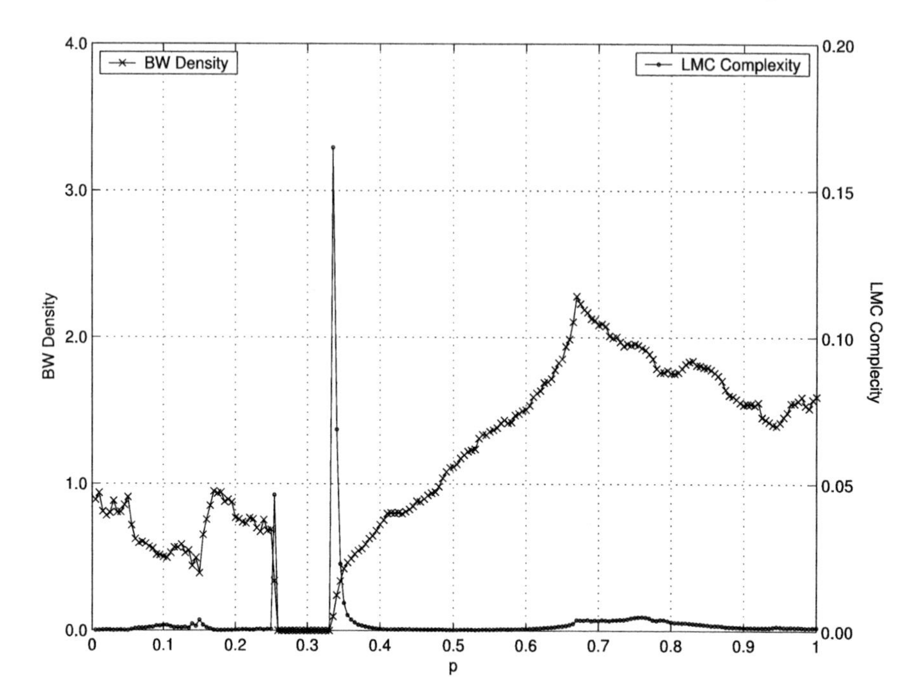

Fig. 4.6 (•) C versus p. Observe the peaks of the LMC complexity located just on the borders of the absorbent region $0.258 < p < 0.335$, where $\beta = 0$ (×). (Computations have been performed with $\Delta p = 0.005$ for a lattice of 10000 sites after a transient of 5000 iterations and a running of other 2000 iterations)

region loses its influence on the global dynamics and less structured and more random patterns than before are obtained (Fig. 4.5a–d). For $p = 0$ we have no coupling of the maps, and each map generates so called fully developed chaos, where the invariant measure is well-known to be symmetric around 0.5. From this we conclude that $\beta(p = 0) = 1$. Let us observe that this symmetrical behavior of the invariant measure is broken for small p, and β decreases slightly in the vicinity of $p = 0$.

If the LMC complexities are quantified as function of p, our *intuition* is confirmed. The method proposed in [7] to calculate C is now adapted to the case of two-dimensional patterns. First, we let the system evolve until the asymptotic regime is attained. This transient is discarded. Then, for each time n, we map the whole lattice in a binary sequence: 0 if $x_i^n < 0.5$ and 1 if $x_i^n > 0.5$, for $i = 1, \ldots, L$. This L-binary string is analyzed by blocks of n_o bits, where n_o can be considered the scale of observation. For this scale, there are 2^{n_o} possible states but only some of them are accessible. These accessible states as well as their probabilities are found in the L-binary string. Next, the magnitudes H, D, D_{JS}, C and C_{JS} are directly calculated for this particular time n by applying the formulas (4.3), (4.7). We repeat this process for a set of successive time units $(n, n+1, \ldots, n+m)$. The mean values of H, D, D_{JS}, C and C_{JS} for these m time units are finally obtained and plotted in Figs. 4.6, 4.7.

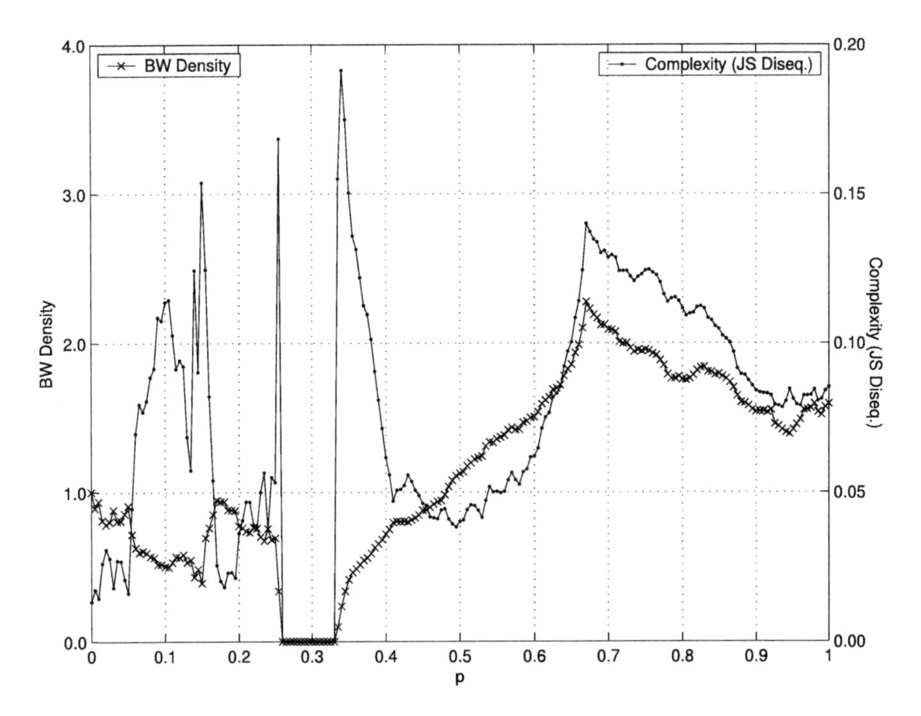

Fig. 4.7 $(\cdot)$ C_{JS} versus p. The peaks of this modified LMC complexity are also evident just on the borders of the absorbent region $0.258 < p < 0.335$, where $\beta = 0$ $(\times)$. (Computations have been performed with $\Delta p = 0.005$ for a lattice of 10000 sites after a transient of 5000 iterations and a running of other 2000 iterations)

Figures 4.6, 4.7 show the result for the case of $n_o = 10$. Let us observe that the highest C and C_{JS} are reached when the dynamics displays spatio-temporal intermittency, that is, the *most complex patterns* are obtained for those values of p that are located on the borders of the absorbent region $0.258 < p < 0.335$. Thus the plot of C and C_{JS} versus p shows two tight peaks around the values $p = 0.256$ and $p = 0.34$ (Figs. 4.6, 4.7). Let us remark that the LMC complexity C can be neglected far from the absorbent region. Contrarily to this behavior, the magnitude C_{JS} also shows high peaks in some other sharp transition of β located in the region $0 < p < 25$, and an intriguing correlation with the *black/white* statistics in the region $0.4 < p < 1$. All these facts as well as the stability study of the different dynamical regions of system (4.8) are not the object of the present writing but they could deserve some attention in a further inspection.

If the detection of complexity in the two-dimensional case requires to identify some sharp change when comparing different patterns, those regions in the parameter space where an abrupt transition happens should be explored in order to obtain the most complex patterns. Smoothness seems not to be at the origin of complexity. As well as a selected few distinct molecules among all the possible are in the basis of life [42], discreteness and its spiky appearance could indicate the way towards complexity. As we show in the next section, the distributions with the highest LMC complexity are just those distributions with a spiky-like appearance [19]. In this line, the striking result here exposed confirms the capability of the LMC-like complexities for signaling a transition to complex behavior when regarding two-dimensional patterns [18, 43].

4.1.2 Complexity Versus Time: The Tetrahedral Gas

As before explained, several definitions of complexity, in the general sense of the term, have been presented in the literature. These can be classified according to their calculation procedure into two broad and loosely defined groups. One of these groups is based on computational science and consists of all definitions based on algorithms or automata to derive the complexity. Examples are the algorithmic complexity [9, 10], the logical depth [12] and the ε-machine complexity [16]. These definitions have been shown to be very useful in describing symbolic dynamics of chaotic maps, but they have the disadvantage of being very difficult to calculate. Another broad group consists of those complexities based on the measure of entropy or entropy rate. Among these, we may cite the effective measure complexity [13], the thermodynamic depth [15], the simple measure for complexity [26] and the metric or K–S entropy rate [44, 45]. These definitions have also been very useful in describing symbolic dynamics in maps, the simple measure of complexity having been also applied to some physical situation such as a non-equilibrium Fermi gas [46]. They suffer the disadvantage of either being very difficult to calculate or having a simple relation to the regular entropy.

Other definition types of complexity have been introduced. These are based on quantities that can be calculated directly from the distribution function describing the system. One of these is based on "meta-statistics" [47] and the other on the notion of "disequilibrium" [7]. This latter definition has been referred above as the LMC complexity. These definitions, together with the simple measure for complexity [26], have the great advantage of allowing easy calculations within the context of kinetic theory and of permitting their evaluation in a natural way in terms of statistical mechanics.

As we have shown in the former sections, the disequilibrium-based complexity is easy to calculate and shows some interesting properties [7], but suffers from the main drawback of not being very well behaved as the system size increases, or equivalently, as the distribution function becomes continuous. Feldman and Crutchfield [21] tried to solve this problem by defining another equivalent term for disequilibrium, but ended up with a complexity that was a trivial function of the entropy.

Whether these definitions of complexity are useful in non-equilibrium thermodynamics will depend on how they behave as a function of time. There is a general belief that, although the second law of thermodynamics requires average entropy (or disorder) to increase, this does not in any way forbid local order from arising [48]. The clearest example is seen with life, which can continue to exist and grow in an isolated system for as long as internal resources last. In other words, in an isolated system the entropy must increase, but it should be possible, under certain circumstances, for the complexity to increase.

Here we examine how LMC complexity evolves with time in an isolated system and we show that it indeed has some interesting properties. The disequilibrium-based complexity [7] defined in (4.3) actually tends to be maximal as the entropy increases in a Boltzmann integro-differential equation for a simplified gas.

We proceed to calculate the distributions which maximize and minimize the complexity and its asymptotic behavior, and also introduce the basic concepts underlying the time evolution of LMC complexity in Sect. 4.1.2.1. Later, in Sects. 4.1.2.2 and 4.1.2.3, by means of numerical computations following a restricted version of the Boltzmann equation, we apply this to a special system, which we shall term "tetrahedral gas". Finally, in Sect. 4.1.2.4, the results and conclusions for this system are given, together with their possible applications.

4.1.2.1 Maximum and Minimum Complexity

In this section, we assume that the system can be in one of its N possible accessible states, i. The probability of the system being in state i will be given by the discrete distribution function, $f_i \geq 0$, with the normalization condition $I \equiv \sum_{i=1}^{N} f_i = 1$. The system is defined such that, if isolated, it will reach equilibrium, with all the states having equal probability, $f_e = \frac{1}{N}$. Since we are supposing that H is normalized, $0 \leq H \leq 1$, and $0 \leq D \leq (N-1)/N$, then complexity, C, is also normalized, $0 \leq C \leq 1$.

When an isolated system evolves with time, the complexity cannot have any possible value in a C versus H map as it can be seen in Fig. 4.2, but it must stay within certain bounds, $C_{\max}$ and $C_{\min}$. These are the maximum and minimum values of C for a given H. Since $C = D \cdot H$, finding the extrema of C for constant H is equivalent to finding the extrema of D.

There are two restrictions on D: the normalization, I, and the fixed value of the entropy, H. To find these extrema undetermined Lagrange multipliers are used. Differentiating expressions of D, I and H, we obtain

$$\frac{\partial D}{\partial f_j} = 2(f_j - f_e), \tag{4.10}$$

$$\frac{\partial I}{\partial f_j} = 1, \tag{4.11}$$

$$\frac{\partial H}{\partial f_j} = -\frac{1}{\ln N}(\ln f_j + 1). \tag{4.12}$$

Table 4.1 Probability values, f_j, that give a maximum of disequilibrium, $D_{\rm max}$, for a given H

Number of states with f_j	f_j	Range of f_j
1	$f_{\rm max}$	$\frac{1}{N} \dots 1$
$N-1$	$\frac{1-f_{\rm max}}{N-1}$	$0 \dots \frac{1}{N}$

Table 4.2 Probability values, f_j, that give a minimum of disequilibrium, $D_{\rm min}$, for a given H

Number of states with f_j	f_j	Range of f_j
n	0	0
1	$f_{\rm min}$	$0 \dots \frac{1}{N-n}$
$N-n-1$	$\frac{1-f_{\rm min}}{N-n-1}$	$\frac{1}{N-n} \cdots \frac{1}{N-n-1}$

n can have the values $0, 1, \dots, N-2$

Defining λ_1 and λ_2 as the Lagrange multipliers, we get:

$$2(f_j - f_{\rm e}) + \lambda_1 + \lambda_2(\ln f_j + 1)/\ln N = 0. \tag{4.13}$$

Two new parameters, α and β, which are a linear combinations of the Lagrange multipliers are defined:

$$f_j + \alpha \ln f_j + \beta = 0, \tag{4.14}$$

where the solutions of this equation, f_j, are the values that minimize or maximize the disequilibrium.

In the maximum complexity case there are two solutions, f_j, to (4.14) which are shown in Table 4.1. One of these solutions, $f_{\rm max}$, is given by

$$H = -\frac{1}{\ln N}\left[f_{\rm max} \ln f_{\rm max} + (1 - f_{\rm max}) \ln\left(\frac{1 - f_{\rm max}}{N-1}\right)\right], \tag{4.15}$$

and the other solution by $(1 - f_{\rm max})/(N-1)$.

The maximum disequilibrium, $D_{\rm max}$, for a fixed H is

$$D_{\rm max} = (f_{\rm max} - f_{\rm e})^2 + (N-1)\left(\frac{1 - f_{\rm max}}{N-1} - f_{\rm e}\right)^2, \tag{4.16}$$

and thus, the maximum complexity, which depends only on H, is

$$C_{\rm max}(H) = D_{\rm max} \cdot H. \tag{4.17}$$

The behavior of the maximum value of complexity versus $\ln N$ was computed in [49].

Equivalently, the values, f_j, that give a minimum complexity are shown in Table 4.2. One of the solutions, $f_{\rm min}$, is given by

$$H = -\frac{1}{\ln N}\left[f_{\rm min} \ln f_{\rm min} + (1 - f_{\rm min}) \ln\left(\frac{1 - f_{\rm min}}{N-n-1}\right)\right], \tag{4.18}$$

where n is the number of states with $f_j = 0$ and takes a value in the range $n = 0, 1, \dots, N-2$.

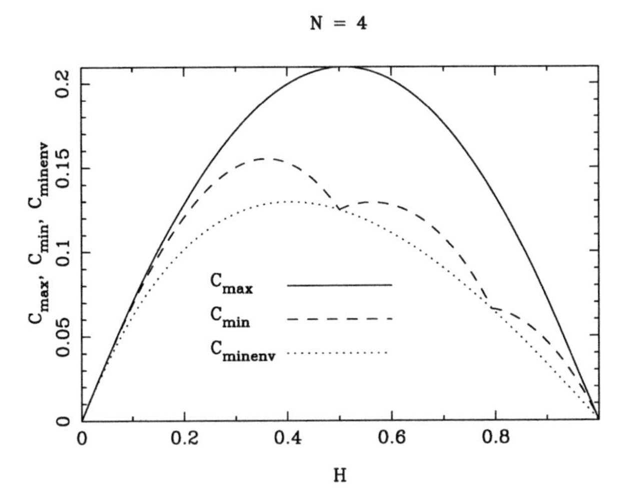

Fig. 4.8 Maximum, minimum, and minimum envelope complexity, C_{max}, C_{min}, and C_{minenv} respectively, as a function of the entropy, H, for a system with $N=4$ accessible states

The resulting minimum disequilibrium, D_{min}, for a given H is,

$$D_{\text{min}} = (f_{\text{min}} - f_{\text{e}})^2 + (N-n-1)\left(\frac{1-f_{\text{min}}}{N-n-1} - f_{\text{e}}\right)^2 + nf_{\text{e}}^2. \tag{4.19}$$

Note that in this case $f_j = 0$ is an additional hidden solution that stems from the positive restriction in the f_i values. To obtain these solutions explicitly we can define x_i such that $f_i \equiv {x_i}^2$. These x_i values do not have the restriction of positivity imposed to f_i and can take a positive or negative value. If we repeat the Lagrange multiplier method with these new variables a new solution arises: $x_j = 0$, or equivalently, $f_j = 0$.

The resulting minimum complexity, which again only depends on H, is

$$C_{\text{min}}(H) = D_{\text{min}} \cdot H. \tag{4.20}$$

As an example, the maximum and minimum of complexity, C_{max} and C_{min}, are plotted as a function of the entropy, H, in Fig. 4.8 for $N=4$. Also, in this figure, it is shown the minimum envelope complexity, $C_{\text{minenv}} = D_{\text{minenv}} \cdot H$, where D_{minenv} is defined below. In Fig. 4.9 the maximum and minimum disequilibrium, D_{max} and D_{min}, versus H are also shown.

As shown in Fig. 4.9 the minimum disequilibrium function is piecewise defined, having several points where its derivative is discontinuous. Each of these function pieces corresponds to a different value of n (Table 4.2). In some circumstances it might be helpful to work with the "envelope" of the minimum disequilibrium function. The function, D_{minenv}, that traverses all the discontinuous derivative points in the D_{min} versus H plot is

$$D_{\text{minenv}} = e^{-H \ln N} - \frac{1}{N}, \tag{4.21}$$

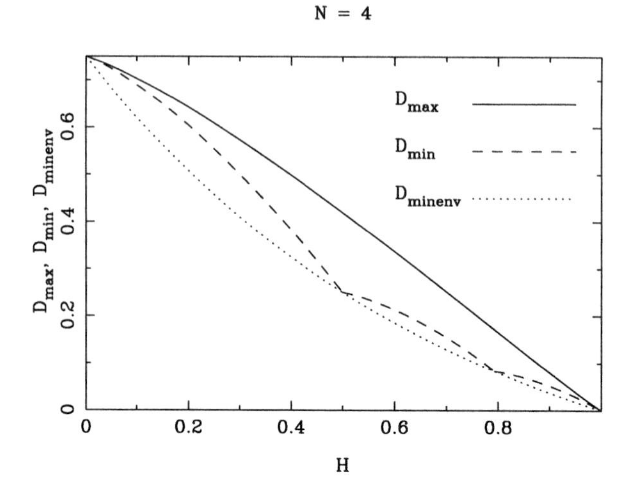

Fig. 4.9 Maximum, minimum, and minimum envelope disequilibrium, $D_{\max}$, $D_{\min}$, and D_{minenv} respectively, as a function of the entropy, H, for a system with $N = 4$ accessible states

and is also shown in Fig. 4.9.

When N tends toward infinity the probability, $f_{\max}$, of the dominant state has a linear dependence with the entropy,

$$\lim_{N\to\infty} f_{\max} = 1 - H, \tag{4.22}$$

and thus the maximum disequilibrium scales as $\lim_{N\to\infty} D_{\max} = (1 - H)^2$. The maximum complexity tends to

$$\lim_{N\to\infty} C_{\max} = H \cdot (1 - H)^2 . \tag{4.23}$$

The limit of the minimum disequilibrium and complexity vanishes, $\lim_{N\to\infty} D_{\text{minenv}} = 0$, and thus

$$\lim_{N\to\infty} C_{\min} = 0. \tag{4.24}$$

In general, in the limit $N \to \infty$, the complexity is not a trivial function of the entropy, in the sense that for a given H there exists a range of complexities between 0 and $C_{\max}$, given by (4.24) and (4.23), respectively.

In particular, in this asymptotic limit, the maximum of $C_{\max}$ is found when $H = 1/3$, or equivalently $f_{\max} = 2/3$, which gives a maximum of the maximum complexity of $C_{\max} = 4/27$. This value was numerically calculated in [49].

4.1.2.2 An out Equilibrium System: The Tetrahedral Gas

We present a simplified example of an ideal gas: the tetrahedral gas. This system is generated by a simplification of the Boltzmann integro-differential equation of an ideal gas. We are interested in studying the disequilibrium time evolution.

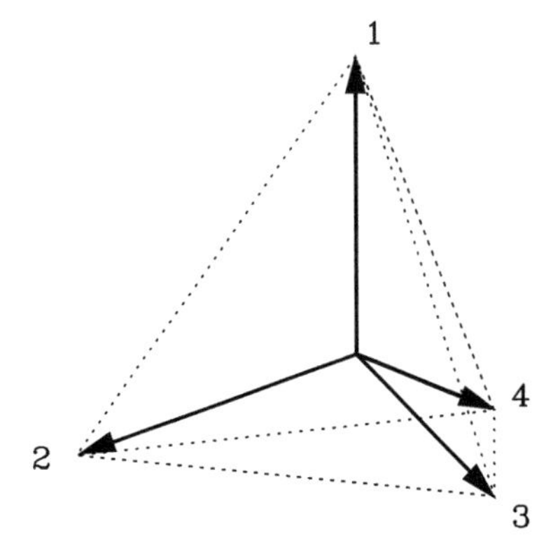

Fig. 4.10 The four possible directions of the velocities of the tetrahedral gas in space. Positive senses are defined as emerging from the center point and with integer numbers 1, 2, 3, 4

The Boltzmann integro-differential equation of an ideal gas with no external forces and no spatial gradients is

$$\frac{\partial f(\mathbf{v};t)}{\partial t} = \int d^3\mathbf{v}_* \int d\Omega_{\mathrm{c.m.}} \sigma(\mathbf{v}_* - \mathbf{v} \rightarrow \mathbf{v}'_* - \mathbf{v}')|\mathbf{v}_* - \mathbf{v}| \times \left[f(\mathbf{v}'_*;t) f(\mathbf{v}';t) - f(\mathbf{v}_*;t) f(\mathbf{v};t) \right], \tag{4.25}$$

where σ represents the cross section of a collision between two particles with initial velocities $\mathbf{v}$ and $\mathbf{v}_*$ and after the collision with velocities $\mathbf{v}'$ and $\mathbf{v}'_*$; and $\Omega_{\mathrm{c.m.}}$ are all the possible dispersion angles of the collision as seen from its center of mass.

In the tetrahedral gas, the particles can travel only in four directions in three-dimensional space and all have the same absolute velocity. These directions are the ones given by joining the center of a tetrahedron with its corners. The directions can be easily viewed by recalling the directions given by a methane molecule, or equivalently, by a caltrop, which is a device with four metal points so arranged that when any three are on the ground the fourth projects upward as a hazard to the hooves of horses or to pneumatic tires (see Fig. 4.10).

By definition, the angle that one direction forms with any other is the same. It can be shown that the angles between different directions, α, satisfy the relationship $\cos\alpha = -1/3$, which gives $\alpha = 109.47°$. The plane formed by any two directions is perpendicular to the plane formed by the remaining two directions.

We assume that the cross-section, σ, is different from zero only when the angle between the velocities of the colliding particles is 109.47°. It is also assumed that this collision makes the two particles leave in the remaining two directions, thus again forming an angle of 109.47°. A consequence of these restrictions is that the modulus of the velocity is always the same no matter how many collisions a particle has undergone and they always stay within the directions of the vertices of the tetrahedron. Furthermore, this type of gas does not break any law of physics and is perfectly valid, although hypothetical.

We label the four directions originating from the center of the caltrop with numbers, **1**, **2**, **3**, **4** (see Fig. 4.10). The velocity components with the same direction but

Table 4.3 Cross sections, σ, for a particle in direction $-\mathbf{1}$ colliding with particles in the other remaining directions of the tetrahedral gas

Collision of particles	Cross section σ
$(-\mathbf{1}, -\mathbf{2}) \rightarrow (\mathbf{3}, \mathbf{4})$	1
$(-\mathbf{1}, -\mathbf{3}) \rightarrow (\mathbf{2}, \mathbf{4})$	1
$(-\mathbf{1}, -\mathbf{4}) \rightarrow (\mathbf{2}, \mathbf{3})$	1
Other collisions	0

opposite sense, or equivalently, directed toward the center of the caltrop, are labeled with negative numbers $-\mathbf{1}, -\mathbf{2}, -\mathbf{3}, -\mathbf{4}$.

In order to formulate the Boltzmann equation for the tetrahedral gas, and because all directions are equivalent, we need only study the different collisions that a particle with one fixed direction can undergo. In particular if we take a particle with direction $-\mathbf{1}$ the result of the collision with another particle with direction $-\mathbf{2}$ are the same two particles traveling in directions $\mathbf{3}$ and $\mathbf{4}$, that is,

$$(-\mathbf{1}, -\mathbf{2}) \rightarrow (\mathbf{3}, \mathbf{4}). \tag{4.26}$$

With this in mind the last bracket of (4.25) is,

$$f_3 f_4 - f_{-1} f_{-2}, \tag{4.27}$$

where f_i denotes the probability of finding a particle in direction $\mathbf{i}$. Note that the dependence on velocity, $\mathbf{v}$, of the continuous velocity distribution function, $f(\mathbf{v}; t)$, of (4.25) is in our case contained in the discrete subindex, i, of the distribution function f_i.

We can proceed in the same manner with the other remaining collisions,

$$\begin{aligned} (-\mathbf{1}, -\mathbf{3}) &\rightarrow (\mathbf{2}, \mathbf{4}), \\ (-\mathbf{1}, -\mathbf{4}) &\rightarrow (\mathbf{2}, \mathbf{3}). \end{aligned} \tag{4.28}$$

When a particle with direction $-\mathbf{1}$ collides with a particle with direction $\mathbf{2}$, they do not form an angle of 109.47°; i.e., they do not collide, they just pass by each other. This is a consequence of the previous assumption for the tetrahedral gas, which establishes a null cross section for angles different from 109.47°. The same can be said for collisions $(-\mathbf{1}, \mathbf{3})$, $(-\mathbf{1}, \mathbf{4})$, and $(-\mathbf{1}, \mathbf{1})$. All these results are summarized in Table 4.3.

Taking all this into account, (4.25) for direction $-\mathbf{1}$ is reduced to a discrete sum,

$$\frac{df_{-1}}{dt} = (f_3 f_4 - f_{-1} f_{-2}) + (f_2 f_4 - f_{-1} f_{-3}) + (f_2 f_3 - f_{-1} f_{-4}), \tag{4.29}$$

where all other factors have been set to unity for simplicity.

The seven remaining equations for the rest of directions can be easily inferred. If we now make $f_i = f_{-i}$ $(i = 1, 2, 3, 4)$ initially, this property is conserved in time. The final four equations defining the evolution of the system are:

$$\begin{aligned}
\frac{df_1}{dt} &= (f_3 f_4 - f_1 f_2) + (f_2 f_4 - f_1 f_3) + (f_2 f_3 - f_1 f_4), \\
\frac{df_2}{dt} &= (f_3 f_4 - f_1 f_2) + (f_1 f_4 - f_2 f_3) + (f_1 f_3 - f_2 f_4), \\
\frac{df_3}{dt} &= (f_2 f_4 - f_3 f_1) + (f_1 f_4 - f_3 f_2) + (f_1 f_2 - f_3 f_4), \\
\frac{df_4}{dt} &= (f_2 f_3 - f_4 f_1) + (f_1 f_3 - f_4 f_2) + (f_1 f_2 - f_3 f_4).
\end{aligned} \tag{4.30}$$

Note that the ideal gas has been reduced to the tetrahedral gas, which is a four-dimensional dynamical system. The velocity distribution function, f_i, corresponds to a probability distribution function with $N = 4$ accessible states that evolve in time.

4.1.2.3 Evolution of the Tetrahedral Gas with Time

To study the time evolution of the complexity, a diagram of C versus time, t, can be used. But, as we know, the second law of thermodynamics states that the entropy grows monotonically with time, that is,

$$\frac{dH}{dt} \geq 0. \tag{4.31}$$

This implies that an equivalent way to study the time evolution of the complexity can be obtained by plotting C versus H. In this way, the entropy substitutes the time axis, since the former increases monotonically with the latter. The conversion from C vs. H to C vs. t diagrams is achieved by stretching or shrinking the entropy axis according to its time evolution. This method is a key point in all this discussion. Note that, in any case, the relationship of H versus t will, in general, not be a simple one [50].

The tetrahedral gas, (4.30), reaches equilibrium when $f_i = 1/N$ for $i = 1, 2, 3, 4$ and $N = 4$. This stationary state, $df_i/dt = 0$, represents the equiprobability towards which the system evolves in time. This is consistent with the definition of disequilibrium in which we assumed that equilibrium was reached at equiprobability, $f_i = f_e$, where $D = 0$.

As the isolated system evolves it gets closer and closer to equilibrium. In this sense, one may intuitively think that the disequilibrium will decrease with time. In fact, it can be analytically shown [19] that, as the system approaches to equilibrium, D tends to zero monotonically with time:

$$\frac{dD}{dt} \leq 0. \tag{4.32}$$

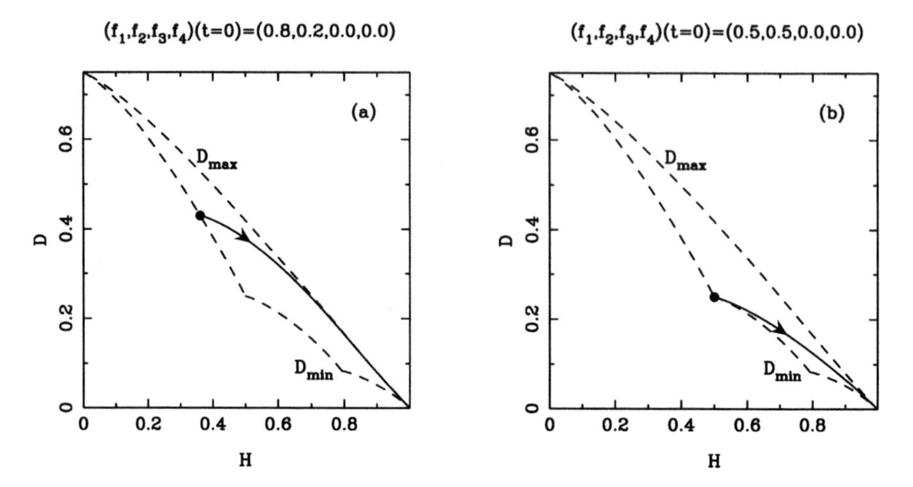

Fig. 4.11 Time evolution of the system in (H, D) phase space for two different initial conditions at time $t = 0$: (**a**) $(f_1, f_2, f_3, f_4) = (0.8, 0.2, 0, 0)$ and (**b**) $(f_1, f_2, f_3, f_4) = (0.5, 0.5, 0, 0)$. The maximum and minimum disequilibrium are shown by *dashed lines*

There are even more restrictions on the evolution of this system. It would be expected that the system approaches equilibrium, $D = 0$, by following the most direct path. To verify this, numerical simulations for several initial conditions have been undertaken. In all of these we observe the additional restriction that D approaches D_{max} on its way to $D = 0$. In fact it appears as an exponential decay of D towards D_{max} in a D versus H plot. As an example, two of these are shown in Fig. 4.11, where Fig. 4.11(a) shows a really strong tendency towards D_{max}. Contrary to intuition, among all the possible paths that the system can follow toward equilibrium, it chooses those closest to D_{max} in particular.

We can also observe this effect in a complexity, C, versus H plot. This is shown for the same two initial conditions in Fig. 4.12.

This additional restriction to the evolution of the system is better viewed by plotting the difference $C_{\text{max}} - C$ versus H. In all the cases analyzed (see two of them in Fig. 4.13) the following condition is observed:

$$\frac{d(C_{\text{max}} - C)}{dt} \leq 0. \tag{4.33}$$

This has been verified numerically and is illustrated in Fig. 4.14, where this time derivative, which always remains negative, is shown as a function of H for a grid of uniformly spaced distribution functions, (f_1, f_2, f_3, f_4), satisfying the normalization condition I. Two system trajectories are also shown for illustrative purposes. The numerical method used to plot this function is explained in [19].

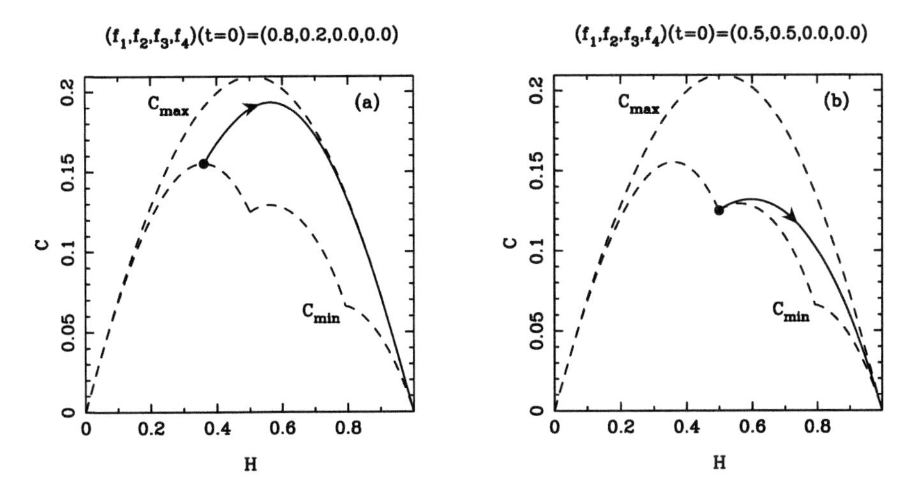

Fig. 4.12 Time evolution of the system in (H, C) phase space for two different initial conditions at time $t = 0$: **(a)** $(f_1, f_2, f_3, f_4) = (0.8, 0.2, 0, 0)$ and **(b)** $(f_1, f_2, f_3, f_4) = (0.5, 0.5, 0, 0)$. The maximum and minimum complexity are shown by *dashed lines*

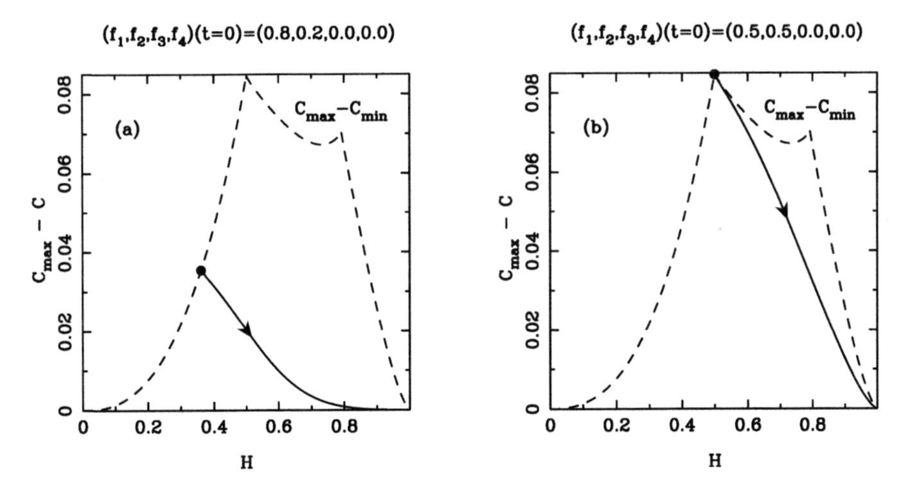

Fig. 4.13 Time evolution of the system in $(H, C_{\max} - C)$ phase space for two different initial conditions at time $t = 0$: **(a)** $(f_1, f_2, f_3, f_4) = (0.8, 0.2, 0, 0)$ and **(b)** $(f_1, f_2, f_3, f_4) = (0.5, 0.5, 0, 0)$. The values $C_{\max} - C_{\min}$ are shown by *dashed lines*

We proceed now to show another interesting property of this system. As shown in Table 4.1, a collection of maximum complexity distributions for $N = 4$ can take the form

$$\begin{aligned} f_1 &= f_{\max}, \\ f_i &= \frac{1 - f_{\max}}{3}, \quad i = 2, 3, 4, \end{aligned} \tag{4.34}$$

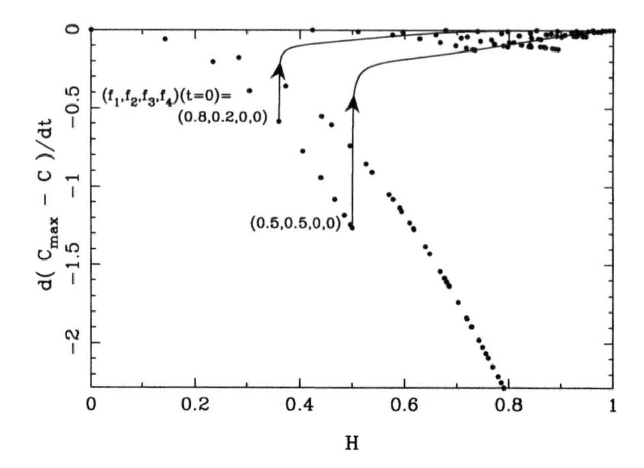

Fig. 4.14 Numerical verification of $d(C_{\max} - C)/dt \leq 0$. This time derivative is shown as a function of H. A grid of uniformly spaced, $\Delta f_i = 0.5$, distribution functions, (f_1, f_2, f_3, f_4), satisfying the normalization condition I, have been used. Two system trajectories for initial conditions, $t = 0$, $(f_1, f_2, f_3, f_4) = (0.8, 0.2, 0, 0)$ and $(f_1, f_2, f_3, f_4) = (0.5, 0.5, 0, 0)$ are also shown for illustrative purposes. It can be seen how the above-mentioned time derivative always remains negative

where $f_{\max}$ runs from $1/N$ (equiprobability distribution) to 1 ("crystal" distribution). The complexity of this collection of distributions covers all possible values of $C_{\max}$.

There is actually a time evolution of the tetrahedral gas, or trajectory of the system, formed by this collection of distributions. Inserting (4.34) in the evolution (4.30), it is found that all equations are compatible with each other and the dynamical equations are reduced to the relation,

$$\frac{df_{\max}}{dt} = \frac{1}{3}(4f_{\max}^2 - 5f_{\max} + 1). \tag{4.35}$$

This trajectory is denoted as the *maximum complexity path.*

Note that the equiprobability or equilibrium, $f_{\max} = 1/4$, is a stable fixed point and the maximum disequilibrium "crystal" distribution, $f_{\max} = 1$, is an unstable fixed point. Thus the maximum complexity path is a heteroclinic connection between the "crystal" and equiprobability distributions.

The maximum complexity path is locally attractive. Let us assume, for instance, the following perturbed trajectory

$$\begin{aligned} f_1 &= f_{\max}, \\ f_2 &= \frac{1 - f_{\max}}{3}, \\ f_3 &= \frac{1 - f_{\max}}{3} + \delta, \\ f_4 &= \frac{1 - f_{\max}}{3} - \delta, \end{aligned} \tag{4.36}$$

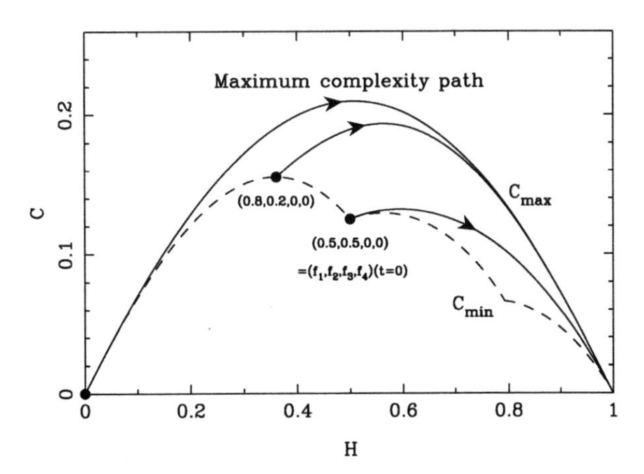

Fig. 4.15 The time evolution of the system for three different initial conditions, $t = 0$, $(f_1, f_2, f_3, f_4) = (0.8, 0.2, 0, 0)$, $(f_1, f_2, f_3, f_4) = (0.5, 0.5, 0, 0)$, and the maximum complexity path are shown. The minimum complexity is shown by *dashed lines*. It can be seen how the system tends to approach the maximum complexity path as it evolves in time toward equilibrium

whose evolution according to (4.30) gives the exponential decay of the perturbation, δ:

$$\frac{d\delta}{dt} \sim -\left(\frac{4 f_{\max} + 2}{3}\right)\delta, \tag{4.37}$$

showing the attractive nature of these trajectories.

4.1.2.4 Conclusions and Further Remarks

In the former section, the time evolution of the LMC complexity, C, has been studied for a simplified model of an isolated ideal gas: the tetrahedral gas. In general, the dynamical behavior of this quantity is bounded between two extremum curves, $C_{\max}$ and $C_{\min}$, when observed in a C versus H phase space. These complexity bounds have been derived and computed. A continuation of this work applied to the study of complexity in gases out of equilibrium can be found in [51, 52].

For the isolated tetrahedral gas two constraints on its dynamics are found. The first, which is analytically demonstrated, is that the disequilibrium, D, decreases monotonically with time until it reaches the value $D = 0$ for the equilibrium state. The second is that the maximum complexity paths, $C_{\max}$, are attractive in phase space. In other words, the complexity of the system tends to equilibrium always approaching those paths. This has been verified numerically, that is, the time derivative of the difference between $C_{\max}$ and C is negative. Figure 4.15 summarizes the dynamical behavior of the tetrahedral gas. The different trajectories starting with arbitrary initial conditions, which represent systems out of equilibrium, evolve towards equilibrium approaching the maximum complexity path.

Whether these properties are useful in real physical systems can need of a further inspection, particularly the macroscopical nature of the disequilibrium in more general systems, such as to the ideal gas following the complete Boltzmann integro–differential equation. Another feature that could deserve attention is the possibility of approximating the evolution of a real physical system trajectory to its maximum complexity path. Note that in general, for a real system, the calculation of the maximum complexity path will not be an easy task.

4.2 The Statistical Complexity in the Continuous Case

As explained in the former sections, the LMC statistical measure of complexity [7] identifies the entropy or information stored in a system and its distance to the equilibrium probability distribution, the disequilibrium, as the two ingredients giving the correct asymptotic properties of a well-behaved measure of complexity. In fact, it vanishes both for completely ordered and for completely random systems. Besides giving the main features of an intuitive notion of complexity, it has been shown that LMC complexity successfully enables us to discern situations regarded as complex in discrete systems out of equilibrium: one instance of phase transitions via intermittency in coupled logistic maps [18] or via stochastic synchronization in cellular automata [43], the dynamical behavior of this quantity in a out-equilibrium gases [19, 51, 52] and other applications in classical statistical mechanics [31, 53].

A possible formula of LMC complexity for continuous systems was suggested in formula (4.4). Anteneodo and Plastino [49] pointed out some peculiarities concerning such an extension for continuous probability distributions. It is the aim of this section to offer a discussion of the extension of LMC complexity for continuous systems and to present a slightly modified extension [54] of expression (4.4) that displays interesting and very striking properties. A further generalization of this work has been done in [55, 56].

In Sect. 4.2.1 the extension of information and disequilibrium concepts for the continuous case are discussed. In Sect. 4.2.2 the LMC measure of complexity is reviewed and possible extensions for continuous systems are suggested. We proceed to present some properties of one of these extensions in Sect. 4.2.3.

4.2.1 Entropy/Information and Disequilibrium

Depending on the necessary conditions to fulfill, the extension of an established formula from the discrete to the continuous case always requires a careful study and in many situations some kind of choice between several possibilities. Next we carry out this process for the entropy and disequilibrium formulas.

4.2.1.1 Entropy or Information

As we know, given a discrete probability distribution $\{p_i\}_{i=1,2,\dots,N}$ satisfying $p_i \geq 0$ and $\sum_{i=1}^{N} p_i = 1$, the *Boltzmann-Gibss-Shannon formula* [4] that accounts for the entropy or information, S, stored in a system is defined by

$$S(\{p_i\}) = -k \sum_{i=1}^{N} p_i \log p_i, \tag{4.38}$$

where k is a positive constant. If we identify H with S, then some properties of this quantity are: (i) *positivity*: $H \geq 0$ for any arbitrary set $\{p_i\}$, (ii) *concavity*: H is concave for arbitrary $\{p_i\}$ and reaches the extremal value for equiprobability ($p_i = 1/N$ $\forall i$), (iii) *additivity*: $H(A \cup B) = H(A) + H(B)$ where A and B are two independent systems, and (iv) *continuity*: H is continuous for each of its arguments. And vice versa, it has been shown that the only function of $\{p_i\}$ verifying the latter properties is given by (4.38) [4, 57]. For an isolated system, the *irreversibility* property is also verified, that is, the time derivative of H is positive, $dH/dt \geq 0$, reaching the equality only for equilibrium.

Calculation of H for a continuous probability distribution $p(x)$, with support on $[-L, L]$ and $\int_{-L}^{L} p(x)\,dx = 1$, can be performed by dividing the interval $[-L, L]$ in small equal-length pieces $\Delta x = x_i - x_{i-1}$, $i = 1, \dots, n$, with $x_0 = -L$ and $x_n = L$, and by considering the approximated discrete distribution $\{p_i\} = \{p(\bar{x}_i)\Delta x\}$, $i = 1, \dots, n$, with $\bar{x}_i$ a point in the segment $[x_{i-1}, x_i]$. It gives us

$$\begin{aligned} H^* &= H(\{p_i\}) \\ &= -k \sum_{i=1}^{n} p(\bar{x}_i) \log p(\bar{x}_i) \Delta x - k \sum_{i=1}^{n} p(\bar{x}_i) \log(\Delta x) \Delta x. \end{aligned} \tag{4.39}$$

The second adding term of H^* in the expression (4.39) grows as $\log n$ when n goes to infinity. Therefore it seems reasonable to take just the first and finite adding term of H^* as the extension of H to the continuous case: $H(p(x))$. It characterizes with a finite number the information contained in a continuous distribution $p(x)$. In the limit $n \to \infty$, we obtain

$$\begin{aligned} H(p(x)) &= \lim_{n\to\infty} \left[-k \sum_{i=1}^{n} p(\bar{x}_i) \log p(\bar{x}_i) \Delta x \right] \\ &= -k \int_{-L}^{L} p(x) \log p(x)\, dx. \end{aligned} \tag{4.40}$$

If $p(x) \geq 1$ in some region, the entropy defined by (4.40) can become negative. Although this situation is mathematically possible and coherent, it is unfounded from a physical point of view. See [58] for a discussion on this point. Let $f(p,q)$ be a probability distribution in phase space with coordinates (p,q), $f \geq 0$ and $dp\,dq$ having the dimension of an action. In this case the volume element is $dp\,dq/h$ with h the Planck constant. Suppose that $H(f) < 0$. Because of $\int (dp\,dq/h) f = 1$, the extent of the region where $f > 1$ must be smaller than h. Hence a negative classical entropy arises if one tries to localize a particle in phase space in a region smaller than h, that is, if the uncertainty relation is violated. In consequence, not every classical probability distribution can be observed in nature. The condition $H(f) = 0$ could give us the minimal width that is physically allowed for the distribution and so the maximal localization of the system under study. This *cutting* property has been used in the calculations performed in [53].

4.2.1.2 Disequilibrium

Given a discrete probability distribution $\{p_i\}_{i=1,2,\ldots,N}$ satisfying $p_i \geq 0$ and $\sum_{i=1}^{N} p_i = 1$, its *Disequilibrium*, D, can be defined as the quadratic distance of the actual probability distribution $\{p_i\}$ to equiprobability:

$$D(\{p_i\}) = \sum_{i=1}^{N} \left(p_i - \frac{1}{N} \right)^2 . \tag{4.41}$$

D is maximal for fully regular systems and vanishes for completely random ones.

In the continuous case with support on the interval $[-L, L]$, the rectangular function $p(x) = 1/(2L)$, with $-L < x < L$, is the natural extension of the equiprobability distribution of the discrete case. The disequilibrium could be defined as

$$D^* = \int_{-L}^{L} \left(p(x) - \frac{1}{2L} \right)^2 dx = \int_{-L}^{L} p^2(x)\, dx - \frac{1}{2L} . \tag{4.42}$$

If we redefine D omitting the constant adding term in D^*, the disequilibrium reads now:

$$D(p(x)) = \int_{-L}^{L} p^2(x)\, dx . \tag{4.43}$$

$D > 0$ for every distribution and it is minimal for the rectangular function which represents the equipartition. D does also tend to infinity when the width of $p(x)$ narrows strongly and becomes extremely peaked.

4.2.2 *The Continuous Version $\hat{C}$ of the LMC Complexity*

As shown in the previous sections, LMC complexity has been successfully calculated in different systems out of equilibrium. However, Feldman and Crutchfield [21] presented as a main drawback that C vanishes and it is not an extensive variable for finite-memory regular Markov chains when the system size increases. This is not the general behavior of C in the thermodynamic limit as it has been suggested by Calbet and López-Ruiz [19]. On the one hand, when $N \to \infty$ and $k = 1/\log N$, LMC complexity is not a trivial function of the entropy, in the sense that for a given

H there exists a range of complexities between 0 and $C_{\max}(H)$, where $C_{\max}$ is given by expression (4.23).

Observe that in this case H is normalized, $0 < H < 1$, because $k = 1/\log N$. On the other hand, non-extensivity cannot be considered as an obstacle since it is nowadays well known that there exists a variety of physical systems for which the classical statistical mechanics seems to be inadequate and for which an alternative non-extensive thermodynamics is being hailed as a possible basis of a theoretical framework appropriate to deal with them [59].

According to the discussion in Sect. 4.2.1, the expression of C for the case of a continuum number of states, x, with support on the interval $[-L, L]$ and $\int_{-L}^{L} p(x)\,dx = 1$, is defined by

$$\begin{aligned} C(p(x)) &= H(p(x)) \cdot D(p(x)) \\ &= \left(-k \int_{-L}^{L} p(x) \log p(x)\,dx\right) \cdot \left(\int_{-L}^{L} p^2(x)\,dx\right). \end{aligned} \tag{4.44}$$

Hence, C can become negative. Obviously, $C < 0$ implies $H < 0$. Although this situation is coherent from a mathematical point of view, it is not physically possible. Hence a negative entropy means to localize a system in phase space into a region smaller than h (Planck constant) and this would imply to violate the uncertainty principle (see discussion of Sect. 4.2.1.1). Then a distribution can broaden without any limit but it cannot become extremely peaked. The condition $H = 0$ could indicate the minimal width that $p(x)$ is allowed to have. Similarly to the discrete case, C is positive for any situation and vanishes both for an extreme localization and for the most widely delocalization embodied by the equiprobability distribution. Thus, LMC complexity can be straightforwardly calculated for any continuous distribution by (4.44). Anyway, the positivity of C for every distribution in the continuous case can be recovered by taking the exponential of S [60] and redefining H according to this exponential, i.e. $H = e^S$. To maintain the same nomenclature than in the precedent text we continue to identify H with S and we introduce the symbol $\hat{H} = e^H$. Then the new expression of the statistical measure of complexity C is identified as $\hat{C}$ in the rest of this section and is given by [54]

$$\hat{C}(p(x)) = \hat{H}(p(x)) \cdot D(p(x)) = e^{H(p(x))} \cdot D(p(x)). \tag{4.45}$$

In addition to the positivity, $\hat{C}$ encloses other interesting properties that we describe in the next section.

4.2.3 Properties of $\hat{C}$

The quantity $\hat{C}$ given by (4.45) has been presented as one of the possible extensions of the LMC complexity for continuous systems [54]. We proceed now to present some of the properties that characterize such a complexity indicator.

4.2.3.1 Invariance under Translations and Rescaling Transformations

If $p(x)$ is a density function defined on the real axis $\mathbf{R}$, $\int_{\mathbf{R}} p(x)\,dx = 1$, and $\alpha > 0$ and β are two real numbers, we denote by $p_{\alpha,\beta}(x)$ the new probability distribution obtained by the action of a β-translation and an α-rescaling transformation on $p(x)$,

$$p_{\alpha,\beta}(x) = \alpha p(\alpha(x - \beta)). \tag{4.46}$$

When $\alpha < 1$, $p_{\alpha,\beta}(x)$ broadens whereas if $\alpha > 1$ it becomes more peaked. Observe that $p_{\alpha,\beta}(x)$ is also a density function. After making the change of variable $y = \alpha(x - \beta)$ we obtain

$$\int_{\mathbf{R}} p_{\alpha,\beta}(x)\,dx = \int_{\mathbf{R}} \alpha p(\alpha(x - \beta))\,dx = \int_{\mathbf{R}} p(y)\,dy = 1. \tag{4.47}$$

The behaviour of H under the transformation given by (4.46) is the following:

$$\begin{aligned} H(p_{\alpha,\beta}) &= -\int_{\mathbf{R}} p_{\alpha,\beta}(x) \log p_{\alpha,\beta}(x)\,dx = -\int_{\mathbf{R}} p(y) \log(\alpha p(y))\,dy \\ &= -\int_{\mathbf{R}} p(y) \log p(y)\,dy - \log\alpha \int_{\mathbf{R}} p(y)\,dy \\ &= H(p) - \log\alpha. \end{aligned} \tag{4.48}$$

Then,

$$\hat{H}(p_{\alpha,\beta}) = e^{H(p_{\alpha,\beta})} = \frac{\hat{H}(p)}{\alpha}. \tag{4.49}$$

It is straightforward to see that $D(p_{\alpha,\beta}) = \alpha D(p)$, and to conclude that

$$\hat{C}(p_{\alpha,\beta}) = \hat{H}(p_{\alpha,\beta}) \cdot D(p_{\alpha,\beta}) = \frac{\hat{H}(p)}{\alpha} \alpha D(p) = \hat{C}(p). \tag{4.50}$$

Observe that translations and rescaling transformations keep also the shape of the distributions. Then it could be reasonable to denominate the invariant quantity $\hat{C}$ as the *shape complexity* of the family formed by a distribution $p(x)$ and its transformed $p_{\alpha,\beta}(x)$. Hence, for instance, the rectangular $\Pi(x)$, the isosceles-triangle shaped $\Lambda(x)$, the Gaussian $\Gamma(x)$, or the exponential $\Xi(x)$ distributions continue to belong to the same Π, Λ, Γ or Ξ family, respectively, after applying the transformations defined by (4.46). Calculation of $\hat{C}$ on these distribution families gives us

$$\hat{C}(\Pi) = 1, \tag{4.51}$$

$$\hat{C}(\Lambda) = \frac{2}{3}\sqrt{e} \approx 1.0991, \tag{4.52}$$

$$\hat{C}(\Gamma) = \sqrt{\frac{e}{2}} \approx 1.1658, \tag{4.53}$$

$$\hat{C}(\Xi) = \frac{e}{2} \approx 1.3591. \tag{4.54}$$

Remark that the family of rectangular distributions has a smaller $\hat{C}$ than the rest of distributions. This fact is true for every distribution and it will be proved in Sect. 4.2.3.4.

4.2.3.2 Invariance under Replication

Lloyd and Pagels [15] recommend that a complexity measure should remain essentially unchanged under replication. We show now that $\hat{C}$ is replicant invariant, that is, the shape complexity of m replicas of a given distribution is equal to the shape complexity of the original one.

Suppose $p(x)$ a compactly supported density function, $\int_{-\infty}^{\infty} p(x)\,dx = 1$. Take n copies $p_m(x)$, $m = 1,\ldots,n$, of $p(x)$,

$$p_m(x) = \frac{1}{\sqrt{n}} p(\sqrt{n}(x-\lambda_m)), \quad 1 \le m \le n, \tag{4.55}$$

where the supports of all the $p_m(x)$, centered at $\lambda'_m s$ points, $m = 1,\ldots,n$, are all disjoint. Observe that $\int_{-\infty}^{\infty} p_m(x)\,dx = \frac{1}{n}$, what make the union

$$q(x) = \sum_{i=1}^{n} p_m(x) \tag{4.56}$$

to be also a normalized probability distribution, $\int_{-\infty}^{\infty} q(x)\,dx = 1$. For every $p_m(x)$, a straightforward calculation shows that

$$H(p_m) = \frac{1}{n} H(p) + \frac{1}{n} \log \sqrt{n}, \tag{4.57}$$

$$D(p_m) = \frac{1}{n\sqrt{n}} D(p). \tag{4.58}$$

Taking into account that the m replicas are supported on disjoint intervals on $\mathbf{R}$, we obtain

$$H(q) = H(p) + \log \sqrt{n}, \tag{4.59}$$

$$D(q) = \frac{1}{\sqrt{n}} D(p). \tag{4.60}$$

Then,

$$\hat{C}(q) = \hat{C}(p), \tag{4.61}$$

what completes the proof of the replicant invariance of $\hat{C}$.

4.2.3.3 Near-Continuity

Continuity is a desirable property of an indicator of complexity. For a given scale of observation, similar systems should have a similar complexity. In the continuous case, similarity between density functions defined on a common support suggests that they take close values almost everywhere. More strictly speaking, let δ be a positive real number. It will be said that two density functions $f(x)$ and $g(x)$ defined on the interval $I \in \mathbf{R}$ are *δ-neighboring functions* on I if the Lebesgue measure of the points $x \in I$ verifying $|f(x) - g(x)| \ge \delta$ is zero. A real map T defined on density functions on I will be called *near-continuous* if for any $\varepsilon > 0$ there exists $\delta(\varepsilon) > 0$ such that if $f(x)$ and $g(x)$ are δ-neighboring functions on I then $|T(f)-T(g)| < \varepsilon$.

It can be shown that the information H, the disequilibrium D and the shape complexity $\hat{C}$ are near-continuous maps on the space of density functions defined on a compact support. We must stress at this point the importance of the compactness condition of the support in order to have near-continuity. Take, for instance, the density function defined on the interval $[-1, L]$,

$$g_{\delta,L}(x) = \begin{cases} 1-\delta & \text{if } -1 \le x \le 0, \\ \frac{\delta}{L} & \text{if } 0 \le x \le L, \\ 0 & \text{otherwise,} \end{cases} \tag{4.62}$$

with $0 < \delta < 1$ and $L > 1$. If we calculate H and D for this distribution we obtain

$$H(g_{\delta,L}) = -(1-\delta)\log(1-\delta) - \delta \log\left(\frac{\delta}{L}\right), \tag{4.63}$$

$$D(g_{\delta,L}) = (1-\delta)^2 + \frac{\delta^2}{L}. \tag{4.64}$$

Consider also the rectangular density function

$$\chi_{[-1,0]}(x) = \begin{cases} 1 & \text{if } -1 \le x \le 0, \\ 0 & \text{otherwise.} \end{cases} \tag{4.65}$$

If $0 < \delta < \bar{\delta} < 1$, $g_{\delta,L}(x)$ and $\chi_{[-1,0]}(x)$ are $\bar{\delta}$-neighboring functions. When $\delta \to 0$, we have that $\lim_{\delta\to 0} g_{\delta,L}(x) = \chi_{[-1,0]}(x)$. In this limit process the support is maintained and near-continuity manifests itself as following,

$$\left[\lim_{\delta\to 0} \hat{C}(g_{\delta,L})\right] = \hat{C}(\chi_{[-1,0]}) = 1. \tag{4.66}$$

But if we allow the support L to become infinitely large, the compactness condition is not verified and, although $\lim_{L\to\infty} g_{\delta,L}(x)$ and $\chi_{[-1,0]}(x)$ are $\bar{\delta}$-neighboring distributions, we have that

$$\left[\left(\lim_{L\to\infty} \hat{C}(g_{\delta,L})\right) \to \infty\right] \neq \hat{C}(\chi_{[-1,0]}) = 1. \tag{4.67}$$

Then near-continuity in the map $\hat{C}$ is lost due to the non-compactness of the support when $L \to \infty$. This example suggests that the shape complexity $\hat{C}$ is near-continuous on compact supports and this property will be rigorously proved elsewhere.

4.2.3.4 The Minimal Shape Complexity

If we calculate $\hat{C}$ on the example given by (4.62), we can verify that the shape complexity can be as large as wanted. Take, for instance, $\delta = \frac{1}{2}$. The measure $\hat{C}$ reads now

$$\hat{C}(g_{\delta=\frac{1}{2},L}) = \frac{1}{2}\sqrt{L}\left(1 + \frac{1}{L}\right). \tag{4.68}$$

Thus $\hat{C}$ becomes infinitely large after taking the limits $L \to 0$ or $L \to \infty$. Remark that even in the case $g_{\delta,L}$ has a finite support, $\hat{C}$ is not upper bounded. The density functions, $g_{(\delta=\frac{1}{2}),(L\to 0)}$ and $g_{(\delta=\frac{1}{2}),(L\to\infty)}$, of infinitely increasing complexity have two zones with different probabilities. In the case $L \to 0$ there is a narrow zone where probability rises to infinity and in the case $L \to \infty$ there exists an increasingly large zone where probability tends to zero. Both kind of density functions show a similar pattern to distributions of maximal LMC complexity in the discrete case, where there is an state of dominating probability and the rest of states have the same probability.

The minimal $\hat{C}$ given by (4.68) is found when $L = 1$, that is, when $g_{\delta,L}$ becomes the rectangular density function $\chi_{[-1,1]}$. In fact, the value $\hat{C} = 1$ is the minimum of possible shape complexities and it is reached only on the rectangular distributions. We sketch now some steps that prove this result.

Suppose

$$f = \sum_{k=1}^{n} \lambda_k \chi_{E_k} \tag{4.69}$$

to be a density function consisting of several rectangular pieces E_k, $k = 1, \ldots, n$, on disjoint intervals. If μ_k is the Lebesgue measure of E_k, calculation of $\hat{C}$ gives

$$\hat{C}(f) = \prod_{k=1}^{n} \left(\lambda_k^{-\lambda_k \mu_k}\right) \cdot \left(\sum_{k=1}^{n} \lambda_k^2 \mu_k\right). \tag{4.70}$$

Lagrange multipliers method is used to find the real vector $(\mu_1, \mu_2, \ldots, \mu_n; \lambda_1, \lambda_2, \ldots, \lambda_n)$ that makes extremal the quantity $\hat{C}(f)$ under the condition $\sum_{k=1}^{n} \lambda_k \mu_k = 1$. This is equivalent to studying the extrema of $\log \hat{C}(f)$. We define the function $z(\lambda_k, \mu_k) = \log \hat{C}(f) + \alpha(\sum_{k=1}^{n} \lambda_k \mu_k - 1)$, then

$$z(\lambda_k, \mu_k) = -\sum_{k=1}^{n} \mu_k \lambda_k \log \lambda_k + \log\left(\sum_{k=1}^{n} \mu_k \lambda_k^2\right) + \alpha\left(\sum_{k=1}^{n} \lambda_k \mu_k - 1\right). \tag{4.71}$$

Differentiating this expression and making the result equal to zero we obtain

$$\frac{\partial z(\lambda_k, \mu_k)}{\partial \lambda_k} = -\mu_k \log \lambda_k - \mu_k + \frac{2\lambda_k \mu_k}{\sum_{j=1}^{n} \mu_j \lambda_j^2} + \alpha \mu_k = 0, \tag{4.72}$$

$$\frac{\partial z(\lambda_k, \mu_k)}{\partial \mu_k} = -\lambda_k \log \lambda_k + \frac{\lambda_k^2}{\sum_{j=1}^{n} \mu_j \lambda_j^2} + \alpha \lambda_k = 0. \tag{4.73}$$

Dividing (4.72) by μ_k and (4.73) by λ_k we get

$$\frac{2\lambda_k}{\sum_{j=1}^{n} \mu_j \lambda_j^2} + \alpha - 1 = \log \lambda_k, \tag{4.74}$$

$$\frac{\lambda_k}{\sum_{j=1}^{n} \mu_j \lambda_j^2} + \alpha = \log \lambda_k. \tag{4.75}$$

Solving these two equations for every λ_k we have

$$\lambda_k = \sum_{j=1}^{n} \mu_j \lambda_j^2 \quad \text{for all } k. \tag{4.76}$$

Therefore f is a rectangular function taking the same value λ for every interval E_k, that is, f is the rectangular density function

$$f = \lambda \cdot \chi_L \quad \text{with } \lambda = \frac{1}{\sum_{i=1}^{n} \mu_i} = \frac{1}{L}, \tag{4.77}$$

where L is the Lebesgue measure of the support.

Then $\hat{C}(f) = 1$ is the minimal value for a density function composed of several rectangular pieces because, as we know for the example given by (4.68), $\hat{C}(f)$ is not upper bounded for this kind of distributions.

Furthermore, for every compactly supported density function g and for every $\varepsilon > 0$, it can be shown that near-continuity of $\hat{C}$ allows to find a δ-neighboring density function f of the type given by expression (4.69) verifying $|\hat{C}(f) - \hat{C}(g)| < \varepsilon$. The arbitrariness of the election of ε brings us to conclude that $\hat{C}(g) \geq 1$ for every probability distribution g. Thus, we can conclude that the minimal value of $\hat{C}$ is 1 and it is reached only by the rectangular density functions.

4.3 Fisher-Shannon Information Product. Some Applications

4.3.1 Fisher-Shannon Information: Definition and Properties

The description of electronic properties by means of information measures was introduced into quantum chemistry by the pioneering works [61–65]. In particular Shannon entropy [66] and Fisher information [67] have attracted special attention in atomic and molecular physics. (See e.g. [68–97].) It is known that these two information measures give complementary descriptions of the concentration and uncertainty of the probability density: S_ρ (I_ρ) can be seen as a global (local) measure of spreading. In this context, the Fisher-Shannon information product was found as a link between these information measures to improve the characterization of a probability density function in terms of information measures [77].

The single-electron density, the basic variable of the density functional theory [98] of D-dimensional many-electron systems is given by

$$\rho(\mathbf{r}) = \int |\Psi(\mathbf{r}, \mathbf{r}_2, \ldots, \mathbf{r}_N)|^2 d^D\mathbf{r}_2 \ldots d^D\mathbf{r}_N \tag{4.78}$$

where $\Psi(\mathbf{r}_1, \ldots, \mathbf{r}_N)$ denotes the normalized wavefunction of the N-electron system and $\rho(\mathbf{r})$ is normalized to unity. The spreading of this quantity is best measured by the Shannon information entropy

$$S_\rho = -\int \rho(\mathbf{r}) \ln \rho(\mathbf{r}) d^D\mathbf{r}, \tag{4.79}$$

or equivalently by the Shannon entropy power [60, 66]

$$J_\rho \equiv \frac{1}{2\pi e} e^{\frac{2}{D} S_\rho}. \tag{4.80}$$

On the other hand the Fisher information [60, 67] of $\rho(\mathbf{r})$ is given by

$$I_\rho = \int \frac{|\nabla \rho(\mathbf{r})|^2}{\rho(\mathbf{r})} d^D\mathbf{r}. \tag{4.81}$$

The sharpness, concentration or delocalization of the electronic cloud is measured by both quantities. It is known that these two information measures give complementary descriptions of the smoothness and uncertainty of the electron localization: S_ρ and I_ρ are global and local measures of smoothness, respectively [60–67, 77].

For completeness let us point out that the aforementioned information measures, which refer to an unity-normalized density $\rho_1(\mathbf{r}) \equiv \rho(\mathbf{r})$, are related to the corresponding measures of the N-normalized density $\rho_N(\mathbf{r})$ by

$$S_{\rho_N} = -N \ln N + N S_\rho \quad \text{and} \quad I_{\rho_N} = N I_\rho \tag{4.82}$$

for the Shannon and Fisher quantities, respectively.

The information product concept P_ρ was originally defined in [77] as

$$P_\rho \equiv \frac{1}{D} J_\rho I_\rho, \tag{4.83}$$

and it was applied in the study of electronic properties of quantum systems during last years. (See, e.g. [77, 90, 93, 94, 96, 99–101].) Next we will put forward some mathematical properties which have been obtained in [77, 82, 102, 103] for the Fisher-Shannon information product P_ρ.

4.3.1.1 Scaling Property

The Fisher information and the Shannon entropy power transform as

$$I_{\rho_\gamma} = \gamma^{D-1} I_\rho; \qquad J_{\rho_\gamma} = \gamma^{-(D-1)} J_\rho \tag{4.84}$$

under scaling of the probability density $\rho(\mathbf{r})$ by a real scalar factor γ; i.e. when $\rho_\gamma(\mathbf{r}) = \gamma^D \rho(\gamma \mathbf{r})$. This indicates that they are homogeneous density functionals of degrees 2 and -2, respectively. Consequently, the information product $P_\rho = \frac{1}{D} J_\rho I_\rho$ is invariant under this scaling transformation, i.e.

$$P_{\rho_\gamma} = P_\rho. \tag{4.85}$$

4.3.1.2 Uncertainty Properties

The Fisher information I_ρ and the Shannon entropy power J_ρ satisfy the uncertainty relationship [60]

$$\frac{1}{D} J_\rho I_\rho \geq 1. \tag{4.86}$$

Remark that when one of the involved quantities decreases near to zero, the other has to increase to a large value. Moreover, it is closely linked to the uncertainty relation $\langle r^2\rangle\langle p^2\rangle \geq \frac{D^2}{4}$, where $\langle r^2\rangle$ is defined in terms of the charge position density $\rho(\mathbf{r})$ as $\langle r^2\rangle = \int r^2\rho(\mathbf{r})d^D\mathbf{r}$, and $\langle p^2\rangle$ is given in terms of the momentum density $\Pi(\mathbf{p})$ in an analogous way, where $\Pi(\mathbf{p})$ is defined by means of the Fourier transform of $\Psi(\mathbf{r}_1,\ldots,\mathbf{r}_N)$, $\Phi(\mathbf{p}_1,\ldots,\mathbf{p}_N)$, as

$$\Pi(\mathbf{p}) = \int |\Phi(\mathbf{p},\mathbf{p}_2,\ldots,\mathbf{p}_N)|^2 d^D\mathbf{p}_2\ldots d^D\mathbf{p}_N. \tag{4.87}$$

The Fisher information has been used as a measure of uncertainty in quantum physics. (See e.g. [82, 103–112].) It has been shown to fulfill the Stam inequalities [113]

$$I_\rho \leq 4\langle p^2\rangle; \qquad I_\pi \leq 4\langle r^2\rangle, \tag{4.88}$$

and the Cramer-Rao inequalities [60, 102, 112, 114, 115]

$$I_\rho \geq \frac{D^2}{\langle r^2\rangle}; \qquad I_\pi \geq \frac{D^2}{\langle p^2\rangle} \tag{4.89}$$

for the general single-particle systems. The multiplication of each pair of these inequalities produces

$$\frac{D^4}{\langle r^2\rangle\langle p^2\rangle} \leq I_\rho I_\pi \leq 16\langle r^2\rangle\langle p^2\rangle, \tag{4.90}$$

valid for ground and excited states of general systems, which shows the close connection between the Heisenberg-like uncertainty product and the product of the position and momentum Fisher informations.

Indeed, taken into account $1/D\langle r^2\rangle \geq J_\rho$ [116] one has that

$$\frac{4}{D^2}\langle p^2\rangle\langle r^2\rangle \geq \frac{1}{D}J_\rho I_\rho \geq 1 \tag{4.91}$$

and

$$\frac{4}{D^2}\langle p^2\rangle\langle r^2\rangle \geq \sqrt{P_\rho P_\pi} \geq 1. \tag{4.92}$$

It is straightforward to show that the equality limit of these two inequalities is reached for Gaussian densities.

An special case is given by a single-particle in a central potential. In this framework an uncertainty Fisher information relation was obtained in [103]:

$$I_\rho I_\pi \geq 4D^2\left[1 - \frac{(2l+1)|m|}{2l(l+1)}\right]^2 \tag{4.93}$$

and Fisher information in position space was derived in [82] as

$$I_\rho = 4\langle p^2\rangle - 2(2l+1)|m|\langle r^{-2}\rangle \tag{4.94}$$

where l and m are the orbital and magnetic quantum numbers. Taking into account the duality of the position and momentum spaces as well as the separability of the wavefunction, one can express the Fisher information of the momentum distribution density as

$$I_{\pi} = 4\langle r^2 \rangle - 2(2l+1)|m|\langle p^{-2} \rangle. \tag{4.95}$$

On the other hand, the radial expectation values $\langle p^2 \rangle$ and $\langle r^{-2} \rangle$ ($\langle r^2 \rangle$ and $\langle p^{-2} \rangle$) are related [82, 103] by

$$\langle p^2 \rangle \geq l(l+1)\langle r^{-2} \rangle, \tag{4.96}$$

$$\langle r^2 \rangle \geq l(l+1)\langle p^{-2} \rangle, \tag{4.97}$$

and combining above expressions the fisher uncertainty-like relation (4.93) is obtained.

4.3.1.3 Nonadditivity Properties

The superadditivity of the Fisher information and the subadditivity of the Shannon information of a probability density, can be used to prove [77] that

$$I_W \geq N I_{\rho}, \tag{4.98}$$

$$S_W \leq N S_{\rho}, \tag{4.99}$$

where

$$I_W = \int \frac{|\nabla |\Psi(\mathbf{r}_1, \ldots, \mathbf{r}_N)|^2|^2}{|\Psi(\mathbf{r}_1, \ldots, \mathbf{r}_N)|^2} d\mathbf{r}_1 \ldots d\mathbf{r}_N \tag{4.100}$$

and

$$S_W = \int |\Psi(\mathbf{r}_1, \ldots, \mathbf{r}_N)|^2 \ln |\Psi(\mathbf{r}_1, \ldots, \mathbf{r}_N)|^2 d\mathbf{r}_1 \ldots d\mathbf{r}_N \tag{4.101}$$

for general N-fermion systems in three dimensions. The D-dimensional generalization is obvious. We will show the proof below.

Let $\rho(\mathbf{r})$ a probability density on $\mathbf{R}^t$, that is, $\rho(\mathbf{r})$ non-negative and $\int \rho(\mathbf{r}) d\mathbf{r} = 1$. We will suppose that Fisher information and Shannon information of $\rho(\mathbf{r})$ exits. Corresponding to any orthogonal decomposition $\mathbf{R}^t = \mathbf{R}^r \oplus \mathbf{R}^s$, $t = r + s$, the marginal densities are given by:

$$\rho_1(\mathbf{x}) = \int_{\mathbf{R}^r} \rho(\mathbf{x}, \mathbf{y}) d^r \mathbf{y}, \qquad \rho_2(\mathbf{y}) = \int_{\mathbf{R}^s} \rho(\mathbf{x}, \mathbf{y}) d^s \mathbf{x} \tag{4.102}$$

then [117]

$$I_{\rho} \geq I_{\rho_1} + I_{\rho_2} \tag{4.103}$$

property which is known as superadditivity of Fisher information, and

$$S_{\rho} \leq S_{\rho_1} + S_{\rho_2} \tag{4.104}$$

which is known as subadditivity of Shannon information. Both inequalities saturate when $\rho(\mathbf{x}, \mathbf{y}) = \rho_1(\mathbf{x})\rho_2(\mathbf{y})$ [117].

On the other hand, let us consider an N-fermion system and denote the ith-electron density by

$$\rho_i \equiv \rho(\mathbf{r}_i) = \int |\Psi(\mathbf{r}_1, \ldots, \mathbf{r}_i, \ldots, \mathbf{r}_N)|^2 d\mathbf{r}_1 \ldots d\mathbf{r}_{i-1} d\mathbf{r}_{i+1} \ldots d\mathbf{r}_N, \qquad (4.105)$$

for $i = 1, \ldots, N$. Then, taken into account that the wavefunction is antisymmetric and (4.103) and (4.104), the wavefunction Fisher information fulfills

$$I_W = \int \frac{|\nabla|\Psi(\mathbf{r}_1, \ldots, \mathbf{r}_N)|^2|^2}{|\Psi(\mathbf{r}_1, \ldots, \mathbf{r}_N)|^2} d\mathbf{r}_1 \ldots d\mathbf{r}_N \geq \sum_{i=1}^{N} I_{\rho_i} = N I_\rho, \qquad (4.106)$$

and the wavefunction Shannon information fulfills:

$$S_W = \int |\Psi(\mathbf{r}_1, \ldots, \mathbf{r}_N)|^2 \ln |\Psi(\mathbf{r}_1, \ldots, \mathbf{r}_N)|^2 d\mathbf{r}_1 \ldots d\mathbf{r}_N \leq \sum_{i=1}^{N} S_{\rho_i} = N S_\rho. \qquad (4.107)$$

Inequalities (4.106) and (4.107) are equalities when $|\Psi(\mathbf{r}_1, \ldots, \mathbf{r}_N)|^2 = \rho(\mathbf{r}_1) \ldots \rho(\mathbf{r}_N)$.

These properties have allowed us to generalize the following uncertainty relationships:

- The Stam's uncertainty relation for wave functions normalized to unity [77, 113] is generalized via the inequality (4.106) by

$$N I_\rho \leq I_W \leq 4N \langle p^2 \rangle \qquad (4.108)$$

- The Shannon information uncertainty relation for wave functions normalized to unity [116] is generalized via inequality (4.107) by

$$\begin{aligned} 3N(1 + \ln \pi) &\leq -\int |\Psi(\mathbf{r}_1, \ldots, \mathbf{r}_N)|^2 \ln |\Psi(\mathbf{r}_1, \ldots, \mathbf{r}_N)|^2 d\mathbf{r}_1 \ldots d\mathbf{r}_N \\ &\quad - \int |\Phi(\mathbf{p}_1, \ldots, \mathbf{p}_N)|^2 \ln |\Phi(\mathbf{p}_1, \ldots, \mathbf{p}_N)|^2 d\mathbf{p}_1 \ldots d\mathbf{p}_N \qquad (4.109) \\ &\leq N(S_\rho + S_\pi) \qquad (4.110) \end{aligned}$$

  where $S_\rho(S_\pi)$ denotes the Shannon information of the single-particle distribution density in position (momentum) space.

4.3.2 *Fisher-Shannon Product as an Electronic Correlation Measure*

The Fisher-Shannon information product was earlier employed [77] as a tool for studying the electron correlation in atomic systems, in particular in two electron isoelectronic series. The application of this indicator to the electronic shell structure of atoms has received a special attention for systems running from on-electron atoms to many-electron ones as those corresponding to the periodic table (see, e.g. [93, 94, 96, 99, 102]).

Many electron systems such as atoms, molecules and clusters show the electron correlation phenomenon. This feature has been characterized in terms of the correlation energy [118], which gives the difference between the exact non-relativistic energy and the Hartree-Fock approximation, as well as by some statistical correlation coefficients [119], which asses radial and angular correlation in both the position and momentum density distributions. Some information-theoretic measures of the electron correlation in many electron systems have been proposed during last years [77, 120–130]. Here we will focus on the Fisher-Shannon Information Product as a measure of electron correlation.

The Fisher-Shannon Information Product has been studied in two types of two-electron systems [77] which differ in the Coulomb- and oscillator-like form of the electron-nucleus interaction. The Hamiltonian of such a system is

$$H = -\frac{1}{2}\nabla_1^2 - \frac{1}{2}\nabla_2^2 + V(r_1) + V(r_2) + \frac{1}{|\mathbf{r}_1 - \mathbf{r}_2|}, \tag{4.111}$$

where $V(r_i)$ denotes the electron-nucleus interaction of the ith-electron. $V(r_i) = Z/r_i$ for He-like ions (Z being the nuclear charge) and $V(r_i) = \frac{1}{2}\omega r_i^2$ for the Hooke atoms. The Hooke atom is especially well suited for the understanding of correlation phenomena because of its amenability to analytical treatment.

4.3.2.1 He-Like Ions

In the bare coulomb field case (BCF), i.e. without Coulombic interelectronic interaction in the Hamiltonian, the ground state wave function of He(Z) is a single Slater determinant and the charge density is a hydrogenlike one, so $J_{\rho_Z} = \frac{e}{2\pi^{1/3}} \frac{1}{Z^2}$ and $I_{\rho_Z} = 4Z^2$, so $P_{BCF} = K_{BCF}$ with $K_{BCF} \simeq 1.237333$. To consider the inclusion of electronic interaction we will work with the 204-terms Hylleraas type functions of Koga et al. [131] for the ground states of H^-, He, Li^+, Be^{2+}, B^{3+}, and Ne^{8+} ($Z = 1$–5, 10).

In Fig. 4.16 we have compared the dependence of the information product P_{ρ_Z} on the nuclear charge Z for He-like ions with the bare coulomb field information product. It is apparent the monotonic decrease of P_{ρ_Z} when Z increased, asymptotically approaching the bare or no-correlation value $P_{BCF} = K_{BCF}$ and showing that the electron correlation effect gradually decreases with respect to the electron-nucleus interaction when the nuclear charge of the system is raised up.

4.3.2.2 Hooke's Atoms

For the bare oscillator-field case (BOF), it is known that $J_{\rho_\omega} = 1/(2\omega)$ and $I_{\rho_\omega} = 6\omega$, so that the information product $P_{BOF} = 1$. On the other hand the Schrödinger equation of the entire Hooke atom can be solved analytically for an infinite set of oscillator frequencies [132]. The use of relative and center of mass coordinates allows

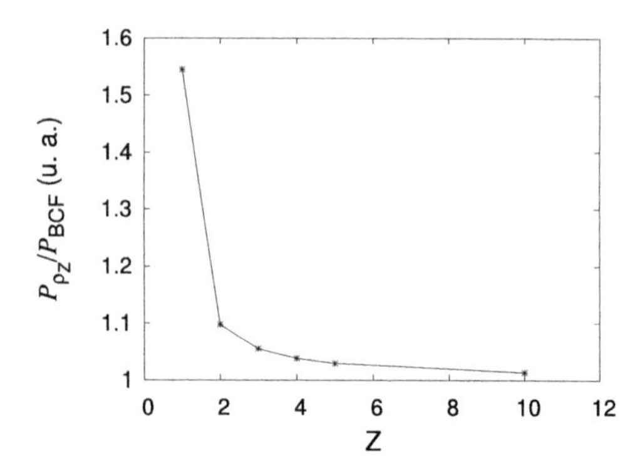

Fig. 4.16 The ratio $P_{\rho Z}/P_{BCF}$ of the information product for the He-like ions and the information product for bare two-electron atoms as a function of the nuclear charge Z. The points correspond to the values of He(Z) ions with $Z = 1$–5 and 10. The *solid line* has been drawn only to guide the eye

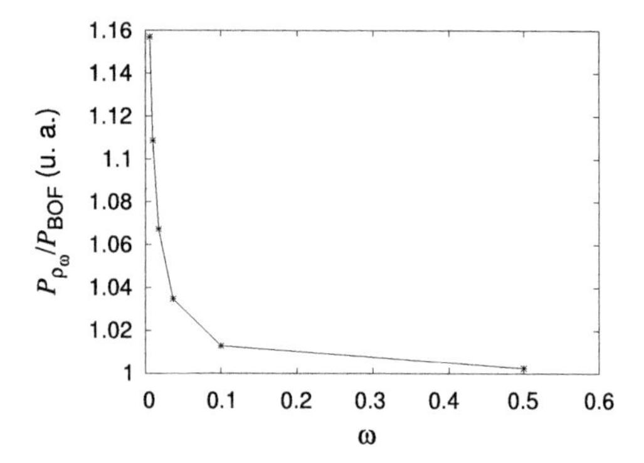

Fig. 4.17 The information product P_{ρ_ω}/P_{BOF} for the Hooke atoms with the oscillator strength $\omega = 0.5$, 0.1, 0.03653727, 0.01734620, 0.009578420, and 0.005841700 and the bare oscillator field information product P_{BOF}. The *solid line* has been drawn only to guide the eye

the Hamiltonian to be separable so that the total wavefunction for singlet states is given by $\Psi(\mathbf{r_1}, \sigma_1, \mathbf{r_2}, \sigma_2) = \xi(\mathbf{R})\Phi(\mathbf{u})\tau(\sigma_1, \sigma_2)$, where $\tau(\sigma_1, \sigma_2)$ is the singlet spin wave function, $\xi(\mathbf{R})$ and $\Phi(\mathbf{u})$ being the solutions of the Schrödinger equations

$$\left(-\frac{1}{4}\nabla_R^2 + \omega R^2\right)\xi(\mathbf{R}) = E_R\xi(\mathbf{R}), \tag{4.112}$$

$$\left(-\nabla_u^2 + \frac{1}{4}\omega u^2 + \frac{1}{u}\right)\Phi(\mathbf{u}) = E_u\Phi(\mathbf{u}), \tag{4.113}$$

respectively, and the total energy $E = E_R + E_u$.

The computed results for the Fisher information and entropy power of these systems are shown in Fig. 4.17 for several ω values, (namely, 0.5, 0.1, 0.03653727, 0.01734620, 0.009578420, and 0.005841700). For these particular values the ground state solution can be obtained [132] as

$$\xi(\mathbf{R}) = \left(\frac{2\omega}{\pi}\right)^{3/4} e^{-\omega R^2} \quad \text{and} \quad \Phi(\mathbf{u}) = e^{-\frac{\omega r^2}{4}} Q_n(r) \tag{4.114}$$

where $Q_n(r)$ is a polynomial whose coefficients can be determined analytically.

Cioslowski et al. [133] quantify the domains of the weakly correlated regime of this system which corresponds to the values of ω greater than $\omega_c \simeq 4.011624 \times 10^{-2}$, and the strongly correlated regime that encompasses the values of ω smaller than ω_c.

In Fig. 4.17 we have drawn P_{ρ_ω}/P_{BOF} as a function of the oscillator electron-nucleus strength ω. It is apparent that the value of the electron density functional P_{ρ_ω}/P_{BOF} (dots) is always bigger than unity, when the electron-electron repulsion becomes very small with respect to the oscillator electron-nucleus interaction, the points approach to the value 1, indicating the decrease of the relative importance of electron correlation when the strength ω is increased.

4.3.3 Fisher Information for a Single Particle in a Central Potential

As another application, let us consider the Fisher information in the position space (for momentum space is analogous) of a single-particle system in a central potential $V(r)$, defined by

$$I_\rho = \int \frac{|\nabla \rho(\mathbf{r})|^2}{\rho(\mathbf{r})} d\mathbf{r} \tag{4.115}$$

where $\rho(\mathbf{r}) = |\psi(\mathbf{r})|^2$ and where $\psi(\mathbf{r})$ is the bound solutions of the Schrödinger equation

$$\left[-\frac{1}{2}\nabla^2 + V(r)\right]\psi(\mathbf{r}) = E\psi(\mathbf{r}). \tag{4.116}$$

For bounded states the solution of above equation is given by

$$\psi_{nlm}(\mathbf{r}) = R_{nl}(r)Y_{lm}(\Omega) \tag{4.117}$$

where $R_{nl}(r)$ is the radial part of the function and $Y_{lm}(\Omega)$ is the spherical harmonic of order l that is given by

$$Y_{lm}(\Omega) = \frac{1}{\sqrt{2\pi}} e^{im\phi} \Theta_{lm}(\cos\theta) \quad (-l \le m \le l \text{ and } 0 \le \theta \le \pi,\ 0 \le \phi \le 2\pi) \tag{4.118}$$

where $\Theta_{lm}(x)$ are given in terms of the associated Legendre functions of the first kind $P_l^m(x)$:

$$\Theta_{lm}(x) = \sqrt{\frac{2l+1}{2}\frac{(l-m)!}{(l+m)!}} P_l^m(x). \tag{4.119}$$

So the Fisher information for a single particle in a central potential is given by

$$\begin{aligned} I_{\rho_{nlm}} &= 4\int |\nabla \rho_{nlm}^{1/2}(\mathbf{r})|^2 \\ &= \int \left[\Theta_{lm}^2(\theta)\left(\frac{\partial R_{nl}^2(r)}{\partial r}\right)^2 + \frac{1}{r^2}R_{nl}^2(r)\left(\frac{\partial \Theta_{lm}(\theta)}{\partial \theta}\right)^2\right] d\mathbf{r}, \end{aligned} \tag{4.120}$$

on the other hand the kinetic energy is given by:

$$\langle p^2\rangle_{nlm} = \int |\nabla\psi_{nlm}(\mathbf{r})|^2 = \int \left[\left(\frac{\partial R_{nl}(r)}{\partial r}\right)^2 |Y_{lm}(\Omega)|^2\right] d\mathbf{r}$$
$$+ \int \left[\frac{1}{r^2} R_{nl}^2(r) \left(\frac{\partial \Theta_{lm}(\theta)}{\partial\theta}\right)^2 + \frac{1}{r^2}\frac{1}{\sin^2\theta} R_{nl}^2(r)\Theta_{lm}^2(\theta)m^2\right] d\mathbf{r} \tag{4.121}$$

thus

$$I_{\rho_{nlm}} = 4\langle p^2\rangle_{nlm} - 2\langle r^{-2}\rangle_{nlm}(2l+1)|m|. \tag{4.122}$$

4.3.3.1 Hydrogen Atom

For this system the potential is $V(r) = -1/r$ and the expectation values $\langle p^2\rangle_{nlm} = \frac{1}{n^2}$ and $\langle r^{-2}\rangle_{nlm} = \frac{2}{(2l+1)n^3}$ thus

$$I_{\rho_{nlm}} = \frac{4}{n^2}\left(1 - \frac{|m|}{n}\right). \tag{4.123}$$

4.3.3.2 Isotropic Harmonic Oscillator

In this case the potential is $V(r) = \frac{1}{2}\omega^2 r^2$ and the expectation values $\langle p^2\rangle_{nlm} = \omega(2n + l + \frac{3}{2})$ and $\langle r^{-2}\rangle_{nlm} = \frac{\omega}{(2l+1)}$

$$I_{\rho_{nlm}} = 4\omega\left(2n + l + \frac{3}{2} - |m|\right). \tag{4.124}$$

4.4 Applications to Quantum Systems

4.4.1 Formulas in Position and Momentum Spaces

Here, we summarize the formulas and the nomenclature that will use in all this section.

The measure of complexity C has been defined as

$$C = H \cdot D, \tag{4.125}$$

where H represents the information content of the system and D gives an idea of how much concentrated is its spatial distribution.

The simple exponential Shannon entropy, in the position and momentum spaces, takes the form, respectively,

$$H_r = e^{S_r}, \qquad H_p = e^{S_p}, \tag{4.126}$$

where S_r and S_p are the Shannon information entropies,

$$S_r = -\int \rho(\mathbf{r}) \log \rho(\mathbf{r})\, d\mathbf{r}, \qquad S_p = -\int \gamma(\mathbf{p}) \log \gamma(\mathbf{p})\, d\mathbf{p}, \tag{4.127}$$

and $\rho(\mathbf{r})$ and $\gamma(\mathbf{p})$ are the densities normalized to 1 of the quantum system in position and momentum spaces, respectively.

The disequilibrium is:

$$D_r = \int \rho^2(\mathbf{r})\, d\mathbf{r}, \qquad D_p = \int \gamma^2(\mathbf{p})\, d\mathbf{p}. \tag{4.128}$$

In this manner, the final expressions for C in position and momentum spaces are:

$$C_r = H_r \cdot D_r, \qquad C_p = H_p \cdot D_p. \tag{4.129}$$

Second, the Fisher-Shannon information, P, in the position and momentum spaces, is given respectively by

$$P_r = J_r \cdot I_r, \qquad P_p = J_p \cdot I_p, \tag{4.130}$$

where the first factor

$$J_r = \frac{1}{2\pi e} e^{2S_r/3}, \qquad J_p = \frac{1}{2\pi e} e^{2S_p/3}, \tag{4.131}$$

is a version of the exponential Shannon entropy, and the second factor

$$I_r = \int \frac{[\nabla \rho(\mathbf{r})]^2}{\rho(\mathbf{r})} d\mathbf{r}, \qquad I_p = \int \frac{[\nabla \gamma(\mathbf{p})]^2}{\gamma(\mathbf{p})} d\mathbf{p}, \tag{4.132}$$

is the Fisher information measure, that quantifies the narrowness of the probability density.

4.4.2 The H-atom

The atom can be considered a complex system. Its structure is determined through the well established equations of Quantum Mechanics [134, 135]. Depending on the set of quantum numbers defining the state of the atom, different conformations are available to it. As a consequence, if the wave function of the atomic state is known, the probability densities in the position and the momentum spaces are obtained, and from them, the different statistical magnitudes such as Shannon and Fisher informations, different indicators of complexity, etc., can be calculated.

These quantities enlighten new details of the hierarchical organization of the atomic states. In fact, states with the same energy can display, for instance, different values of complexity. This is the behavior shown by the simplest atomic system, that is, the hydrogen atom (H-atom). Now, we present the calculations for this system [94].

The non-relativistic wave functions of the H-atom in position space ($\mathbf{r} = (r, \Omega)$, with r the radial distance and Ω the solid angle) are:

$$\Psi_{n,l,m}(\mathbf{r}) = R_{n,l}(r) Y_{l,m}(\Omega), \tag{4.133}$$

where $R_{n,l}(r)$ is the radial part and $Y_{l,m}(\Omega)$ is the spherical harmonic of the atomic state determined by the quantum numbers (n,l,m). The radial part is expressed as [135]

$$R_{n,l}(r) = \frac{2}{n^2}\left[\frac{(n-l-1)!}{(n+l)!}\right]^{1/2}\left(\frac{2r}{n}\right)^l e^{-\frac{r}{n}} L_{n-l-1}^{2l+1}\left(\frac{2r}{n}\right), \tag{4.134}$$

being $L_\alpha^\beta(t)$ the associated Laguerre polynomials. Atomic units are used here.

The same functions in momentum space ($\mathbf{p} = (p, \hat{\Omega})$, with p the momentum modulus and $\hat{\Omega}$ the solid angle) are:

$$\hat{\Psi}_{n,l,m}(\mathbf{p}) = \hat{R}_{n,l}(p) Y_{l,m}(\hat{\Omega}), \tag{4.135}$$

where the radial part $\hat{R}_{n,l}(p)$ is now given by the expression [136]

$$\hat{R}_{n,l}(p) = \left[\frac{2}{\pi}\frac{(n-l-1)!}{(n+l)!}\right]^{1/2} n^2 2^{2l+2} l! \frac{n^l p^l}{(n^2p^2+1)^{l+2}} C_{n-l-1}^{l+1}\left(\frac{n^2p^2-1}{n^2p^2+1}\right), \tag{4.136}$$

with $C_\alpha^\beta(t)$ the Gegenbauer polynomials.

Taking the former expressions, the probability density in position and momentum spaces,

$$\rho(\mathbf{r}) = |\Psi_{n,l,m}(\mathbf{r})|^2, \qquad \gamma(\mathbf{p}) = |\hat{\Psi}_{n,l,m}(\mathbf{p})|^2, \tag{4.137}$$

can be explicitly calculated. From these densities, the statistical complexity and the Fisher-Shannon information are computed.

C_r and C_p (see expression (4.129)) are plotted in Fig. 4.18 as function of the modulus of the third component m of the orbital angular momentum l for different pairs of (n,l) values. The range of the quantum numbers is: $n \geq 1$, $0 \leq l \leq n-1$, and $-l \leq m \leq l$. Figure 4.18(a) shows C_r for $n = 15$ and Fig. 4.18(b) shows C_r for $n = 30$. In both figures, it can be observed that C_r splits in different sets of discrete points. Each one of these sets is associated to a different l value. It is worth to note that the set with the minimum values of C_r corresponds just to the highest l, that is, $l = n-1$. The same behavior can be observed in Figs. 4.18(c) and 4.18(d) for C_p.

Figure 4.19 shows the calculation of P_r and P_p (see expression (4.130)) as function of the modulus of the third component m for different pairs of (n,l) values. The second factor, I_r or I_p, of this indicator can be analytically obtained in both spaces (position and momentum). The results are [82]:

$$I_r = \frac{4}{n^2}\left(1 - \frac{|m|}{n}\right), \tag{4.138}$$

$$I_p = 2n^2\left\{5n^2 + 1 - 3l(l+1) - (8n - 3(2l+1))|m|\right\}. \tag{4.139}$$

In Fig. 4.19(a), P_r is plotted for $n = 15$, and P_r is plotted for $n = 30$ in Fig. 4.19(b). Here P_r also splits in different sets of discrete points, showing a behavior parallel to the above signaled for C (Fig. 4.18). Each one of these sets is also related with a different l value. It must be remarked again that the set with the minimum values of P_r corresponds just to the highest l. In Figs. 4.19(c) and 4.19(d), the same behavior can be observed for P_p.

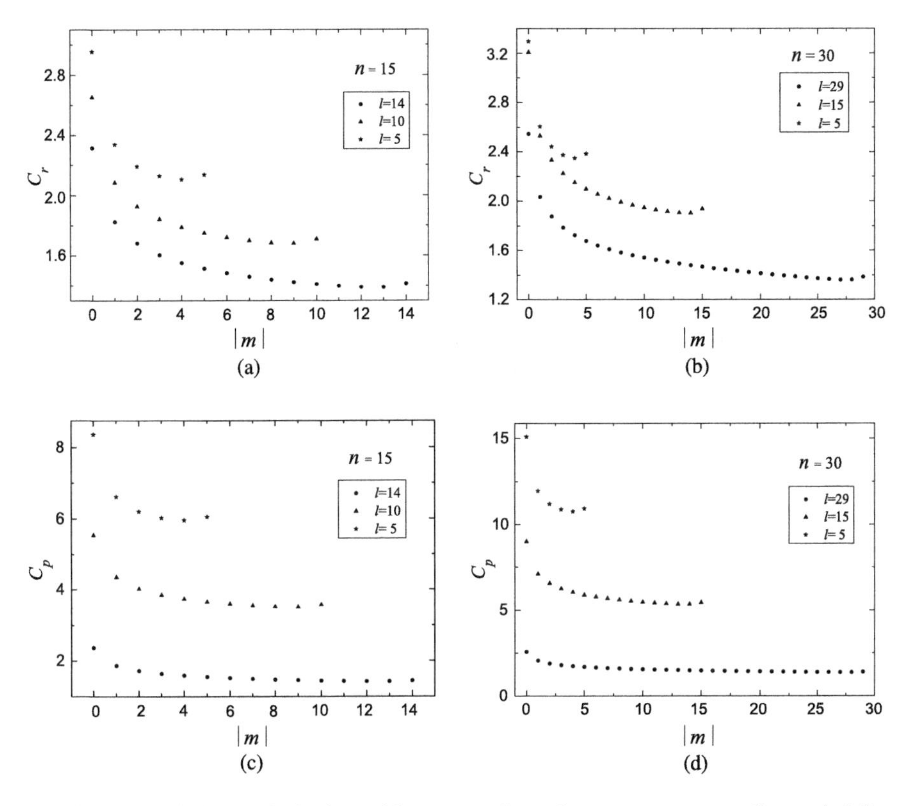

Fig. 4.18 Statistical complexity in position space, C_r, and momentum space, C_p, vs. $|m|$ for different (n, l) values in the hydrogen atom. C_r for (**a**) $n = 15$ and (**b**) $n = 30$. C_p for (**c**) $n = 15$ and (**d**) $n = 30$. All values are in atomic units

Then, it is put in evidence that, for a fixed level of energy n, these statistical magnitudes take their minimum values for the highest allowed orbital angular momentum, $l = n - 1$. It is worth to remember at this point that the mean radius of an $(n, l = n - 1)$ orbital, $\langle r \rangle_{n,l}$, is given by [137]

$$\langle r \rangle_{n,l=n-1} = n^2 \left(1 + \frac{1}{2n} \right), \tag{4.140}$$

that tends, when n is very large, to the radius of the nth energy level, $r_{Bohr} = n^2$, of the Bohr atom. The radial part of this particular wave function, that describes the electron in the $(n, l = n - 1)$ orbital, has no nodes. In fact, if we take the standard deviation, $(\Delta r) = \langle (r - \langle r \rangle)^2 \rangle^{1/2}$, of this wave function, $(\Delta r) = n\sqrt{2n + 1}/2$, the ratio $(\Delta r)/\langle r \rangle$ becomes $1/\sqrt{2n}$ for large n. This means that the spatial configuration of this atomic state is like a spherical shell that converges to a semiclassical Bohr-like orbit when n tends to infinity. These highly excited H-atoms are referred as Rydberg atoms, that have been intensively studied [138] for its importance in areas as astrophysics, plasma physics, quantum optics, etc., and also in studies of the classical limit of quantum mechanics [139].

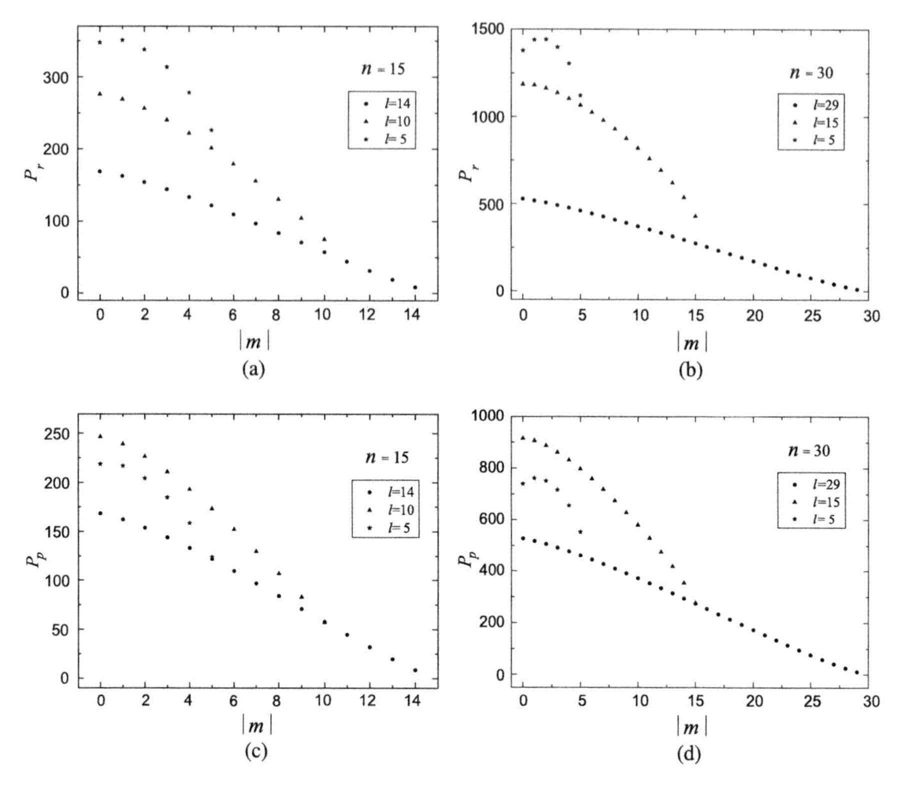

Fig. 4.19 Fisher-Shannon information in position space, P_r, and momentum space, P_p, vs. $|m|$ for different (n, l) values in the hydrogen atom. P_r for (**a**) $n = 15$ and (**b**) $n = 30$. P_p for (**c**) $n = 15$ and (**d**) $n = 30$. All values are in atomic units

We conclude this section by remarking that the minimum values of these statistical measures calculated from the quantum wave functions of the H-atom enhance our intuition by selecting just those orbitals that for a large principal quantum number converge to the Bohr-like orbits in the pre-quantum image. Therefore, these results show that insights on the structural conformation of quantum systems can be inferred from these magnitudes, as it can also be seen in the next sections.

4.4.3 The Quantum Harmonic Oscillator

As suggested in the previous section, a variational process on the statistical measures calculated in the H-atom could select just those orbitals that in the pre-quantum image are the Bohr-like orbits. Now, we show that a similar behavior for the statistical complexity and Fisher-Shannon information is also found in the case of the isotropic quantum harmonic oscillator [93].

We recall the three-dimensional non-relativistic wave functions of this system when the potential energy is written as $V(r) = \lambda^2 r^2/2$, with λ a positive real constant expressing the potential strength. In the same way as in the H-atom (4.133), these wave functions in position space ($\mathbf{r} = (r, \Omega)$, with r the radial distance and Ω the solid angle) are:

$$\Psi_{n,l,m}(\mathbf{r}) = R_{n,l}(r) Y_{l,m}(\Omega), \tag{4.141}$$

where $R_{n,l}(r)$ is the radial part and $Y_{l,m}(\Omega)$ is the spherical harmonic of the quantum state determined by the quantum numbers (n, l, m). Atomic units are used here. The radial part is expressed as [140]

$$R_{n,l}(r) = \left[\frac{2n!\lambda^{l+3/2}}{\Gamma(n+l+3/2)} \right]^{1/2} r^l e^{-\frac{\lambda}{2}r^2} L_n^{l+1/2}(\lambda r^2), \tag{4.142}$$

where $L_\alpha^\beta(t)$ are the associated Laguerre polynomials. The levels of energy are given by

$$E_{n,l} = \lambda(2n + l + 3/2) = \lambda(e_{n,l} + 3/2), \tag{4.143}$$

where $n = 0, 1, 2, \ldots$ and $l = 0, 1, 2, \ldots$. Let us observe that $e_{n,l} = 2n + l$. Thus, different pairs of (n, l) can give the same $e_{n,l}$, and then the same energy $E_{n,l}$.

The wave functions in momentum space ($\mathbf{p} = (p, \hat{\Omega})$, with p the momentum modulus and $\hat{\Omega}$ the solid angle) present the same form as in the H-atom (4.135):

$$\hat{\Psi}_{n,l,m}(\mathbf{p}) = \hat{R}_{n,l}(p) Y_{l,m}(\hat{\Omega}), \tag{4.144}$$

where the radial part $\hat{R}_{n,l}(p)$ is now given by the expression [140]

$$\hat{R}_{n,l}(p) = \left[\frac{2n!\lambda^{-l-3/2}}{\Gamma(n+l+3/2)} \right]^{1/2} p^l e^{-\frac{p^2}{2\lambda}} L_n^{l+1/2}(p^2/\lambda). \tag{4.145}$$

Taking the former expressions, the probability density in position and momentum spaces,

$$\rho_\lambda(\mathbf{r}) = |\Psi_{n,l,m}(\mathbf{r})|^2, \qquad \gamma_\lambda(\mathbf{p}) = |\hat{\Psi}_{n,l,m}(\mathbf{p})|^2, \tag{4.146}$$

can be explicitly calculated. From these densities, the statistical complexity (see expression (4.129)) and the Fisher-Shannon information (see expression (4.130)) are computed. It is shown in Sect. 4.4.3.1 that these quantities are independent of λ, the potential strength, and also that they are the same in both position and momentum spaces, i.e. $C_r = C_p$ and $P_r = P_p$.

In Fig. 4.20, C_r (or C_p) is plotted as function of the modulus of the third component m, $-l \leq m \leq l$, of the orbital angular momentum l for different l values with a fixed energy. That is, according to expression (4.143), the quantity $e_{n,l} = 2n + l$ is constant in each figure. Figure 4.20(a) shows C_r for $e_{n,l} = 15$ and Fig. 4.20(b) shows C_r for $e_{n,l} = 30$. In both figures, it can be observed that C_r splits in different sets of discrete points. Each one of these sets is associated to a different l value. It is worth noting that the set with the minimum values of C_r corresponds just to the highest l, that is, $l = 15$ in Fig. 4.20(a) and $l = 30$ in Fig. 4.20(b).

Figure 4.21 shows P as function of the modulus of the third component m for different pairs of $(e_{n,l} = 2n + l, l)$ values. The second factor, I_r or I_p, of this indicator can be analytically obtained in both spaces (position and momentum) [82]:

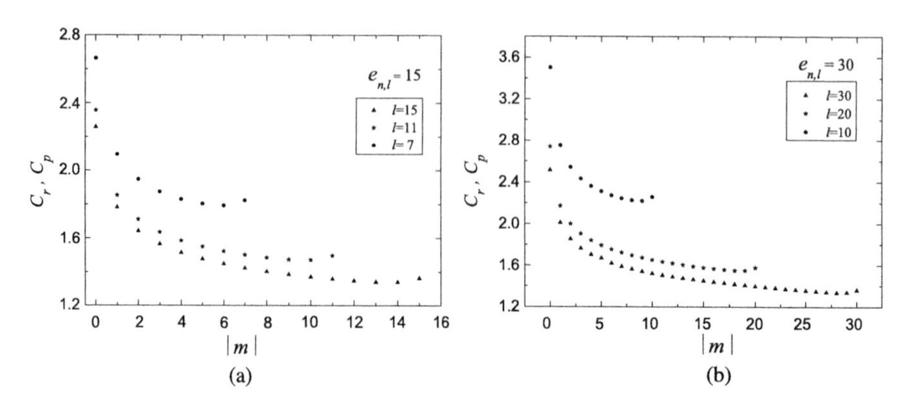

Fig. 4.20 Statistical complexity in position space, C_r, and momentum space, C_p, vs. $|m|$ for different energy $e_{n,l}$-values in the quantum isotropic harmonic oscillator for (**a**) $e_{n,l} = 15$ and (**b**) $e_{n,l} = 30$. Recall that $C_r = C_p$. All values are in atomic units

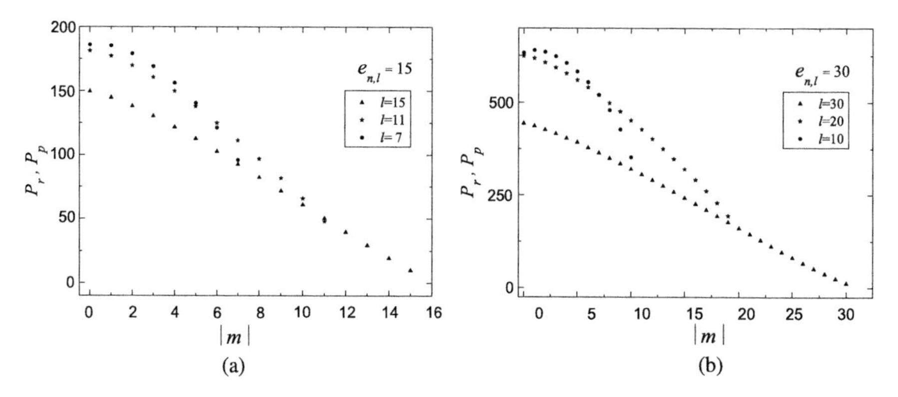

Fig. 4.21 Fisher-Shannon information in position space, P_r, and momentum space, P_p, vs. $|m|$ for different energy $e_{n,l}$-values in the quantum isotropic harmonic oscillator for (**a**) $e_{n,l} = 15$ and (**b**) $e_{n,l} = 30$. Recall that $P_r = P_p$. All values are in atomic units

$$I_r = 4(2n + l + 3/2 - |m|)\lambda, \tag{4.147}$$

$$I_p = 4(2n + l + 3/2 - |m|)\lambda^{-1}. \tag{4.148}$$

Let us note that I_r and I_p depend on λ, although the final result for P_r and P_p are non λ-dependent (see Sect. 4.4.3.1). In Fig. 4.21(a), P_r (or P_p) is plotted for $e_{n,l} = 15$, and P_r is plotted for $e_{n,l} = 30$ in Fig. 4.21(b). Here, P_r also splits in different sets of discrete points, showing a behavior similar to that of C in Fig. 4.20. Each one of these sets is related with a different l value, and the set with the minimum values of P_r also corresponds just to the highest l, that is, $l = 15$ and $l = 30$, respectively.

As in the H-atom, we also see here that, for a fixed level of energy, let us say $e_{n,l} = 2n + l$, these statistical quantities take their minimum values for the highest allowed orbital angular momentum, $l = e_{n,l}$. It is worth remembering at this point that the radial part of this particular wave function, that describes the quantum system in the $(n = 0, l = e_{n,l})$ orbital, has no nodes. This means that the spatial configuration of this state is, in some way, a spherical-like shell. In Sect. 4.4.3.2, the mean radius of this shell, $\langle r \rangle_{n,l,m}$, is found for the case $(n = 0, l = e_{n,l}, m)$. This is:

$$\langle r \rangle_{n=0,l=e_{n,l},m} \equiv \langle r \rangle_{n=0,l=e_{n,l}} \simeq \sqrt{\lambda^{-1}(e_{n,l}+1)}\left(1+\Theta(e_{n,l}^{-1})\right), \tag{4.149}$$

that tends, when $e_{n,l} \gg 1$, to the radius of the Nth energy level, $r_N = \sqrt{\lambda^{-1}(N+1)}$, taking $N = e_{n,l}$ in the Bohr-like picture of the harmonic oscillator (see Sect. 4.4.3.2).

Then, we can remark again that the minimum values of the statistical measures calculated from the wave functions of the quantum isotropic harmonic oscillator also select just those orbitals that in the pre-quantum image are the Bohr-like orbits.

4.4.3.1 Invariance of C and P under Rescaling Transformations

Here, it is shown that the statistical complexities C_r and C_p are equal and independent of the strength potential, λ, for the case of the quantum isotropic harmonic oscillator. Also, the same behavior is displayed by P_r and P_p.

For a fixed set of quantum numbers, (n, l, m), let us define the normalized probability density $\hat{\rho}(\mathbf{t})$:

$$\hat{\rho}(\mathbf{t}) = \frac{2n!}{\Gamma(n+l+3/2)} t^{2l} e^{-t^2} \left[L_n^{l+1/2}(t^2)\right]^2 |Y_{l,m}(\Omega)|^2. \tag{4.150}$$

From expressions (4.141), (4.142) and (4.146), it can be obtained that

$$\rho_\lambda(\mathbf{r}) = \lambda^{3/2}\hat{\rho}(\lambda^{1/2}\mathbf{r}), \tag{4.151}$$

where ρ_λ is the normalized probability density of expression (4.146). Now, it is straightforward to find that

$$H_r(\rho_\lambda) = \lambda^{-3/2} H(\hat{\rho}), \tag{4.152}$$

and that

$$D_r(\rho_\lambda) = \lambda^{3/2} D(\hat{\rho}). \tag{4.153}$$

Then,

$$C_r(\rho_\lambda) = C(\hat{\rho}), \tag{4.154}$$

and the non λ-dependence of C_r is shown.

To show that C_r and C_p are equal, let us note that, from expressions (4.144), (4.145) and (4.146), the normalized probability density $\gamma_\lambda(\mathbf{p})$ for the same set of quantum numbers (n, l, m) can be written as

$$\gamma_\lambda(\mathbf{p}) = \lambda^{-3/2}\hat{\rho}(\lambda^{-1/2}\mathbf{p}). \tag{4.155}$$

Now, it is found that

$$H_p(\gamma_\lambda) = \lambda^{3/2} H(\hat{\rho}), \tag{4.156}$$

and that

$$D_p(\gamma_\lambda) = \lambda^{-3/2} D(\hat{\rho}). \tag{4.157}$$

Then,

$$C_p(\gamma_\lambda) = C(\hat{\rho}), \tag{4.158}$$

and the equality of C_r and C_p, and their non λ-dependence are shown.

Similarly, from expressions (4.130), (4.131), (4.147) and (4.148), it can be found that $P_r = P_p$, and that these quantities are also non λ-dependent.

4.4.3.2 Bohr-Like Orbits in the Quantum Isotropic Harmonic Oscillator

Here, the mean radius of the orbital with the lowest complexity is calculated as function of the energy. Also, the radii of the orbits in the Bohr picture are obtained.

The general expression of the mean radius of a state represented by the wave function $\Psi_{n,l,m}$ is given by

$$\langle r \rangle_{n,l,m} \equiv \langle r \rangle_{n,l} = \frac{n!}{\Gamma(n+l+3/2)} \frac{1}{\lambda^{1/2}} \int_0^\infty t^{l+1} e^{-t} \left[L_n^{l+1/2}(t) \right]^2 dt. \tag{4.159}$$

For the case of minimum complexity (see Figs. 4.20 or 4.21), the state has the quantum numbers ($n = 0, l = e_{n,l}$). The last expression (4.159) becomes:

$$\langle r \rangle_{n=0,l=e_{n,l}} = \frac{(e_{n,l}+1)!}{\Gamma(e_{n,l}+3/2)\lambda^{1/2}}, \tag{4.160}$$

that, in the limit $e_{n,l} \gg 1$, simplifies to expression (4.149):

$$\langle r \rangle_{n=0,l=e_{n,l}\gg 1} \simeq \sqrt{\lambda^{-1}(e_{n,l}+1)} \left(1 + \Theta(e_{n,l}^{-1})\right). \tag{4.161}$$

Now we obtain the radius of an orbit in the Bohr-like image of the isotropic harmonic oscillator. Let us recall that this image establishes the quantization of the energy through the quantization of the classical orbital angular momentum. So, the energy E of a particle of mass m moving with velocity v on a circular orbit of radius r under the harmonic potential $V(r) = m\lambda^2 r^2/2$ is:

$$E = \frac{1}{2} m\lambda^2 r^2 + \frac{1}{2} m v^2. \tag{4.162}$$

The circular orbit is maintained by the central force through the equation:

$$\frac{mv^2}{r} = m\lambda^2 r. \tag{4.163}$$

The angular momentum takes discrete values according to the condition

$$mvr = (N+1)\hbar \quad (N = 0, 1, 2, \ldots). \tag{4.164}$$

Combining the last three equations (4.162)–(4.164), and taking atomic units, $m = \hbar = 1$, the radius r_N of a Bohr-like orbit for this system is obtained

$$r_N = \sqrt{\lambda^{-1}(N+1)} \quad (N = 0, 1, 2, \ldots). \tag{4.165}$$

Let us observe that this expression coincides with the quantum mechanical radius given by expression (4.161) when $e_{n,l} = N$ for $N \gg 1$.

4.4.4 The Square Well

Statistical complexity has been calculated in different atomic systems, such as in the H atom (Sect. 4.4.2) and in the quantum harmonic oscillator (Sect. 4.4.3). The behavior of this statistical magnitude in comparison with that of the energy displays some differences. Among other applications, the energy has a clear physical meaning [134] and it can be used to find the equilibrium states of a system. In the same way, it has also been shown that the complexity can give some insight about the equilibrium configuration in the ground state of the H_2^+ molecule [100]. In this case, Montgomery and Sen have reported that the minimum of the statistical complexity as a function of the internuclear distance for this molecule is an accurate result comparable with that obtained with the minimization of the energy. This fact could suggest that energy and complexity are two magnitudes strongly correlated for any quantum system. But this is not the general case. See, for example, the behavior of both magnitudes in the previous sections for the H-atom and for the quantum isotropic harmonic oscillator. In both systems, the degeneration of the energy is split by the statistical complexity, in such a way that the minimum of complexity for each level of energy is taken on the wave function with the maximum orbital angular momentum. Therefore, energy and complexity are two independent variables.

In this section, we wonder if there exists a quantum system where degeneration of the complexity can be split by the energy. The answer will be affirmative [141]. We show it in two steps. First, a new type of invariance by replication for the statistical complexity is established, and, second, it is seen that the energy eigenstates of the quantum infinite square well fulfill the requirements of this kind of invariance. From there, it is revealed that the degeneration of complexity in this quantum system is broken by the energy.

Different types of replication can be defined on a given probability density. One of them was established in [54]. Here, a similar kind of replication is presented, in such a manner that the complexity C of m replicas of a given distribution is equal to the complexity of the original one. Thus, if $\mathbf{R}$ represents the support of the density function $p(x)$, with $\int_{\mathbf{R}} p(x)\,dx = 1$, take n copies $p_m(x)$, $m = 1, \ldots, n$, of $p(x)$,

$$p_m(x) = p(n(x - \lambda_m)), \quad 1 \le m \le n, \tag{4.166}$$

where the supports of all the $p_m(x)$, centered at $\lambda_m' s$ points, $m = 1, \ldots, n$, are all disjoint. Observe that $\int_{\mathbf{R}} p_m(x)\,dx = \frac{1}{n}$, what makes the replicas union

$$q_n(x) = \sum_{i=1}^{n} p_m(x) \tag{4.167}$$

to be also a normalized probability distribution, $\int_{\mathbf{R}} q_n(x)\,dx = 1$. For every $p_m(x)$, a straightforward calculation shows that the Shannon entropy is

$$S(p_m) = \frac{1}{n} S(p), \tag{4.168}$$

and the disequilibrium is

$$D(p_m) = \frac{1}{n} D(p). \tag{4.169}$$

Taking into account that the m replicas are supported on disjoint intervals on $\mathbf{R}$, we obtain

$$S(q_n) = S(p), \tag{4.170}$$

$$D(q_n) = D(p). \tag{4.171}$$

Then, the statistical complexity ($C = e^S \cdot D$) is

$$C(q_n) = C(p), \tag{4.172}$$

and this type of invariance by replication for C is shown.

Let us see now that the probability density of the eigenstates of the energy in the quantum infinite square well display this type of invariance. The wave functions representing these states for a particle in a box, that is confined in the one-dimensional interval $[0, L]$, are given by [142]

$$\varphi_k(x) = \sqrt{\frac{2}{L}} \sin\left(\frac{k\pi x}{L}\right), \quad k = 1, 2, \ldots. \tag{4.173}$$

Taking $p(x)$ as the probability density of the fundamental state ($k = 1$),

$$p(x) = |\varphi_1(x)|^2, \tag{4.174}$$

the probability density of the kth excited state,

$$q_k(x) = |\varphi_k(x)|^2, \tag{4.175}$$

can be interpreted as the union of k replicas of the fundamental state density, $p(x)$, in the k disjoint intervals $[(m-1)L/k, mL/k]$, with $m = 1, 2, \ldots, k$. That is, we find expression (4.167), $q_k(x) = \sum_{i=1}^{k} p_m(x)$, with

$$p_m(x) = \frac{2}{L} \sin^2\left(\frac{k\pi x}{L} - \pi(m-1)\right), \quad m = 1, 2, \ldots, k, \tag{4.176}$$

where in this case the λ_m's of expression (4.166) are taken as $(m-1)L/k$. Therefore, we conclude that the complexity is degenerated for all the energy eigenstates of the quantum infinite square well. Its value can be exactly calculated. Considering that L is the natural length unit in this problem, we obtain

$$C(p) = C(q_k) = \frac{3}{e} = 1.1036\ldots. \tag{4.177}$$

In the general case of a particle in a d-dimensional box of width L in each dimension, it can also be verified that complexity is degenerated for all its energy eigenstates with a constant value given by $C = (3/e)^d$.

Here we have shown that, in the same way that the complexity breaks the energy degeneration in the H-atom and in the quantum isotropic harmonic oscillator, the contrary behavior is also possible. In particular, the complexity is constant for the whole energy spectrum of the d-dimensional quantum infinite square well. This result is due to the same functional form displayed by all the energy eigenstates of this system. Therefore, it suggests that the study of the statistical complexity in a quantum system allows to infer some properties on its structural conformation.

4.4.5 The Periodic Table

The use of these statistical magnitudes to study the electronic structure of atoms is another interesting application [64, 89, 143–148]. The basic ingredient to calculate these statistical indicators is the electron probability density, $\rho(\mathbf{r})$, that can be obtained from the numerically derived Hartree-Fock atomic wave function in the non-relativistic case [143, 144], and from the Dirac-Fock atomic wave function in the relativistic case [145]. The behavior of these statistical quantifiers with the atomic number Z has revealed a connection with physical measures, such as the ionization potential and the static dipole polarizability [89]. All of them, theoretical and physical magnitudes, are capable of unveiling the shell structure of atoms, specifically the closure of shells in the noble gases. Also, it has been observed that statistical complexity fluctuates around an average value that is non-decreasing as the atomic number Z increases in the non-relativistic case [144, 145]. This average value becomes increasing in the relativistic case [145]. This trend has also been confirmed when the atomic electron density is obtained with a different approach [149]. In another context where the main interactions have a gravitational origin, as it is the case of a white dwarf, it has also been observed that complexity grows as a function of the star mass, from the low-mass non-relativistic case to the extreme relativistic limit. In particular, complexity for the white dwarf reaches a maximum finite value in the Chandrasekhar limit as it was calculated by Sañudo and López-Ruiz [150].

An alternative method to calculate the statistical magnitudes can be used when the atom is seen as a discrete hierarchical organization. The atomic shell structure can also be captured by the fractional occupation probabilities of electrons in the different atomic orbitals. This set of probabilities is here employed to evaluate all these quantifiers for the non-relativistic (NR) and relativistic (R) cases. In the NR case, a non-decreasing trend in complexity as Z increases is obtained and also the closure of shells for some noble gases is observed [96, 151].

For the NR case, each electron shell of the atom is given by $(nl)^w$ [152], where n denotes the principal quantum number, l the orbital angular momentum ($0 \le l \le n-1$) and w is the number of electrons in the shell ($0 \le w \le 2(2l+1)$). For the R case, due to the spin-orbit interaction, each shell is split, in general, in two shells [153]: $(nlj_-)^{w_-}$, $(nlj_+)^{w_+}$, where $j_\pm = l \pm 1/2$ (for $l = 0$ only one value of j is possible, $j = j_+ = 1/2$) and $0 \le w_\pm \le 2j_\pm + 1$. As an example, we explicitly give the electron configuration of Ar($Z = 18$) in both cases,

$$\mathrm{Ar}(NR) : (1\mathrm{s})^2(2\mathrm{s})^2(2\mathrm{p})^6(3\mathrm{s})^2(3\mathrm{p})^6, \tag{4.178}$$

$$\mathrm{Ar}(R) : (1\mathrm{s}1/2)^2(2\mathrm{s}1/2)^2(2\mathrm{p}1/2)^2(2\mathrm{p}3/2)^4(3\mathrm{s}1/2)^2(3\mathrm{p}1/2)^2(3\mathrm{p}3/2)^4. \tag{4.179}$$

For each atom, a fractional occupation probability distribution of electrons in atomic orbitals $\{p_k\}$, $k = 1, 2, \ldots, \Pi$, being Π the number of shells of the atom, can be defined. This normalized probability distribution $\{p_k\}$ ($\sum p_k = 1$) is easily calculated by dividing the superscripts $w_\pm$ (number of electrons in each shell) by Z, the total number of electrons in neutral atoms, which is the case we are considering here. The order of shell filling dictated by nature [152] has been chosen. Then, from this probability distribution, the different statistical magnitudes (Shannon entropy, disequilibrium, statistical complexity and Fisher-Shannon entropy) is calculated.

In order to calculate the statistical complexity $C = H \cdot D$, with $H = e^S$, we use the discrete versions of the Shannon entropy S and disequilibrium D:

$$S = -\sum_{k=1}^{\Pi} p_k \log p_k, \tag{4.180}$$

$$D = \sum_{k=1}^{\Pi} (p_k - 1/\Pi)^2. \tag{4.181}$$

To compute the Fisher-Shannon information, $P = J \cdot I$, with $J = \frac{1}{2\pi e} e^{2S/3}$, the discrete version of I is defined as [96, 151]

$$I = \sum_{k=1}^{\Pi} \frac{(p_{k+1} - p_k)^2}{p_k}, \tag{4.182}$$

where $p_{\Pi+1} = 0$ is taken.

The statistical complexity, C, as a function of the atomic number, Z, for the NR and R cases for neutral atoms is given in Figs. 4.22 and 4.23, respectively. It is observed in both figures that this magnitude fluctuates around an increasing average value with Z. This increasing trend recovers the behavior obtained by using the continuous quantum-mechanical wave functions [144, 145]. A shell-like structure is also unveiled in this approach by looking at the minimum values of C taken on the noble gases positions (the dashed lines in the figures) with the exception of Ne($Z = 10$) and Ar($Z = 18$). This behavior can be interpreted as special arrangements in the atomic configuration for the noble gas cases out of the general increasing trend of C with Z.

The Fisher-Shannon entropy, P, as a function of Z, for the NR and R cases in neutral atoms is given in Figs. 4.24 and 4.25, respectively. The shell structure is again displayed in the special atomic arrangements, particularly in the R case (Fig. 4.25) where P takes local maxima for all the noble gases (see the dashed lines on $Z = 2, 10, 18, 36, 54, 86$). The irregular filling (i.f.) of s and d shells [152] is also detected by peaks in the magnitude P, mainly in the R case. In particular, see the elements Cr and Cu (i.f. of 4s and 3d shells); Nb, Mo, Ru, Rh, and Ag (i.f. of 5s and 4d shells); and finally Pt and Au (i.f. of 6s and 5d shells). Pd also has an irregular filling, but P does not display a peak on it because the shell filling in this case does not follow the same procedure as the before elements (the 5s shell is empty and the 5d is full). Finally, the increasing trend of P with Z is clearly observed.

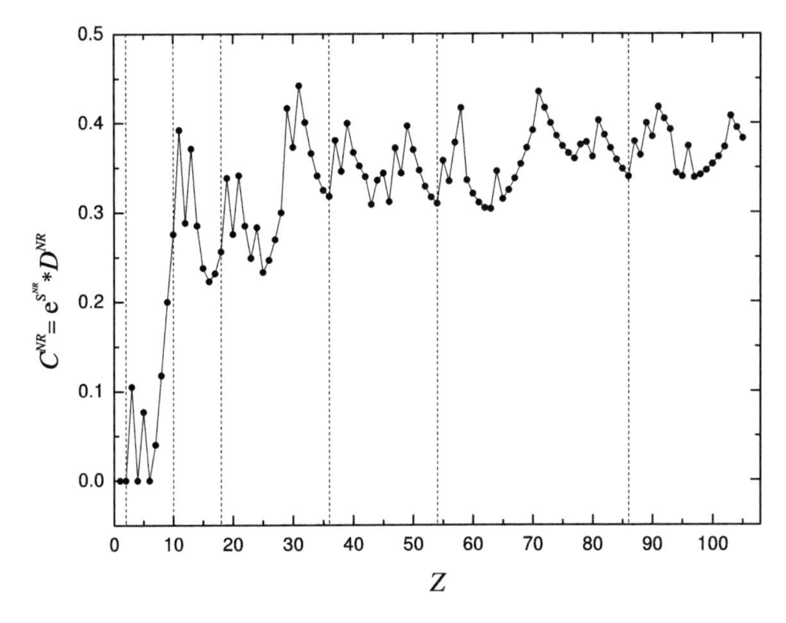

Fig. 4.22 Statistical complexity, C, vs. Z in the non relativistic case (C^{NR}). The *dashed lines* indicate the position of noble gases. For details, see the text

Then, it is found that P, the Fisher-Shannon entropy, in the relativistic case (Fig. 4.25) reflects in a clearer way the increasing trend with Z, the shell structure in noble gases, and the irregular shell filling of some specific elements. The same method that uses the fractional occupation probability distribution is applied in the next section to another many particle system, the atomic nucleus, that has also been described by a shell model.

4.4.6 Magic Numbers in Nuclei

Nucleus is another interesting quantum system that can be described by a shell model [154]. In this picture, just as electrons in atoms, nucleons in nuclei fill in the nuclear shells by following a determined hierarchy. Hence, the fractional occupation probabilities of nucleons in the different nuclear orbitals can capture the nuclear shell structure. This set of probabilities, as explained in the above section, can be used to evaluate the statistical quantifiers for nuclei as a function of the number of nucleons. In this section, by following this method, the calculation of statistical complexity and Fisher-Shannon information for nuclei is presented [155].

The nuclear shell model is developed by choosing an intermediate three-dimensional potential, between an infinite well and the harmonic oscillator, in which nucleons evolve under the Schrödinger equation with an additional spin-orbit interaction [154]. In this model, each nuclear shell is given by $(nlj)^w$, where l denotes

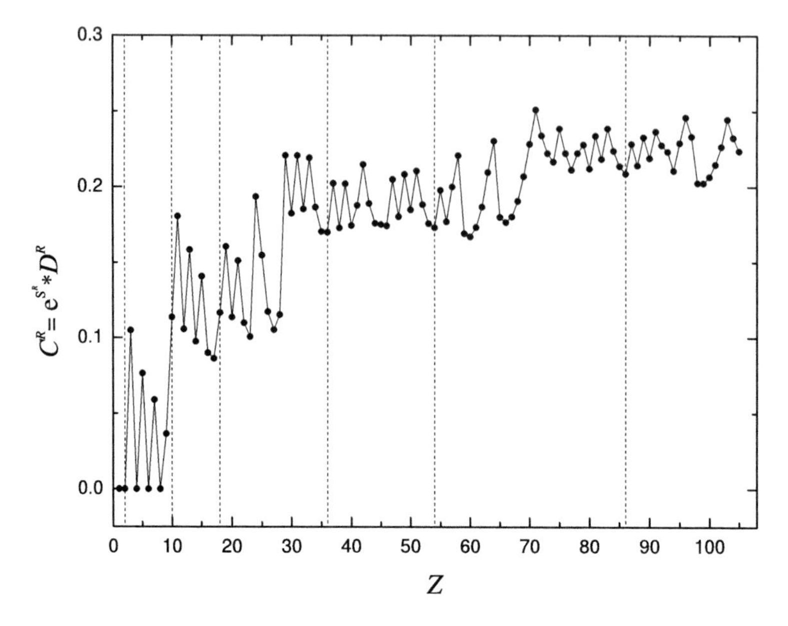

Fig. 4.23 Statistical complexity, C, vs. Z in the relativistic case (C^R). The comments given in Fig. 4.22 are also valid here

the orbital angular momentum ($l = 0, 1, 2, \ldots$), n counts the number of levels with that l value, j can take the values $l + 1/2$ and $l - 1/2$ (for $l = 0$ only one value of j is possible, $j = 1/2$), and w is the number of one-type of nucleons (protons or neutrons) in the shell ($0 \leq w \leq 2j + 1$).

As an example, we explicitly give the shell configuration of a nucleus formed by $Z = 20$ protons or by $N = 20$ neutrons. In both cases, it is obtained [154]:

$$\left\{ \begin{array}{l} (Z = 20) \\ (N = 20) \end{array} \right\} : (1s1/2)^2(1p3/2)^4(1p1/2)^2(1d5/2)^6(2s1/2)^2(1d3/2)^4. \quad (4.183)$$

When one-type of nucleons (protons or neutrons) in the nucleus is considered, a fractional occupation probability distribution of this type of nucleons in nuclear orbitals $\{p_k\}$, $k = 1, 2, \ldots, \Pi$, being Π the number of shells for this type of nucleons, can be defined in the same way as it has been done for electronic calculations in the atom in the previous section. This normalized probability distribution $\{p_k\}$ ($\sum p_k = 1$) is easily found by dividing the superscripts w by the total of the corresponding type of nucleons (Z or N). Then, from this probability distribution, the different statistical magnitudes (Shannon entropy, disequilibrium, statistical complexity and Fisher-Shannon entropy) by following expressions (4.180–4.182) are obtained.

The statistical complexity, C, of nuclei as a function of the number of nucleons, Z or N, is given in Fig. 4.26. Here we can observe that this magnitude fluctuates around an increasing average value with Z or N. This trend is also found for the electronic structure of atoms (see previous section), reinforcing the idea that, in

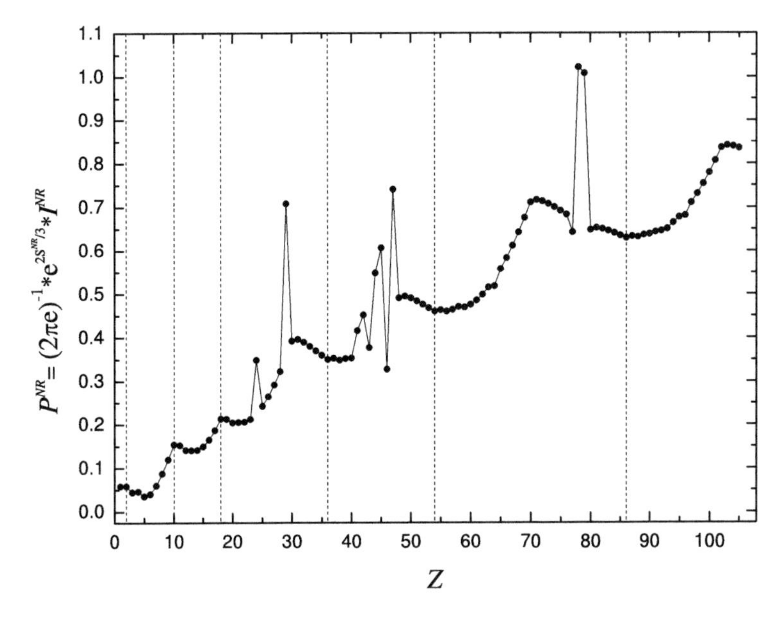

Fig. 4.24 Fisher-Shannon entropy, P, vs. Z, in the non relativistic case (P^{NR}). The *dashed lines* indicate the position of noble gases. For details, see the text

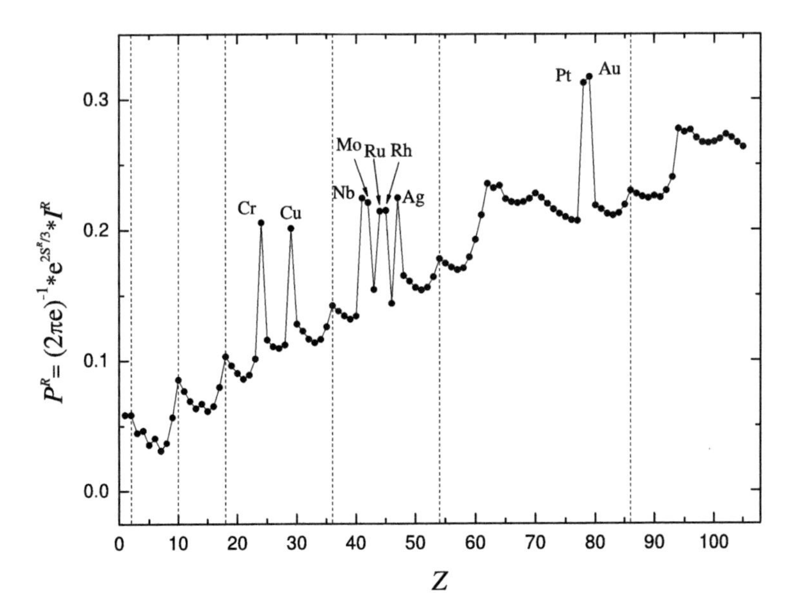

Fig. 4.25 Fisher-Shannon entropy, P, vs. Z, in the relativistic case (P^{R}). The comments given in Fig. 4.24 are also valid here

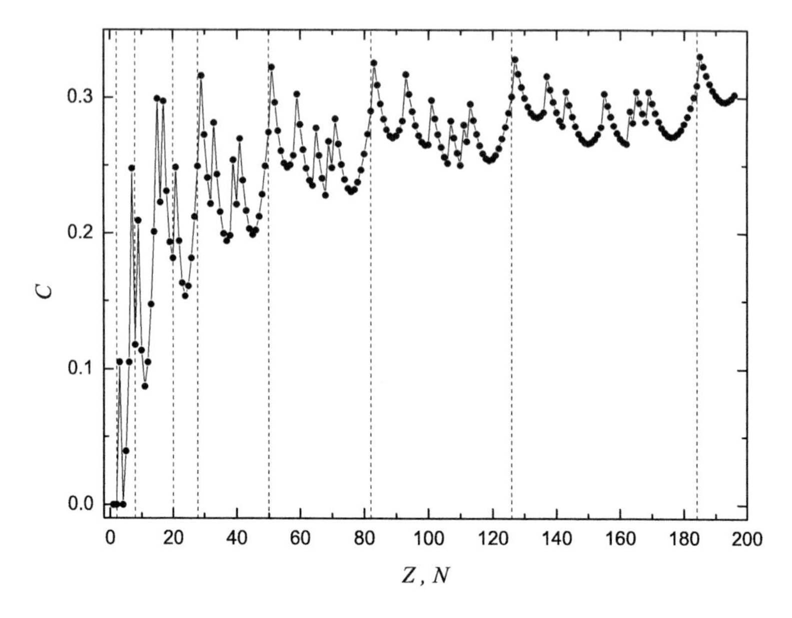

Fig. 4.26 Statistical complexity, C, vs. number of nucleons, Z or N. The *dashed lines* indicate the positions of magic numbers $\{2, 8, 20, 28, 50, 82, 126, 184\}$. For details, see the text

general, complexity increases with the number of units forming a system. However, the shell model supposes that the system encounters certain ordered rearrangements for some specific number of units (electrons or nucleons). This shell-like structure is also unveiled by C in this approach to nuclei. In this case, the extremal values of C are not taken just on the closed shells as happens in the noble gases positions for atoms, if not that they appear to be in the positions one unit less than the closed shells.

The Fisher-Shannon entropy, P, of nuclei as a function of Z or N is given in Fig. 4.27. It presents an increasing trend with Z or N. The spiky behavior of C provoked by the nuclear shell structure becomes smoother for P, that presents peaks (changes in the sign of the derivative) only at a few Z or N, concretely at the numbers $2, 6, 14, 20, 28, 50, 82, 126, 184$. Strikingly, the sequence of magic numbers is $\{2, 8, 20, 28, 50, 82, 126, 184\}$ (represented as dashed vertical lines in the figures). Only the peaks at 6 and 14 disagree with the sequence of magic numbers, what could be justified by saying that statistical indicators work better for high numbers. But in this case, it should be observed that the carbon nucleus, $C^{N=6}_{Z=6}$, and the silicon nucleus, $Si^{N=14}_{Z=14}$, apart from their great importance in nature and industry, they are the stable isotopes with the greatest abundance in the corresponding isotopic series, 98.9% and 92.2%, respectively.

Then, the increasing trend of these statistical magnitudes with Z or N, and the reflect of the shell structure in the spiky behavior of their plots are found when using for their calculation the fractional occupation probability distribution of nucleons, Z or N. It is worth to note that the relevant peaks in the Fisher-Shannon information

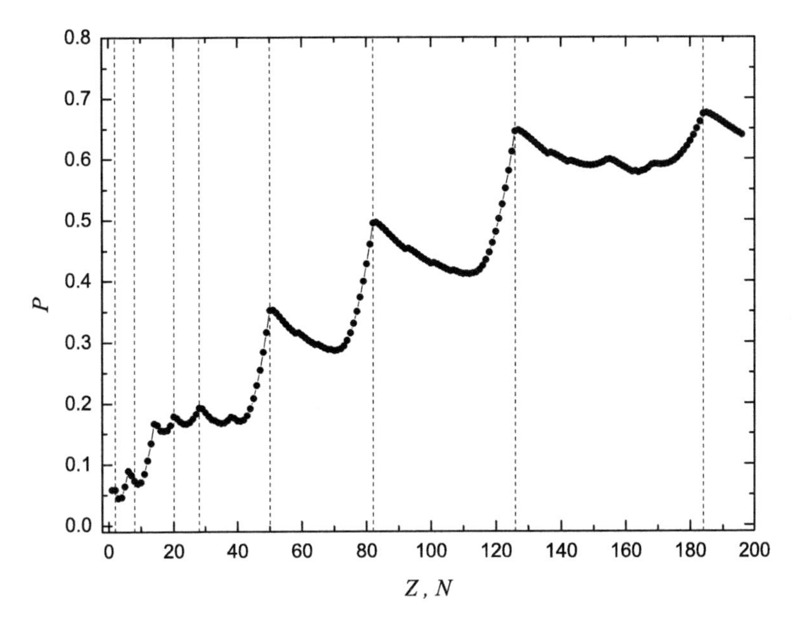

Fig. 4.27 Fisher-Shannon entropy, P, vs. the number of nucleons, Z or N. The *dashed lines* indicate the positions of magic numbers $\{2, 8, 20, 28, 50, 82, 126, 184\}$. For details, see the text

are revealed to be just the series of magic numbers in nuclei. This fact indicates again that these statistical indicators are able to enlighten some structural aspects of quantum many-body systems.

Acknowledgements R.L.-R. thanks Prof. Sen for his invitation to prepare and to present this chapter in this book.

References

1. Hawking S (2000) I think the next century will be the century of complexity. San José Mercury News, Morning Final Edition, January 23
2. Anderson PW (1991) Is complexity physics? Is it science? What is it? Phys Today 9–11, July
3. Parisi G (1993) Statistical physics and biology. Phys World 6:42–47
4. Shannon CE, Weaver W (1949) The mathematical theory of communication. University of Illinois Press, Urbana
5. Nicolis G, Prigogine I (1977) Self-organization in nonequilibrium systems. Wiley, New York
6. López-Ruiz R (1994) On instabilities and complexity. PhD thesis, Universidad de Navarra, Pamplona
7. López-Ruiz R, Mancini HL, Calbet X (1995) A statistical measure of complexity. Phys Lett A 209:321–326
8. Kolmogorov AN (1965) Three approaches to the definition of quantity of information. Probl Inf Transm 1:3–11
9. Chaitin GJ (1966) On the length of programs for computing finite binary sequences. J Assoc Comput Mach 13:547–569

10. Chaitin GJ (1990) Information, randomness & incompleteness. World Scientific, Singapore
11. Lempel A, Ziv J (1976) On the complexity of finite sequences. IEEE Trans Inf Theory 22:75–81
12. Bennett CH (1985) Information, dissipation, and the definition of organization. In: Pines D (ed) Emerging syntheses in science. Santa Fe Institute, Santa Fe, pp 297–313
13. Grassberger P (1986) Toward a quantitative theory of self-generated complexity. Int J Theor Phys 25:907–938
14. Huberman BA, Hogg T (1986) Complexity and adaptation. Physica D 22:376–384
15. Loyd S, Pagels H (1988) Complexity as thermodynamic depth. Ann Phys (NY) 188:186–213
16. Crutchfield JP, Young K (1989) Inferring statistical complexity. Phys Rev Lett 63:105–108
17. Adami C, Cerf NT (2000) Physical complexity of symbolic sequences. Physica D 137:62–69
18. Sánchez JR, López-Ruiz R (2005) A method to discern complexity in two-dimensional patterns generated by coupled map lattices. Physica A 355:633–640
19. Calbet X, López-Ruiz R (2001) Tendency toward maximum complexity in a non-equilibrium isolated system. Phys Rev E 63:066116 (9 pp)
20. Escalona-Morán M, Cosenza MG, López-Ruiz R, García P (2010) Statistical complexity and nontrivial collective behavior in electroencephalographic signals. Int J Bifurc Chaos 20:1723–1729. Special issue on Chaos and dynamics in biological networks, Eds Chávez & Cazelles
21. Feldman DP, Crutchfield JP (1998) Measures of statistical complexity: Why? Phys Lett A 238:244–252
22. Lin J (1991) Divergence measures based on the Shannon entropy. IEEE Trans Inf Theory 37:145–151
23. Martín MT, Plastino A, Rosso OA (2003) Statistical complexity and disequilibrium. Phys Lett A 311:126–132
24. Lamberti W, Martín MT, Plastino A, Rosso OA (2004) Intensive entropic non-triviality measure. Physica A 334:119–131
25. Feng G, Song S, Li P (1998) A statistical measure of complexity in hydrological systems. J Hydrol Eng Chin (Hydrol Eng Soc) 11:14
26. Shiner JS, Davison M, Landsberg PT (1999) Simple measure for complexity. Phys Rev E 59:1459–1464
27. Yu Z, Chen G (2000) Rescaled range and transition matrix analysis of DNA sequences. Commun Theor Phys (Beijing, China) 33:673–678
28. Rosso OA, Martín MT, Plastino A (2003) Tsallis non-extensivity and complexity measures. Physica A 320:497–511
29. Rosso OA, Martín MT, Plastino A (2005) Evidence of self-organization in brain electrical activity using wavelet-based informational tools. Physica A 347:444–464
30. Lovallo M, Lapenna V, Telesca L (2005) Transition matrix analysis of earthquake magnitude sequences. Chaos Solitons Fractals 24:33–43
31. López-Ruiz R (2005) Shannon information, LMC complexity and Rényi entropies: a straightforward approach. Biophys Chem 115:215
32. Pomeau Y, Manneville P (1980) Intermittent transition to turbulence in dissipative dynamical systems. Commun Math Phys 74:189–197
33. Chaté H, Manneville P (1987) Transition to turbulence via spatio-temporal intermittency. Phys Rev Lett 58:112–115
34. Houlrik JM, Webman I, Jensen MH (1990) Mean-field theory and critical behavior of coupled map lattices. Phys Rev A 41:4210–4222
35. Rolf J, Bohr T, Jensen MH (1998) Directed percolation universality in asynchronous evolution of spatiotemporal intermittency. Phys Rev E 57:R2503–R2506
36. Argentina M, Coullet P (1997) Chaotic nucleation of metastable domains. Phys Rev E 56:R2359–R2362
37. Zimmermann MG, Toral R, Piro O, San Miguel M (2000) Stochastic spatiotemporal intermittency and noise-induced transition to an absorbing phase. Phys Rev Lett 85:3612–3615
38. Pomeau Y (1986) Front motion, metastability and subcritical bifurcations in hydrodynamics. Physica D 23:3–11

39. Menon GI, Sinha S, Ray P (2003) Persistence at the onset of spatio-temporal intermittency in coupled map lattices. Europhys Lett 61:27–33
40. López-Ruiz R, Fournier-Prunaret D (2004) Complex behaviour in a discrete logistic model for the symbiotic interaction of two species. Math Biosci Eng 1:307–324
41. López-Ruiz R, Fournier-Prunaret D (2008) Logistic models for symbiosis, predator-prey and competition. In: Encyclopedia of networked and virtual organization, vol II, pp 838–847. Also presented at Conference 'Verhulst 200 on Chaos', abstracts, p 56, Royal Military Academy, Brussels (2004)
42. McKay CP (2004) What is life? PLoS Biol 2:1260–1263
43. Sánchez JR, López-Ruiz R (2005) Detecting synchronization in spatially extended discrete systems by complexity measurements. Discrete Dyn Nat Soc 9:337–342
44. Kolmogorov AN (1958) A new metric invariant of transitive dynamical systems and automorphisms of Lebesgue spaces. Dokl Akad Nauk SSSR 119:861–864
45. Sinai JG (1959) On the concept of entropy of a dynamical system. Dokl Akad Nauk SSSR 124:768–771
46. Landsberg PT, Shiner JS (1998) Disorder and complexity in an ideal non-equilibrium Fermi gas. Phys Lett A 245:228–232
47. Atmanspacher H, Räth C, Wiedermann G (1997) Statistics and meta-statistics in the concept of complexity. Physica A 234:819–829
48. Gell-Mann M (1995) What is complexity. Complexity 1:16–19
49. Anteneodo C, Plastino AR (1996) Some features of the statistical LMC complexity. Phys Lett A 223:348–354
50. Latora V, Baranger M (1999) Kolmogorov-Sinai entropy rate versus physical entropy. Phys Rev Lett 82:520–523
51. Calbet X, López-Ruiz R (2007) Extremum complexity distribution of a monodimensional ideal gas out of equilibrium. Physica A 382:523–530
52. Calbet X, López-Ruiz R (2009) Extremum complexity in the monodimensional ideal gas: the piecewise uniform density distribution approximation. Physica A 388:4364–4378
53. López-Ruiz R (2001) Complexity in some physical systems. Int J Bifurc Chaos 11:2669–2673
54. Catalán RG, Garay J, López-Ruiz R (2002) Features of the extension of a statistical measure of complexity to continuous systems. Phys Rev E 66:011102 (6 pp)
55. Romera E, López-Ruiz R, Sañudo Nagy Á (2009) Generalized statistical complexity and Fisher-Rényi entropy product in the H-atom. Int Rev Phys (IREPHY) 3:207–211
56. López-Ruiz R, Nagy Á, Romera E, Sañudo J (2009) A generalized statistical complexity measure: Applications to quantum systems. J Math Phys 50:123528(10)
57. Khinchin AI (1957) Mathematical foundations of information theory. Dover, New York
58. Wehrl A (1978) General properties of entropy. Rev Mod Phys 50:221–260
59. Tsallis C, Mendes RS, Plastino AR (1998) The role of constraints within generalized nonextensive statistics. Physica A 261:534–554
60. Dembo A, Cover TM, Thomas JA (1991) Information theoretic inequalities. IEEE Trans Inf Theory 37:1501–1518
61. Gadre SR (1984) Information entropy and Thomas-Fermi theory. Phys Rev A 30:620–621
62. Gadre SR, Bendale RD (1985) Maximization of atomic information-entropy sum in configuration and momentum spaces. Int J Quant Chem 28:311–314
63. Gadre SR, Bendale RD (1985) Information entropies in quantum-chemistry. Curr Sci (India) 54:970–977
64. Gadre SR, Sears SB, Chakravorty SJ, Bendale RD (1985) Some novel characteristics of atomic information entropies. Phys Rev A 32:2602–2606
65. Gadre SR, Bendale RD (1987) Rigorous relationships among quantum-mechanical kinetic energy and atomic information entropies: Upper and lower bounds. Phys Rev A 36:1932–1935
66. Shannon CE (1948) A mathematical theory of communication. Bell Syst Tech J 27:379–423
67. Fisher RA (1925) Theory of statistical estimation. Proc Camb Philos Soc 22:700–725
68. Esquivel RO, Rodriguez AL, Sagar RP, Smith VH Jr (1996) Physical interpretation of information entropy: Numerical evidence of the Collins conjecture. Phys Rev A 54:259–265

69. Massen SE, Panos CP (1998) Universal property of the information entropy in atoms, nuclei and atomic clusters. Phys Lett A 246:530–533
70. Massen SE, Panos CP (2001) A link of information entropy and kinetic energy for quantum many-body systems. Phys Lett A 280:65–69
71. Sagar RP, Ramirez JC, Esquivel RO, Ho M Jr (2002) Relationships between Jaynes entropy of the one-particle density matrix and Shannon entropy of the electron densities. J Chem Phys 116:9213–9221
72. Nalewajski RF, Switka E, Michalak A (2002) Information distance analysis of molecular electron densities. Int J Quant Chem 87:198–213
73. Nagy Á (2003) Spin virial theorem in the time-dependent density-functional theory. J Chem Phys 119:9401–9405
74. Massen SE (2003) Application of information entropy to nuclei. Phys Rev C 67:014314 (7 pp)
75. Nalewajski RF (2003) Information principles in the theory of electronic structure. Chem Phys Lett 372:28–34
76. Nalewajski RF (2003) Information principles in the loge theory. Chem Phys Lett 375:196–203
77. Romera E, Dehesa JS (2004) The Fisher-Shannon information plane, an electron correlation tool. J Chem Phys 120:8906–8912
78. Parr RG, Ayers PW, Nalewajski RF (2005) What is an atom in a molecule? J Phys Chem A 109:3957–3959
79. Sen KD (2005) N-derivative of Shannon entropy of shape function for atoms. J Chem Phys 123:074110 (9 pp)
80. Guevara NL, Sagar RP, Esquivel RO (2005) Local correlation measures in atomic systems. J Chem Phys 122:084101
81. Sagar RP, Guevara NL (2005) Local correlation measures in atomic systems. J Chem Phys 123:044108 (10 pp)
82. Romera E, Sánchez-Moreno P, Dehesa JS (2005) The Fisher information of single-particle systems with a central potential. Chem Phys Lett 414:468–472
83. Nagy Á (2006) Fisher information in a two-electron entangled artificial atom. Chem Phys Lett 425:154–156
84. Nagy Á, Sen KD (2006) Atomic Fisher information versus atomic number. Phys Lett A 360:291–293
85. Sagar RP, Guevara NL (2006) Mutual information and electron correlation in momentum space. J Chem Phys 124:134101 (9 pp)
86. Sen KD, Katriel J (2006) Information entropies for eigendensities of homogeneous potentials. J Chem Phys 125:074117 (4 pp)
87. Nagy Á (2007) Fisher information and Steric effect. Chem Phys Lett 449:212–215
88. Liu S (2007) On the relationship between densities of Shannon entropy and Fisher information for atoms and molecules. J Chem Phys 126:191107 (3 pp)
89. Sen KD, Panos CP, Chatzisavas KCh, Moustakidis ChC (2007) Net Fisher information measure versus ionization potential and dipole polarizability in atoms. Phys Lett A 364:286–290
90. Patil SH, Sen KD, Watson NA, Montgomery HE Jr (2007) Characteristic features of net information measures for constrained Coulomb potentials. J Phys B 40:2147–2162
91. Sagar RP, Guevara NL (2008) Relative entropy and atomic structure. J Mol Struct, Theochem 857:72–77
92. Nagy Á, Liu S (2008) Local wave-vector, Shannon and Fisher information. Phys Lett A 372:1654–1656
93. Sañudo, López-Ruiz R (2008) Some features of the statistical complexity, Fisher-Shannon information and Bohr-like orbits in the quantum isotropic harmonic oscillator. J Phys A, Math Theor 41:265303 (7 pp)
94. Sañudo J, López-Ruiz R (2008) Statistical complexity and Fisher-Shannon information in the H-atom. Phys Lett A 372:5283–5286

95. Seo DK, Weng CJ (2008) Orbital interpretation of kinetic energy density and a direct space comparison of chemical bonding in tetrahedral network solids. Phys Chem A 112:7705–7716
96. Sañudo J, López-Ruiz R (2009) Alternative evaluation of statistical indicators in atoms: The non-relativistic and relativistic cases. Phys Lett A 373:2549–2551
97. Nalewajski RF (2009) Entropic descriptors of the chemical bond in H_2: local resolution of stockholder atoms. J Math Chem 45:1041–1054
98. Parr RG, Yang W (1989) Density functional theory of atoms and molecules. Oxford University Press, New York
99. Szabo JB, Sen KD, Nagy Á (2008) The Fisher-Shannon information plane for atoms. Phys Lett A 372:2428–2430
100. Montgomery HE Jr, Sen KD (2008) Statistical complexity and Fisher-Shannon information measure of H_2^+. Phys Lett A 372:2271–2273
101. Mitnik DM, Randazzo J, Gasaneo G (2008) Endohedrally confined helium: Study of mirror collapses. Phys Rev A 78:062501 (10 pp)
102. Romera E (2002) Stam's principle D-dimensional uncertainty-like relationships and some atomic properties. Mol Phys 100:3325–3329
103. Romera E, Sánchez-Moreno P, Dehesa JS (2006) Uncertainty relation for Fisher information of D-dimensional single-particle systems with central potentials. J Math Phys 47:103504 (11 pp)
104. Hall MJW (2000) Quantum properties of classical Fisher information. Phys Rev A 62:012107
105. Hall MJW (2001) Exact uncertainty relations. Phys Rev A 64:052103
106. Hall MJW, Reginatto M (2002) Schrödinger equation from an exact uncertainty principle. J Phys A 35:3289–3302
107. Hall MJW (2004) Prior information: How to circumvent the standard joint-measurement uncertainty relation. Phys Rev A 69:052113
108. Luo S (2000) Quantum Fisher information and uncertainty relations. Lett Math Phys 53:243–251
109. Luo S (2003) Wigner-Yanase skew information and uncertainty relations. Phys Rev Lett 91:180403
110. Luo S, Zhang Z (2004) An informational characterization of Schrödinger's uncertainty relation. J Stat Phys 114:1557–1576
111. Petz D (2003) Covariance and Fisher information in quantum mechanics. J Phys A 35:79–91
112. Romera E, Angulo JC, Dehesa JS (1999) Fisher entropy and uncertainty-like relationships in many-body systems. Phys Rev A 59:4064–4067
113. Stam A (1959) Some inequalities satisfied by the quantities of information of Fisher and Shannon. Inf Control 2:101–112
114. Rao CR (1965) Linear statistical interference and its applications. Wiley, New York
115. Romera E, Dehesa JS (1994) Weizsäcker energy of many-electron systems. Phys Rev A 50:256–266
116. Bialynicki-Birula I, Mycielski J (1975) Uncertainty relations for information entropy in wave mechanics. Commun Math Phys 44:129–132
117. Carlen EA (1991) Superadditivity of Fisher's information and logarithmic Sobolev inequalities. J Funct Anal 101:194–211
118. Fulde P (1995) Electron correlation in molecules and solids. Springer, Berlin
119. Kutzelnigg W, del Re G, Berthier G (1968) Correlation coefficients for electronic wave functions. Phys Rev 172:49–59
120. Grassi A (2008) A relationship between atomic correlation energy and Tsallis entropy. Int J Quant Chem 108:774–778
121. Collins DM (1993) Entropy maximizations on electron-density. Z Naturforsch 48Z:68–74
122. Grassi A, Lombardo GM, March NH, Pucci R (1998) $1/Z$ expansion, correlation energy, and Shannon entropy of heavy atoms in nonrelativistic limit. Int J Quant Chem 69:721–726
123. Mohareji A, Alipour M (2009) Shannon information entropy of fractional occupation probability as an electron correlation measure in atoms and molecules. Chem Phys 360:132–136

124. Guevara NL, Sagar RP, Esquivel RO (2003) Shannon-information entropy sum as a correlation measure in atomic systems. Phys Rev A 67:012507
125. Sagar RP, Laguna HG, Guevara NL (2009) Conditional entropies and position-momentum correlations in atomic systems. Mol Phys 107:2071–2080
126. Ziesche P, Smigh VH Jr, Ho M, Rudin SP, Gersdorfand P, Taut M (1999) The He isoelectronic series and the Hooke's law model: Correlation measures and modifications of Collins' conjecture. J Chem Phys 110:6135–6142
127. Huang Z, Kais S (2005) Entanglement as measure of electron-electron correlation in quantum chemistry calculations. Chem Phys Lett 413:1–5
128. Gottlieb AD, Mauser NJ (2005) New measure of electron correlation. Phys Rev Lett 95:123003 (4 pp)
129. Juhász T, Mazziotti DA (2006) The cumulant two-particle reduced density matrix as a measure of electron correlation and entanglement. J Chem Phys 125:174105 (5 pp)
130. Amovilli C, March NH (2004) Quantum information: Jaynes and Shannon entropies in a two-electron entangled artificial atom. Phys Rev A 69:054302 (4 pp)
131. Koga T, Kasai Y, Thakkar AJ (1993) Accurate algebraic densities and intracules for helium-like ions. Int J Quant Chem 46:689–699
132. Taut M (1993) Two electrons in an external oscillator potential: Particular analytic solutions of a Coulomb correlation problem. Phys Rev A 48:3561–3566
133. Cioslowski J, Pernal K (2000) The ground state of harmonium. J Chem Phys 113:8434–8443
134. Landau LD, Lifshitz LM (1981) Quantum mechanics: non-relativistic theory, 3rd edn, vol 3. Butterworth-Heinemann, Oxford
135. Galindo A, Pascual P (1991) Quantum mechanics I. Springer, Berlin
136. Bethe HA, Salpeter EE (1977) Quantum mechanics of one-and two-electron atoms. Springer, Berlin
137. Eisberg JL (1961) Fundamentals of modern physics. Wiley, New York
138. Lebedev VS, Beigman IL (1998) Physics of highly excited atoms and ions. Springer, Berlin
139. Coffey MW (2003) Semiclassical position entropy for hydrogen-like atoms. J Phys A, Math Gen 36:7441–7448
140. Yánez RJ, van Assche W, Dehesa JS (1994) Position and momentum information entropies of the D-dimensional harmonic oscillator and hydrogen atom. Phys Rev A 50:3065–3079
141. López-Ruiz R, Sañudo J (2009) Complexity invariance by replication in the quantum square well. Open Syst Inf Dyn 16:423–427
142. Cohen-Tannoudji C, Diu B, Laloë F (1977) Quantum mechanics. Wiley, New York. 2 vols
143. Chatzisavvas KCh, Moustakidis ChC, Panos CP (2005) Information entropy, information distances, and complexity in atoms. J Chem Phys 123:174111 (10 pp)
144. Panos CP, Chatzisavvas KCh, Moustakidis ChC, Kyrkou EG (2007) Comparison of SDL and LMC measures of complexity: Atoms as a testbed. Phys Lett A 363:78–83
145. Borgoo A, De Proft F, Geerlings P, Sen KD (2007) Complexity of Dirac-Fock atom increases with atomic number. Chem Phys Lett 444:186–191
146. Angulo JC, Antolín J (2008) Atomic complexity measures in position and momentum spaces. J Chem Phys 128:164109 (7 pp)
147. Romera E, Nagy Á (2008) Rényi information of atoms. Phys Lett A 372:4918–4922
148. Borgoo A, Geerlings P, Sen KD (2008) Electron density and Fisher information of Dirac-Fock atoms. Phys Lett A 372:5106–5109
149. Sañudo J, López-Ruiz R (2008) Complexity in atoms: An approach with a new analytical density. Int Rev Phys (IREPHY) 2:223–230
150. Sañudo J, Pacheco AF (2009) Complexity and white-dwarf structure. Phys Lett A 373:807–810
151. Panos CP, Nikolaidis NS, Chatzisavvas KCh, Tsouros CC (2009) A simple method for the evaluation of the information content and complexity in atoms. A proposal for scalability. Phys Lett A 373:2343–2350
152. Bransden BH, Joachain CJ (2003) Physics of atoms and molecules, 2nd edn. Prentice Hall, London

153. Cowan RD (1981) The theory of atomic structure and spectra. University of California Press, Berkeley
154. Krane KS (1988) Introductory nuclear physics. Wiley, New York
155. López-Ruiz R, Sañudo J (2010) Evidence of magic numbers in nuclei by statistical indicators. Open Syst Inf Dyn 17:279–286

Chapter 5
Entropy and Complexity Analyses of D-dimensional Quantum Systems

J.S. Dehesa, S. López-Rosa, and D. Manzano

Abstract This chapter briefly reviews the present knowledge about the analytic information theory of quantum systems with non-standard dimensionality in the position and momentum spaces. The main concepts of this theory are the power and entropic moments, which are very fertile largely because of their flexibility and multiple interpretations. They are used here to study the most relevant information-theoretic one-element (Fisher, Shannon, Rényi, Tsallis) and some composite two-elements (Fisher-Shannon, LMC shape and Cramér-Rao complexities) measures which describe the spreading measures of the position and momentum probability densities farther beyond the standard deviation. We first apply them to general systems, then to single particle systems in central potentials and, finally, to hydrogenic systems in D-dimensions.

5.1 Introduction

The physics of D-dimensional quantum systems, with D not necessarily equal to 3, is of fundamental interest in field theory [1], superstring theory and quantum cosmology [2–5]; the idea that the universe is trapped on a membrane in some high-dimensional space-time may explain why gravity is so weak, and could be tested by high-energy particle accelerators. The D-dimensional physics is also very relevant to mathematical modelling numerous three-dimensional physical phenomena related to a wide variety of fields ranging from nanotechnology [6–8], quantum computation [9, 10], plasma physics [11], to quantum chemistry and atomic and molecular physics [12–15]. The atomic physics in D dimensions, where the rotation symmetry is $O(D)$, has been shown [1] to simplify as $D \to \infty$ and it can be solved for large D by an expansion in $\frac{1}{D}$ (see. e.g. [12, 16]). Let us only mention its relevance in dimensional scaling or dimensional perturbation methods, where the energy spectrum of three-dimensional systems may be obtained in some situa-

J.S. Dehesa (✉)
Departamento de Física Atómica, Molecular y Nuclear and Instituto Carlos I de Física Teórica y Computacional, Universidad de Granada, 18071 Granada, Spain
e-mail: dehesa@ugr.es

K.D. Sen (ed.), *Statistical Complexity*,
DOI 10.1007/978-90-481-3890-6_5,

tions by generalizing them to some lower or higher dimensions where the particular system is analytically solvable and then by using an interpolation or extrapolation procedure [12, 13]; this new approach, which exploits exact solutions obtainable in the limit of one- or infinite-dimension, has given high accuracy for the electron correlation energy with far less effort than using conventional methods, at least for two-electron systems, as Dudley L. Herschbach (1986 Nobel Prize in Chemistry) and his collaborators have shown. Let us also point out that in chemistry two dimensions are often better than three, since surface-bound reactions can be proved in greater detail than those in a liquid solution as Gerhard Ertl (2007 Nobel Prize in Chemistry) and collaborators have realized in numerous contributions to the field of surface chemistry [17–19]. Moreover, the theory of chaos and fractals [20, 21] has shown an increasing interest in the framework of non-integer dimensions. Fractional dimensionality have also been applied to nanostructured systems, such as excitons in anisotropic or confined quantum well structures to account for the effective medium and the anisotropy of their interactions [6].

This chapter is an introductory tour to the information theory of D-dimensional quantum systems, where the emphasis is put not so much on the conceptual description of the information-theoretic measures (see other chapters of this volume for this purpose) but on their analytical determination. We gather a number of results obtained in the last two decades relative to the measures of spreading and uncertainty of the D-dimensional quantum one- and many-body systems under consideration, which quantify the spatial delocalization of the single-particle density of the system in different and complementary ways. We start in Sect. 5.2 with the power moments and the entropic moments of such a density, which do not only characterize the density itself but also describe numerous fundamental and/or experimental measurable quantities of the system. Then, we provide (i) some variational expressions for the Rényi and Tsallis entropies in terms of a given power moment, which allow us to correlate these measures with other physical quantities; (ii) the uncertainty relations associated to the spreading measures beyond the celebrated Heisenberg relation and (iii) some bounds and inequalities satisfied by the Cramér-Rao, Fisher-Shannon and LMC shape complexities.

In Sect. 5.3 we show more accurate results for single-particle systems with D-dimensional central potentials, such as the recently discovered uncertainty relations relative not only to the power moments but also the Fisher information. In Sect. 5.4, following the reductionist Weisskopf's aphorism "To understand hydrogen is to understand all physics", we study in full detail the information-theoretic properties of D-dimensional hydrogenic systems which include hydrogenic atoms and ions, exotic atoms, antimatter atoms, Rydberg atoms, ions and molecules, excitons and donors in some semiconductors, among others. Moreover, one- and two-dimensional hydrogenic atoms have been used as qubits in the realization of quantum computation. On the other hand, the position and momentum D-dimensional hydrogenic wavefunctions play a relevant role in quantum chemistry for Sturmian approaches and to conform complete orthonormal sets for many-body problems; e.g., for three-body problems (hydrogen molecular ion, helium atom, ...) [14, 22].

Let us also point out that the existence of hydrogenic systems with dimensionality other than 3 have been observed for $D < 3$ [7] and suggested for $D > 3$ [23, 24]. We will realize at the end that we should perhaps paraphrase the previous aphorism as "To understand hydrogen, we must understand all of physics" in Kleppner's words [25]; indeed, there still exist numerous information-theoretic open issues in the hydrogenic problem. Finally some conclusions are given and various open problems are pointed out.

5.2 Information-Theoretic Analysis of Many Particle Systems

Here we study the spatial delocalization or spreading of the single-particle density of a many-particle system by means of its power, logarithmic and entropic moments and the related Rényi and Tsallis entropies and the complexity measures together with their associated uncertainty relations.

5.2.1 Power, Logarithmic and Entropic Moments

A D-dimensional n-particle system is described in quantum mechanics by means of its wavefunction $\Psi(\mathbf{r}_1, \mathbf{r}_2, \ldots, \mathbf{r}_n; \sigma_1, \sigma_2, \ldots, \sigma_n)$, with $\mathbf{r} = (x_1, x_2, \ldots, x_D)$ and where $(\mathbf{r}_i, \sigma_i)$ denotes the position-spin coordinates of the ith-particle, which is assumed to be normalized and antisymmetrized in the pairs $(\mathbf{r}_i, \sigma_i)$. The physical and chemical properties of these systems are controlled [26] by means of the spatial delocalization or spreading of the single particle density $\rho(\mathbf{r})$ defined as

$$\rho(\mathbf{r}) = \sum_{\sigma \in \{-\frac{1}{2}, +\frac{1}{2}\}^D} \int |\Psi(\mathbf{r}, \mathbf{r}_2, \ldots, \mathbf{r}_n; \sigma_1, \sigma_2, \ldots, \sigma_n)|^2 d\mathbf{r}_2 \ldots d\mathbf{r}_n \tag{5.1}$$

which is normalized to unity. The power moments $\{\langle r^\alpha \rangle\}$, the logarithmic moments $\{\langle r^\alpha \ln r \rangle\}$ and the frequency or entropic moments $\{W_\alpha[\rho] \equiv \langle \rho^{\alpha-1} \rangle\}$ provide different ways to characterize the density $\rho(\mathbf{r})$. The symbol $\langle f(\mathbf{r}) \rangle$ denotes the expectation value

$$\langle f(\mathbf{r}) \rangle := \int_{R_D} f(\mathbf{r}) \rho(\mathbf{r}) d\mathbf{r}. \tag{5.2}$$

Similarly, the properties of the system can be analyzed in momentum space by means of the momentum density $\gamma(\mathbf{p})$ defined in terms of the Fourier-transformed $\widehat{\Psi}(\mathbf{p}_1, \mathbf{p}_2, \ldots, \mathbf{p}_n; \sigma_1, \sigma_2, \ldots, \sigma_n)$ as in (5.1). This density can be completely characterized by the three following sets of moments: $\{\langle p^\alpha \rangle\}$, $\{\langle p^\alpha \ln p \rangle\}$ and $\{U_\alpha[\gamma] \equiv \langle \gamma^{\alpha-1} \rangle\}$.

For specific values of α, these moments are physically meaningful and, at times, experimentally accessible. Let us point out here that for $D = 3$: (i) the

Langevin-Pauli diamagnetic susceptibility $\chi = -\alpha_{\text{fs}}^2 \frac{\langle r^2 \rangle}{6}$, α_{fs} being the fine structure constant, (ii) the electron-nucleus attraction constant which is also related to the nuclear magnetic screening constant or diamagnetic screening factor, is given by $E_{eN} = -Z\langle r^{-1}\rangle$, Z being the nuclear charge, (iii) the electron kinetic energy $T_e = \frac{\langle p^2 \rangle}{2}$, (iv) the Breit-Pauli relativistic correction to the kinetic energy $T_{\text{rel}} = -\alpha_{\text{fs}} \frac{\langle p^4 \rangle}{8}$, (v) the height of the peak of the Compton profile $J(0)$ is $h = \frac{\langle p^{-1} \rangle}{2}$. The logarithmic expectation values have been shown to be relevant for atomic systems, not only in an information-theoretic framework [27], but also in physical phenomena such as electron-electron coalescence [28] and elastic electron scattering [29], where e.g. $\langle \ln r \rangle = \frac{d\langle r^\alpha \rangle}{d\alpha}|_{\alpha=0}$ determines the behaviour of the phase shifts at high energy and low angular momenta. Moreover, the entropic moments $\langle \rho^\alpha \rangle$ and the modified moments $\langle r^\alpha \rho \rangle$ have been used to develop ant to interpret all energy components in the density-functional theory of the ground state of atoms [30–32]. Let us point out that the moments $\langle \rho^\alpha \rangle$ describe, up to a proportionality factor, the atomic Thomas-Fermi ($\alpha = \frac{5}{3}$) and the Dirac exchange ($\alpha = \frac{4}{3}$) energies; see [33, 34] for their connection with other atomic density functionals, and [35] for the existence conditions. Finally let us mention that both power and entropic moments have been used as uncertainty measures as we discuss later on.

Remark that $W_1[\rho] = 1$, because $\rho(\mathbf{r})$ is a probability density function and that $W_q[\rho]$ is finite when $q < 0$ only if $\rho(\mathbf{r})$ is of bounded support. Moreover, when $\rho(\mathbf{r})$ is bounded, $W_q[\rho]$ tends to the (Lebesgue) measure of its support $\mu_{\mathcal{L}}\{\mathbf{r} : \rho(\mathbf{r}) > 0\}$ when $q \to 0^+$. In addition, this quantity has the following mathematical properties [36]:

- If $\rho(\mathbf{r})$ is bounded, then $W_q[\rho] < \infty$ for any $q > 1$
- If $W_q[\rho] < \infty$ for some $q < 1$, then $W_{q'}[\rho] < \infty$ for any $q' \in (q, 1)$
- If $\rho(\mathbf{r})$ is of finite support, then $W_{q'}[\rho] < \infty$ for any $q' \in [q, 1)$

The power and entropic moments of D-dimensional systems have been found to be mutually related by means of analytical inequalities of variational origin [37, 38]. The entropic moments of order q of a D-dimensional density $\rho(\mathbf{r})$ are, according to (5.2), given by

$$W_q[\rho] = \langle \rho^{q-1} \rangle = \int_{R_D} [\rho(\mathbf{r})]^q d\mathbf{r}. \tag{5.3}$$

They are bounded from below in terms of the power moments $\langle r^\alpha \rangle$ and $\langle r^\beta \rangle$ as

$$W_q[\rho] \geq F_1(\alpha, \beta, q, D) \left(\frac{\langle r^\beta \rangle^{q(\alpha+D)-D}}{\langle r^\alpha \rangle^{q(\beta+D)-D}} \right)^{\frac{1}{\alpha-\beta}}, \tag{5.4}$$

for $q > 1$ and $\alpha > \beta > \frac{-D(q-1)}{q}$, and as

$$W_q[\rho] \geq F_2(\alpha, \beta, q, D) \left(\frac{\langle r^\alpha \rangle^{-q(\beta+D)-D}}{\langle r^\beta \rangle^{-q(\alpha+D)-D}} \right)^{\frac{1}{\alpha-\beta}}, \tag{5.5}$$

for $q > 1$ and $\beta < \alpha < -\frac{D(q-1)}{q}$, and where $F_i(\alpha, \beta, q, D)$, $i = 1$ and 2, are given by

$$F_1(\alpha,\beta,q,D) = \frac{q^q(\alpha-\beta)^{2q-1}}{[\Omega_D B(\frac{q(\beta+D)-D}{(\alpha-\beta)(q-1)}, \frac{2q-1}{q-1})]^{q-1}} \left\{ \frac{[q(\beta+D)-D]^{q(\beta+D)-D}}{[q(\alpha+D)-D]^{q(\alpha+D)-D}} \right\}^{\frac{1}{\alpha-\beta}} \tag{5.6}$$

and

$$F_2(\alpha,\beta,q,D) = \frac{q^q(\alpha-\beta)^{2q-1}}{[\Omega_D B(\frac{-q(\alpha+D)+D}{(\alpha-\beta)(q-1)}, \frac{2q-1}{q-1})]^{q-1}} \times \left\{ \frac{[-q(\alpha+D)+D]^{-q(\alpha+D)+D}}{[-q(\beta+D)+D]^{-q(\beta+D)+D}} \right\}^{\frac{1}{\alpha-\beta}}, \tag{5.7}$$

where $\Omega_D = \frac{2\pi^{D/2}}{\Gamma(\frac{D}{2})}$ is the surface area of the sphere, and $B(x,y) = \frac{\Gamma(x)\Gamma(y)}{\Gamma(x+y)}$ denotes the beta function of Euler. In addition, the entropic moment $W_q[\rho]$ with $0 < q < 1$ is upper-bounded by means of the expectation values $(\langle r^\alpha \rangle, \langle r^\beta \rangle)$ as

$$W_q[\rho] \leq G(\alpha,\beta,q,D) \left[\langle r^\alpha \rangle^{-q(\beta+D)+D} \langle r^\beta \rangle^{q(\alpha+D)-D} \right]^{\frac{1}{\alpha-\beta}} \tag{5.8}$$

for $\alpha > \frac{D(1-q)}{q} > \beta$, where $G(\alpha, \beta, q, D)$ is given by

$$G(\alpha,\beta,q,D) = \frac{q^q(\alpha-\beta)^{2q-1}}{[\Omega_D B(\frac{q(\alpha+D)-D}{(\alpha-\beta)(1-q)}, \frac{-q(\beta+D)+D}{(\alpha-\beta)(1-q)})]^{q-1}} \times \left\{ \frac{[-q(\beta+D)+D]^{q(\beta+D)-D}}{[q(\alpha+D)-D]^{q(\alpha+D)-D}} \right\}^{\frac{1}{\alpha-\beta}}. \tag{5.9}$$

Expressions similar to the inequalities (5.4)–(5.8) can be derived in momentum space between the qth-order entropic moment $U_q[\gamma] \equiv \langle \gamma^{q-1} \rangle$ and the expectation values $\langle p^\alpha \rangle$ and $\langle p^\beta \rangle$.

It is also possible to obtain the converse inequalities in both reciprocal spaces; that is, expressions which provide lower and upper bounds to a given power moment by means of two entropic moments [33]. It has been found the lower bounds

$$\langle r^k \rangle \geq L_D(a,b,k) \left(\frac{\{W_b[\rho]\}^{a(D+k)-D}}{\{W_a[\rho]\}^{b(D+k)-D}} \right)^{\frac{1}{D(a-b)}} ; \quad k > 0 \tag{5.10}$$

for any $a > b > \frac{D}{D+k}$, and the upper bounds

$$\langle r^k \rangle \leq M_D(a,b,k) \left(\{W_a[\rho]\}^{D-b(D+k)} \{W_b[\rho]\}^{a(D+k)-D} \right)^{\frac{1}{D(a-b)}} ; \quad k < 0 \tag{5.11}$$

where $L_D(a, b, k)$ and $M_D(a, b, k)$ are known functions of the parameters a, b and k [33].

All these inequality-based expressions can be used and have been used to bound and estimate both information-theoretic measures (see Sect. 5.2.2) and fundamental energies. Let us here apply them to bound the Dirac exchange and the exact kinetic energy of a D-dimensional many-electron system. The Dirac exchange energy in the plane-wave approximation (e.g. Thomas-Fermi model) is given by

$$E_{\text{ex}}(D) = -C_D W_{1+\frac{1}{D}}[\rho] \quad \text{with } C_D = \frac{4D}{D^2-1}\left(\frac{D}{2\Omega_D}\right)^{\frac{1}{D}}. \tag{5.12}$$

From (5.8) and (5.12) we obtain that the Dirac exchange energy is bounded from above as

$$E_{\text{ex}}(D) \leq -C_D F_2\left(\alpha, \beta, 1+\frac{1}{D}, D\right)\left[\frac{\langle r^\alpha\rangle^{-\beta(1+\frac{1}{D})-1}}{\langle r^\beta\rangle^{-\alpha(1+\frac{1}{D})-1}}\right]^{\frac{1}{\alpha-\beta}} \tag{5.13}$$

for $\alpha(D+1)+D<0$. In the three-dimensional case, this inequality simplifies as

$$E_{\text{ex}}(D=3) := -C_e W_{\frac{4}{3}}[\rho] \leq -C_e F_2\left(\alpha, \beta, \frac{4}{3}, 3\right)\left[\frac{\langle r^\alpha\rangle^{-\frac{4\beta}{3}-1}}{\langle r^\beta\rangle^{-\frac{4\alpha}{3}-1}}\right]^{\frac{1}{\alpha-\beta}} \tag{5.14}$$

for $\beta < \alpha < -\frac{3}{4}$ and with $C_e = \frac{3}{4}(\frac{3}{\pi})^{1/3} = 0.73856$. This inequality allows us to find very accurate upper bounds to the Dirac exchange energy of atomic systems. Indeed, for $\alpha=-1$ and $\beta=-2$ it gives

$$E_{\text{ex}}(D=3) \leq -\frac{4C_e}{5^{4/3}\pi^{1/3}}\left(\frac{\langle r^{-1}\rangle^5}{\langle r^{-2}\rangle}\right)^{\frac{1}{3}}, \tag{5.15}$$

which gives the best known upper bound to the indirect Coulomb energy of atomic systems. See also [39–41].

Let us now consider the exact kinetic energy T, which is known [42–44] to be lower-bounded by the entropic moment of order $1+\frac{2}{D}$ as

$$T \geq K_D W_{1+\frac{2}{D}}[\rho], \quad \text{with } K_D = \frac{2\pi D}{D+2}\left[\Gamma\left(\frac{D}{2}+1\right)\right]^{\frac{2}{D}}. \tag{5.16}$$

From (5.4), (5.6) and (5.16) we find [45] the following family of lower bounds:

$$T \geq C_D(\alpha,\beta)\left(\frac{\langle r^\beta\rangle^{\alpha(1+\frac{2}{D})+2}}{\langle r^\alpha\rangle^{\beta(1+\frac{2}{D})+2}}\right)^{\frac{1}{\alpha-\beta}} \tag{5.17}$$

with $C_D(\alpha,\beta) = K_D F_1(\alpha, \beta, q=1+\frac{2}{D}, D)$ for $\alpha > \beta > -\frac{2D}{D-2}$. This expression extends and generalizes to D dimensions many other related three-dimensional ones (see [45] for a detailed discussion). In particular, for $\alpha=2$ and $\beta=0$ we have the bound

$$T \geq \frac{D^2(D!)^{\frac{2}{D}}}{2(D+1)^2}\frac{1}{\langle r^2\rangle}; \quad D \geq 1 \tag{5.18}$$

in terms of the mean square radius, and for $\alpha = 0$ and $\beta = -1$ we find the bound

$$T \geq \frac{(D!)^{\frac{2}{D}}(D-2)}{2^{1+\frac{2}{D}}}\langle r^{-1}\rangle^2; \quad D \geq 2. \tag{5.19}$$

Then, for three-dimensional systems we have the bounds

$$T \geq \frac{9 \cdot 3^{2/3}}{16 \cdot 2^{1/3}} \frac{1}{\langle r^2\rangle} \simeq \frac{0.92867}{\langle r^2\rangle} \tag{5.20}$$

and

$$T \geq \frac{1}{2 \cdot 3^{1/3}}\langle r^{-1}\rangle^2 \simeq 0.34668\langle r^{-1}\rangle^2. \tag{5.21}$$

On the other hand, from the Pitt-Beckner inequality it is possible to find [45, 46] the following inequality between the expectation values $\langle p^\alpha\rangle$ and $\langle r^{-\alpha}\rangle$ of D-dimensional systems:

$$\langle p^\alpha\rangle \geq 2^\alpha \left[\frac{\Gamma(D+\frac{\alpha}{4})}{\Gamma(D-\frac{\alpha}{4})}\right]^2 \langle r^{-\alpha}\rangle; \quad 0 \leq \alpha < D, \tag{5.22}$$

which for $\alpha = 2$ allows us to obtain another bound for the kinetic energy; namely,

$$T \geq \frac{(D-2)^2}{8}\langle r^{-2}\rangle; \quad D \geq 2. \tag{5.23}$$

Let us also quote here the Daubechies' lower bound [47] of $\langle p^\alpha\rangle$ in terms of the entropic moments $W_q[\rho]$, not yet sufficiently well explored, of a D-dimensional many-electron system:

$$\langle p^\alpha\rangle \geq C_{\alpha,D} W_{1+\frac{\alpha}{D}}[\rho], \tag{5.24}$$

where

$$C_{\alpha,D} = \frac{[2k_d K_2(D,\alpha)]^{-\frac{\alpha}{D}}}{1+\frac{\alpha}{D}}, \quad \text{with } k_d = \left[D2^{D-1}\pi^{D/2}\Gamma\left(\frac{D}{2}\right)\right]^{-1}, \tag{5.25}$$

and

$$K_2(D,\alpha) = \Gamma\left(\frac{D}{\alpha}\right) \inf_{a>0}\left[a^{-\frac{D}{\alpha}}\left(\int_a^\infty due^{-u}(u-a)n^{-1}\right)^{-1}\right] \tag{5.26}$$

which, taking into account (5.12), allows us to correlate the mean momentum expectation value with the Dirac exchange energy of the system.

Finally, it is interesting to mention that the kinetic energy can also be accurately bounded from below by means of the logarithmic moments of $\rho(\mathbf{r})$ in various forms [45].

5.2.2 *Shannon, Rényi and Tsallis Entropies*

Here we will gather some known results relative to the bounds to the Shannon, Rényi and Tsallis entropies of D-dimensional systems in terms of the power, logarithmic and/or entropic moments of the single-particle density $\rho(\mathbf{r})$, and various mutual inequalities.

The Rényi entropies are defined [48] in terms of the entropic moments $W_q[\rho]$ as

$$R_q[\rho] := \frac{1}{1-q} \ln W_q[\rho] = \frac{1}{1-q} \ln \int_{R_D} [\rho(\mathbf{r})]^q d\mathbf{r}, \tag{5.27}$$

and the Tsallis entropies [49] (also called Havrda-Charvát entropies [50]) are given by

$$T_q[\rho] := \frac{1}{q-1}[1 - W_q[\rho]] = \frac{1}{q-1}\left\{1 - \int_{R_D} [\rho(\mathbf{r})]^q d\mathbf{r}\right\} \tag{5.28}$$

with $q > 0$ and $q \neq 1$. Moreover, it can be shown that

$$T_q[\rho] = -D_q W_x[\rho]|_{x=1} \tag{5.29}$$

where $D_q f(x) = (f(qx) - f(x))/(qx - x)$ is the Jackson derivative of the function $f(x)$.

When q tends to 1, both entropies reduce to the (Boltzmann-Gibbs) Shannon entropy

$$S[\rho] := -\int_{R_D} \rho(\mathbf{r}) \ln \rho(\mathbf{r}) d\mathbf{r}. \tag{5.30}$$

It is interesting to remark that these quantities are global measures of spreading of the density $\rho(\mathbf{r})$ because they are power (Rényi, Tsallis) and logarithmic (Shannon) functionals of $\rho(\mathbf{r})$. The Shannon entropy is the only one satisfying all the hypotheses of Shannon theorem [51] as well as some other important criteria [52]. It can be expressed [33] in terms of the entropic moments $W_q[\rho]$ as the limit

$$S[\rho] = -\lim_{q \to 1} \frac{dW_q[\rho]}{dq}. \tag{5.31}$$

Moreover, the Shannon entropy becomes the well-known thermodynamical entropy in the case of a thermal ensemble. It is worth nothing that, unlike the more familiar entropy $-\sum_i p_i \ln p_i$ (also due to Shannon) of a probability on a discrete sample space, in the continuous case $S[\rho]$ can have any values in $[-\infty, \infty]$, and it can also be undefined (i.e. of the form $\infty - \infty$). Any sharp peaks in $\rho(\mathbf{r})$ will tend to make $S[\rho]$ negative, whereas positive values for $S[\rho]$ are provoked by a slowly decaying tail; hence the Shannon entropy $S[\rho]$ is a measure of how localized the density $\rho(\mathbf{r})$ is [53].

The Rényi entropies are additive while the Tsallis entropies are non-negative, extremal at equiprobability, concave for $q > 0$ and pseudoadditive (i.e. $T_q[\rho_1 \otimes \rho_2] = T_q[\rho_1] + T_q[\rho_2] + (1-q)T_q[\rho_1]T_q[\rho_2]$). Remark that the Tsallis expression may be seen as a linearization of the Rényi expression with respect to $W_q[\rho]$. Let us also point out that the second order Tsallis entropy, i.e.

$$T_2[\rho] = 1 - W_2[\rho] = 1 - \int_{R_D} [\rho(\mathbf{r})]^2 d\mathbf{r} \tag{5.32}$$

is the simplest entropy, providing a good alternative of the Shannon entropy in many cases. In fact, it is more than that; it refers directly to the experimental results of mutually complementary measurements, so opposite to the Shannon entropy which is applicable when the measurements exhibit a preexisting symmetry [54]. It is called linear or linearized entropy [55], having been used not only as an impurity measure of the quantum state but also as measures of decoherence, entanglement, complexity and mixedness of three-dimensional quantum systems.

Extending previous tridimensional results [27], it was variationally shown that the Shannon entropy $S[\rho]$ has the following class of upper bounds [56]

$$S[\rho] \leq A_D(\alpha,\beta) + \beta \ln\langle r^\alpha\rangle + (D-\alpha\beta)\langle \ln r\rangle; \quad \forall \beta > 0,\ \alpha > -D \tag{5.33}$$

in terms of the expectation values $\langle r^\alpha\rangle$ and $\langle \ln r\rangle$, with

$$A_D(\alpha,\beta) = \beta + \ln \frac{\Omega_D \Gamma(\beta)}{|\alpha|\beta^\beta}. \tag{5.34}$$

For the particular case $\beta = \frac{D}{\alpha}$, this inequality provides the following upper bound (see also [57]) in terms of radial expectation values $\langle r^\alpha\rangle$:

$$S[\rho] \leq A_D\left(\alpha, \frac{D}{\alpha}\right) + \frac{D}{\alpha}\ln\langle r^\alpha\rangle; \quad \forall \alpha > 0. \tag{5.35}$$

Some instances of this expression are the upper bounds

$$S[\rho] \leq D + \ln\left(\frac{\Omega_D (D-1)!}{D^D}\langle r\rangle^D\right) \tag{5.36}$$

for $\alpha = 1$, and

$$S[\rho] \leq \frac{D}{2}[1 + \ln(2\pi\langle r^2\rangle)] \tag{5.37}$$

for $\alpha = 2$, in terms of the mean radius or centroid $\langle r\rangle$ and the second-order central moment $\langle r^2\rangle$, respectively. Moreover, for a given $\langle r^2\rangle$ the Shannon entropy is maximum for a Gaussian density of covariance matrix $R = \frac{\langle r^2\rangle}{D} I$, where I is the identity matrix.

A second type of variational upper bound to $S[\rho]$ has been also found [56] to depend on the mean logarithmic radius $\langle \ln r\rangle$ and the logarithmic uncertainty $\Delta(\ln r) = (\langle(\ln r)^2\rangle - \langle \ln r\rangle^2)^{1/2}$; it has the expression

$$S[\rho] \leq B_D + \ln \Delta(\ln r) + D\langle \ln r\rangle, \tag{5.38}$$

with

$$B_D = \frac{1}{2} + \ln\left(\sqrt{2\pi}\,\Omega_D\right); \quad \Omega_D = \frac{2\pi^{D/2}}{\Gamma(\frac{D}{2})}. \tag{5.39}$$

Moreover, Angulo et al. [33] have found a class of upper bounds in terms of the mean logarithmic radius $\langle \ln r\rangle$ and the entropic moment $W_q[\rho]$:

$$S[\rho] \leq \frac{1}{1-\beta(q-1)}[B_D(q,\beta) + \beta \ln W_q[\rho] + D\langle \ln r\rangle], \quad 1 \leq q \leq 1 + \frac{1}{\beta}, \tag{5.40}$$

where

$$B_D(q,\beta) \equiv q\beta + \ln \frac{\Omega_D \Gamma(\beta)}{D|q-1|(q\beta)^\beta}. \tag{5.41}$$

This interesting inequality allows us to correlate the Shannon entropy with the Dirac exchange K_0 and Thomas-Fermi T_0 energies

$$K_0 = \frac{3^{4/3}}{4\pi^{1/3}} W_{\frac{4}{3}}[\rho]; \qquad T_0 = \frac{3^{5/3}}{10} W_{\frac{5}{3}}[\rho] \tag{5.42}$$

in the three-dimensional case, obtaining

$$S_\rho \le C_1 + \frac{3}{2} \ln K_0 + \frac{9}{2} \langle \ln r \rangle \tag{5.43}$$

for $q = \frac{4}{3}$ and $\beta = 1$, and

$$S_\rho \le C_2 + \frac{6}{5} \ln T_0 + \frac{27}{5} \langle \ln r \rangle \tag{5.44}$$

for $q = \frac{5}{3}$ and $\beta = \frac{2}{3}$, respectively. A numerical Hartree-Fock study for all ground state neutral atoms of the periodic table has shown [33] that these two inequalities are quite accurate. Greater accuracy can be obtained for other choices of the parameter β for each specific atom.

It is interesting to point out that the D-dimensional maximum entropy problem inherent to the previous bounds (5.18)–(5.26) has a unique solution [57] whenever it exists. The existence conditions are given in [57]. Moreover, the usefulness of this maxent problem to explain the main characteristics of the periodic table, such as periodicity and shell structure, is discussed in [58].

Let us now tackle the maximum-Rényi-entropy or maxrent problem which provides variational upper bounds to the Rényi entropy $R_q[\rho]$ defined by (5.27). When the only constraint is a radial expectation value $\langle r^k \rangle$, it is possible to find the upper bounds given by

$$R_q[\rho] \le \frac{1}{1-q} \ln\left\{L_1(q,k,D) \langle r^k \rangle^{-\frac{D}{k}(q-1)}\right\} \tag{5.45}$$

in terms of $\langle r^k \rangle$ with $k = 1, 2, \ldots$, and the upper bounds given by

$$R_q[\rho] \le \frac{1}{1-q} \ln\left\{L_2(q,k,D) \langle r^{-k} \rangle^{-\frac{D}{k}(q-1)}\right\} \tag{5.46}$$

in terms of $\langle r^{-k} \rangle$ with $k = 1, 2, \ldots$ but subject to the condition $k < \frac{D}{q}(q-1)$. The functions $L_i(q,k,D)$, $i = 1$ and 2, have the expressions

$$L_1(q,k,D) = \frac{qk}{D(q-1)+kq} \left\{ \frac{k\Gamma(D/2)\left[\frac{D(q-1)}{D(q-1)+kq}\right]^{\frac{D}{k}}}{2\pi^{\frac{D}{2}} B\left(\frac{q}{q-1}, \frac{D}{k}\right)} \right\}^{q-1} \tag{5.47}$$

and

$$L_2(q,k,D) = \frac{qk}{D(q-1)-kq} \left\{ \frac{k\Gamma(D/2)\left[\frac{D(q-1)-kq}{D(q-1)}\right]^{\frac{D}{k}}}{2\pi^{\frac{D}{2}} B\left(\frac{D}{k} - \frac{1}{q-1}, \frac{q}{q-1}\right)} \right\}^{q-1}. \tag{5.48}$$

We have used the variational bounds [38] to the entropic moments $W_\alpha[\rho]$ with a single expectation value $\langle r^k \rangle$ as constraint. These bounds (5.45)–(5.48) extend and generalize similar bounds obtained in the one-dimensional [59, 60] and three-dimensional [61] cases used in various contexts, ranging from finances to atomic physics.

The maximum-Tsallis-entropy or maxtent problem with the constraint $\langle r^k \rangle$ follows in a similar manner. We find the lower bounds

$$1 + (1-q)T_q[\rho] \geq L_1(q,k,D)\langle r^k \rangle^{-\frac{D}{k}(q-1)} \tag{5.49}$$

in terms of $\langle r^k \rangle$ with $k = 1, 2, \ldots$, and the lower bounds

$$1 + (1-q)T_q[\rho] \geq L_2(q,k,D)\langle r^{-k} \rangle^{\frac{D}{k}(q-1)} \tag{5.50}$$

in terms of $\langle r^{-k} \rangle$ with $k = 1, 2, \ldots$ and $k < \frac{-D(q-1)}{q}$. The functions $L_i(q,k,D)$ are given by (5.47) and (5.48), respectively. The maxtent problem is discussed in [57] in the three-dimensional case and its usefulness to interpret various physical phenomena of the periodic table have been recently shown in [58].

Finally, let us comment that expressions similar to the position inequalities (5.33)–(5.50) are also valid in momentum space for the corresponding quantities in this space.

5.2.3 Fisher Information

In this section we will gather a number of results, necessarily of inequality type, about the Fisher information of D dimensional systems in terms of the power and entropic moments of the single-particle density $\rho(\mathbf{r})$.

The (translationally invariant) Fisher information of the D-dimensional density $\rho(\mathbf{r})$ is defined by

$$I[\rho] := \int_{R_D} \rho(\mathbf{r}) |\nabla_D \ln \rho(\mathbf{r})|^2 d\mathbf{r} = 4 \int_{R_D} \left| \nabla_D \sqrt{\rho(\mathbf{r})} \right|^2 d\mathbf{r} \tag{5.51}$$

where ∇_D denotes the D-dimensional gradient. The corresponding quantity for the momentum-space probability density $\gamma(\mathbf{p})$ will be denoted by $I[\gamma]$.

This concept was firstly introduced for one-dimensional random variables in statistical estimation [62] but nowadays it is playing a increasing role in numerous fields [63], in particular, for many-electron systems, partially because of its formal resemblance with kinetic [63–67] and Weiszäcker [26, 68] energies. The Fisher information, contrarily to the Rényi, Shannon and Tsallis entropies, is a local measure of spreading of the density $\rho(\mathbf{r})$ because it is a gradient functional of $\rho(\mathbf{r})$. The higher this quantity is, the more localized is the density, the smaller is the uncertainty and the higher is the accuracy in estimating the localization of the particle. It has, however, an intrinsic connection with Shannon's entropy via the de Bruijn identity [69, 70] as well as a simple connection with the precision (variance $V[\rho]$) of the experiments by means of the celebrated Cramér-Rao inequality [69–71]

$$I[\rho] \times V[\rho] \geq D^2. \tag{5.52}$$

The notion of Fisher information has been shown to be very fertile to identify, characterize and interpret numerous phenomena and processes in atomic and molecular physics such as e.g., correlation properties in atoms, spectral avoided crossings of atoms in external fields [72], the periodicity and shell structure in the periodic table of chemical elements [73] and the transition state and other stationary points in chemical reactions [74]. Moreover, it has been used for the variational characterization of quantum equations of motion [63] as well as to rederive the classical thermodynamics without requiring the usual concept of Boltzmann's entropy [75].

Extending previous three-dimensional bounds to Fisher information of atomic systems [68, 76], it has been found by use of Redheffer-Weyl's inequality that the Fisher information of D dimensional systems is bounded from below [73] as

$$I[\rho] \geq (\beta + D - 1)^2 \frac{\langle r^{\beta-1} \rangle}{\langle r^{2\beta} \rangle}, \quad \text{for } \beta \geq \max\{-D+1, -1\} \tag{5.53}$$

and

$$I[\rho] \geq \langle r^{-2} \rangle \left[(D-2)^2 + \frac{(\beta+1)^2 \langle r^{\beta-1} \rangle^2}{\langle r^{2\beta} \rangle \langle r^{-2} \rangle - \langle r^{\beta-1} \rangle^2} \right], \quad \beta \geq -1 \tag{5.54}$$

in terms of the radial expectation values $\langle r^{\alpha} \rangle$, and as

$$I[\rho] \geq \langle r^{-2} \rangle \left[(D-2)^2 + \frac{\langle r^{-2} \rangle^2}{\langle r^{-2} (\ln r)^2 \rangle \langle r^{-2} \rangle - \langle r^{-2} \ln r \rangle^2} \right] \tag{5.55}$$

in terms of the radial and logarithmic expectation values.

On the other hand, Frieden's variational principle of extreme physical information [63] and, in particular, the minimization problem of the Fisher information (minfin problem) allows us [57] to find that the D-dimensional density $\rho_F(\mathbf{r}) \equiv g(\mathbf{r})$ which minimizes the Fisher information (5.51) subject to the m constraints $\langle f_k(\mathbf{r}) \rangle$, fulfills the differential equation

$$\left[\frac{\nabla_D g(\mathbf{r})}{g(\mathbf{r})} \right]^2 + 2\nabla_D \left[\frac{\nabla_D g(\mathbf{r})}{g(\mathbf{r})} \right] + \lambda_0 + \sum_{k=1}^{m} \lambda_k f_k(\mathbf{r}) = 0. \tag{5.56}$$

The general solution of this equation is very difficult, even for a single constraint of the type $\langle r^{\alpha} \rangle$. In the case of $D = 3$, this problem has been partially solved [68] for a single constraint $\langle r^{\alpha} \rangle$ with $\alpha = -1$ and 2, and for two constraints of the types (r^{-1}, r^{-2}) and (r^{-2}, r^2). Let us here give the lower bound

$$I[\rho] \geq 4 \langle r^{-1} \rangle. \tag{5.57}$$

This inequality is saturated by the exponential minimized density

$$g(r) = \frac{1}{\pi} \langle r^{-1} \rangle^3 \exp(-2 \langle r^{-1} \rangle r) \tag{5.58}$$

which has very interesting information-theoretic properties [57]. These properties have allowed us [58] to nicely interpret the shell structure of the atomic systems. The extension of these three-dimensional bounds to D dimension is a yet unsolved problem.

5.2.4 Uncertainty Relations

In this Section we shall give the uncertainty relations corresponding to the power logarithmic and entropic moments together with those associated to the Rényi, Shannon and Tsallis entropies and the Fisher information. These relations are different mathematical formulations of the quantum-mechanical uncertainty principle which describes a characteristic feature of quantum mechanics and states the limitations to perform measurements on a system without disturbing it. Moreover, since the two canonically conjugate observables involved in the uncertainty relations here considered—position and momentum—do not commute, both observables cannot be precisely determined in any quantum state of the system.

The D-dimensional position-momentum uncertainty relation is the Heisenberg-like inequality [77]

$$\langle r^{D/\alpha}\rangle^{\alpha}\langle p^{D/\beta}\rangle^{\beta} \geq \alpha^{\alpha}\beta^{\beta}\frac{\Gamma^2(1+\frac{D}{2})}{\Gamma(1+\alpha)\Gamma(1+\beta)}e^{D-\alpha-\beta}; \quad \alpha>0, \beta>0 \qquad (5.59)$$

obtained by using information-theoretic methods.

For $\alpha=\beta=D/2$, this expression simplifies to the familiar D-dimensional form of Heisenberg inequality

$$\langle r^2\rangle\langle p^2\rangle \geq \frac{D^2}{4} \qquad (5.60)$$

which shows that the more accurately the position is known, the less accurately is the momentum determined, and vice versa.

For completeness let us quote here that (5.59) for $D=3$ can be cast in the form

$$\langle r^a\rangle^{1/a}\langle p^b\rangle^{1/b} \geq \left(\frac{\pi ab}{16\Gamma(\frac{3}{a})\Gamma(\frac{3}{b})}\right)^{1/3}\left(\frac{3}{a}\right)^{\frac{1}{a}}\left(\frac{3}{b}\right)^{\frac{1}{b}}e^{1-\frac{1}{a}-\frac{1}{b}}; \quad a>0,\ b>0 \qquad (5.61)$$

which for the specially interesting case $a=b>0$ takes the form

$$\langle r^a\rangle\langle p^a\rangle \geq \left\{\left(\frac{27\pi}{16a\Gamma(\frac{3}{a})}\right)^{\frac{1}{3}}\left(\frac{ae}{3}\right)^{1-\frac{2}{a}}\right\}^a, \quad a>0. \qquad (5.62)$$

Note that for $a=2$ this inequality reduces to (5.60) with $D=3$.

For alternative forms of these two inequalities in terms of variances or even in terms of moments around a point other than the origin, see (2.9) et sequel of [78].

In addition, from the Pitt-Beckner inequality (5.22) for $(\langle p^a\rangle, \langle r^{-a}\rangle)$ and its version for $(\langle p^{-a}\rangle, \langle r^a\rangle)$ given by

$$\langle p^{-a}\rangle \leq 2^{-a}\left[\frac{\Gamma(\frac{D-a}{4})}{\Gamma(\frac{D+a}{4})}\right]^2\langle r^a\rangle, \quad 0\leq a<D \qquad (5.63)$$

one obtains the uncertainty inequality

$$\langle r^a\rangle^{\frac{1}{a}}\langle p^{-a}\rangle^{-\frac{1}{a}} \le 2\left[\frac{\Gamma(\frac{D+a}{4})}{\Gamma(\frac{D-a}{4})}\right]^{\frac{2}{a}}, \quad 0\le a<D \tag{5.64}$$

which generalizes various uncertainty relations of similar type found in the literature [73, 76]. See also [79] for some slight improvements in the $D=3$ case. Moreover, since we may exchange roles of r and p in (5.61), we can write

$$\langle r^{-a}\rangle \le \frac{1}{2^a}\left[\frac{\Gamma(\frac{D-a}{4})}{\Gamma(\frac{D+a}{4})}\right]^2 \langle p^a\rangle, \quad 0\le a\le D. \tag{5.65}$$

Then, from (5.61) and (5.63) we have another uncertainty relation given by

$$\langle r^{-a}\rangle\langle p^{-a}\rangle \le \frac{1}{2^{2a}}\left[\frac{\Gamma(\frac{D-a}{4})}{\Gamma(\frac{D+a}{4})}\right]^4 \langle r^a\rangle\langle p^a\rangle, \tag{5.66}$$

valid for $0\le a<D$. Similar uncertainty relations can be obtained for the modified moments $\langle r^\alpha\rho\rangle$ and $\langle p^\alpha\gamma\rangle$ by means of the results of Sect. 3 of Folland and Sitaran [78].

There are quantitative formulations of the uncertainty principle more stringent than (5.59). With the same procedure used to obtain the relation (5.59) but using logarithmic moments instead of power moments, one obtains the *logarithmic uncertainty inequality* [27]

$$\Delta(\ln r)\Delta(\ln p) \ge \frac{\Gamma^2(\frac{D}{2})}{8\pi}\exp[D-1-D(\langle\ln r\rangle+\langle\ln p\rangle)] \tag{5.67}$$

where

$$\Delta(\ln r) \equiv (\langle(\ln r)^2\rangle - \langle\ln r\rangle^2)^{1/2} \tag{5.68}$$

denotes the logarithmic uncertainty. For $D=3$ we have

$$\Delta(\ln r)\Delta(\ln p) \ge \frac{1}{32}\exp[2-3(\langle\ln r\rangle+\langle\ln p\rangle)]. \tag{5.69}$$

Moreover, the logarithmic sum involved in (5.67) satisfies Beckner's uncertainty relation [80]

$$\langle\ln r\rangle + \langle\ln p\rangle \ge \psi\left(\frac{D}{4}\right) + \ln 2 \tag{5.70}$$

where $\psi(x) = \Gamma'(x)/\Gamma(x)$ is the Psi or digamma function. This expression can alternatively be obtained by taking the limit $a\to 0$ in the uncertainty product $\langle r^a\rangle\langle p^{-a}\rangle$ with the values (5.22) and (5.69) for $\langle r^a\rangle$ and $\langle p^{-a}\rangle$, respectively.

Another generalization of the (power-moments-based) Heisenberg-like uncertainty relation was obtained by Rajagopal [81] by use of the entropic moments in position and momentum spaces. He extended to D dimensions and improve the one-dimensional results of Maassen-Uffink [82], obtaining the following *entropic-moment-based uncertainty relation* (see also [83, 84]):

$$\{W_{\alpha+1}[\rho]\}^{-\frac{1}{\alpha}} \times \{W_{\beta+1}[\gamma]\}^{-\frac{1}{\beta}} \geq \left[\frac{\pi(1+2\alpha)^{1+\frac{1}{2\alpha}}}{1+\alpha}\right]^{D} \tag{5.71}$$

which is valid for $\alpha \geq -\frac{1}{2}$ and $\beta = -\frac{\alpha}{1+2\alpha}$. For $\alpha = \beta = 0$ it reduces to the (Shannon-entropy-based) *entropic uncertainty relation*

$$S[\rho] + S[\gamma] \geq D(1 + \ln \pi) \tag{5.72}$$

first derived by Hirschman [85] and later improved independently by Beckner [80] and Bialynicki-Birula and Mycielski [53]. This expression indicates that the total uncertainty in position and momentum cannot be decreased beyond the value given by (5.72). The entropic uncertainty relation can be recast into the form

$$J[\rho] \times J[\gamma] \geq \frac{1}{4} \tag{5.73}$$

where the position Shannon entropic power is defined by

$$J[\rho] := \frac{1}{2\pi e} e^{\frac{2}{D} S[\rho]} \tag{5.74}$$

and similarly for the momentum Shannon entropy power $J[\gamma]$. Let us point out here that the Shannon-entropy sum $S[\rho] + S[\gamma]$ has shown its usefulness for numerous physical issues (e.g., to study the correlation energy of atomic systems [86]), having been postulated a new entropy maximization principle based in it by Gadre [87]. This author and his collaborators [88] have numerically shown some interesting properties of this entropy sum for atoms in a Hartree-Fock framework: they obtain a minimum value for the ground state which is scale invariant while the individual entropies are not.

The expressions (5.28) and (5.71) have allowed Rajagopal [81] to obtain the *Tsallis-entropy-based uncertainty relation* as

$$\{1 + (1-p)T_p[\rho]\}^{\frac{-1}{2p}} \times \{1 + (1-q)T_q[\gamma]\}^{\frac{1}{2q}} \geq \left(\frac{q}{\pi}\right)^{\frac{D}{4q}} \left(\frac{p}{\pi}\right)^{\frac{-D}{4p}} \tag{5.75}$$

with $\frac{1}{p} + \frac{1}{q} = 2$. Note that in the limit of (p,q) going to unity, this expression transforms into the entropic uncertainty relation (5.70). Finally, let us quote here the *Rényi-entropy-based uncertainty relation* found by Bialynicki-Birula [89] and independently by Zozor and Vignat [90]. They have obtained it in the one-dimensional case as

$$R_{\alpha-1}[\rho] + R_{\beta-1}[\gamma] \geq -\frac{1}{2(1-\alpha)} \ln \frac{\alpha}{\pi} - \frac{1}{2(1-\beta)} \ln \frac{\beta}{\pi}, \tag{5.76}$$

for $\alpha > \beta$. The extension to D dimensions has been recently found by Zozor et al. [91].

Finally, let us discuss the uncertainty relations which involve the Fisher informations. Since the fifties [92] it is known that the Stam inequalities

$$I[\rho] \leq 4\langle p^2 \rangle; \qquad I[\gamma] \leq 4\langle r^2 \rangle, \tag{5.77}$$

link the position (momentum) Fisher information and the momentum (position) radial expectation value $\langle p^2 \rangle$ (respectively $\langle r^2 \rangle$). See also [73] for its generalization to finite many-electron systems. Recently, the *Fisher-information-based uncertainty relations*

$$I[\rho] \times I[\gamma] \geq 4D^2 \tag{5.78}$$

has been found [93] to hold not only for one-dimensional [94] but also for D-dimensional real-valued wavefunctions. For further information see Sect. 5.3, where the lower bound $4D^2$ is further improved for all wavefunctions of central potentials.

5.2.5 Complexity Measures

Here we consider the Cramér-Rao, Fisher-Shannon and LMC shape complexities of a D-dimensional quantum system. We will gather their known analytical results, without making any emphasis on their meaning and physical and chemical significances. In the latter case, see other contributions of this volume (particularly [96]). These information-theoretic measures of spreading of the quantum-mechanical probability density characterizing the physical state of our system, are composed by two single quantities of local (Fisher information) and/or global (variance, Shannon entropy) types. They have a common property: vanishing for the two extreme probability densities which corresponds to perfect order and maximum disorder.

The LMC shape complexity $C_{LMC}[\rho]$ is defined [97–99] as the product

$$C_{LMC}[\rho] := D[\rho] \exp(S[\rho]), \tag{5.79}$$

where

$$D[\rho] \equiv W_2[\rho] = \int \rho^2(\mathbf{r}) d\mathbf{r} \tag{5.80}$$

is the second-order entropic moment (5.3), heretoforth the disequilibrium because it quantifies the departure of $\rho(\mathbf{r})$ from equiprobability, and $S[\rho]$ denotes the Shannon entropy (5.30) which measures the randomness or global spreading of the distribution.

The Fisher-Shannon complexity is given by the product

$$C_{FS}[\rho] := I[\rho] \times J[\rho] \tag{5.81}$$

where $I[\rho]$ and $J[\rho]$ denote the (local) Fisher information (5.51) and the (global) Shannon entropy power (5.74) of $\rho(\mathbf{r})$. The former ingredient measures the local internal disorder of the distribution, quantifying the concentration of the quantum-mechanical probability cloud around the maxima of $\rho(\mathbf{r})$. The Shannon entropy power $J[\rho]$ is another global measure of the total spreading of the density all over its domain of definition.

The Cramér-Rao complexity $C_{CR}[\rho]$ is defined as

$$C_{CR}[\rho] := I[\rho] \times V[\rho] \tag{5.82}$$

which has a local ingredient, the Fisher information, and a global spreading measure of the density $\rho(\mathbf{r})$, the variance (which measures the concentration of the probability cloud around the centroid). So, contrary to the two previous complexities, this quantity does depend on a specific point of the definition domain of $\rho(\mathbf{r})$; namely, its centroid.

It has been shown (see [96] and [71] for further details) that these complexities have the following lower bounds

$$C_{LMC}[\rho] \geq 1, \tag{5.83}$$

$$C_{FS}[\rho] \geq D, \tag{5.84}$$

$$C_{CR}[\rho] \geq D^2 \tag{5.85}$$

in position space, and similar bounds in momentum space for the corresponding momentum complexities. These inequalities can be improved either by taking into account some known data (as e.g. some power moments) or by referring to specific quantum systems (as e.g. the D-dimensional particle in a box [100] and the D-dimensional hydrogenic system [71]). In particular, let us quote here that the Cramér-Rao inequality (5.85) has been improved by

$$C_{CR}[\rho] \geq D^2 + (D-1)\langle r^{-1}\rangle[(D-1)\langle r^2\rangle\langle r^{-1}\rangle - 2D\langle r\rangle] \tag{5.86}$$

in terms of the radial expectation values $\langle r^\alpha\rangle$ with $\alpha = -1$, 1 and 2. Moreover, the three complexity measures satisfy some (not yet fully accomplished) uncertainty relations. Let us point out here the corresponding relation for the LMC shape complexity [71]:

$$C_{LMC}[\rho] \times C_{LMC}[\gamma] \geq \frac{e^D \Gamma^2(\frac{D}{2})(D+\alpha)(D+\beta)}{\langle r^\alpha\rangle^{D/\alpha}\langle p^\beta\rangle^{D/\beta}} \times \left(\frac{D}{D+2\alpha}\right)^{1+\frac{D}{\alpha}} \left(\frac{D}{D+2\beta}\right)^{1+\frac{D}{\beta}} \tag{5.87}$$

for $\alpha, \beta > -\frac{D}{2}$. See [71, 96] for further details.

Just recently, a generalized statistical measure based on Rényi entropies has been introduced and characterized by a detailed study of its mathematical properties [101]. It extends the LMC shape complexity previously discussed.

5.3 Entropy Analysis of Quantum Systems with Central Potentials

Here we survey the recent work on the spreading and uncertainty measures (power moments, Shannon entropy, Fisher Information) of single-particle systems moving in spherically symmetric potentials by means of entropic ideas and methods extracted from information theory. The associated uncertainty relations are also discussed.

We start with some basic characteristics of the D-dimensional problem of a spinless particle in a central potential $V_D(r)$. The quantum-mechanical wavefunctions have, in atomic units, the form $\Psi_D(\mathbf{r},t) = \psi_D(\mathbf{r})\exp(-iE_Dt)$, where (E_D, ψ_D) denote the physical eigenfunctions of the Schrödinger equation [102, 103]

$$\left[-\frac{1}{2}\nabla_D^2 + V_D(r)\right]\psi_D(\mathbf{r}) = E_D\psi_D(\mathbf{r}) \tag{5.88}$$

where the D-dimensional position vector $\mathbf{r} = (r, \theta_1, \theta_2, \ldots, \theta_{D-1}) \equiv (r, \Omega_{D-1})$ in polar hyperspherical coordinates, where r denotes the hyperradius of the particle, and the Laplacian operator is expressed as

$$\nabla_D^2 = \frac{1}{r^{D-1}}\frac{\partial}{\partial r}r^{D-1}\frac{\partial}{\partial r} - \frac{\Lambda_{D-1}^2}{r^2}, \tag{5.89}$$

Λ_{D-1}^2 being the D-dimensional generalization of the squared angular momentum operator which only depends on the $D-1$ angular coordinates Ω_{D-1} of the hypersphere in the form

$$\Lambda_{D-1}^2 = -\sum_{i=1}^{D-1}\frac{(\sin\theta_i)^{i+1-D}}{(\prod_{i=j-1}^{i-1}\sin\theta_j)^2}\frac{\partial}{\partial\theta_i}\left[(\sin\theta_i)^{D-1}\frac{\partial}{\partial\theta_i}\right]. \tag{5.90}$$

This operator is known to fulfill the eigenvalue equation [104, 105]

$$\Lambda_{D-1}^2\mathcal{Y}_{l,\{\mu\}}(\Omega_{D-1}) = l(l+D-2)\mathcal{Y}_{l,\{\mu\}}(\Omega_{D-1}) \tag{5.91}$$

where $\mathcal{Y}$-symbol describes the hyperspherical harmonics characterized by the $D-1$ hyperangular quantum numbers $(l \equiv \mu_1, \mu_2, \mu_3, \ldots, \mu_{D-1} \equiv m) \equiv (l, \{\mu\})$, which are natural numbers with values $l = 0, 1, 2, \ldots$, and $l \equiv \mu_1 \geq \mu_2 \geq \mu \geq \cdots \geq \mu_{D-2} \geq |\mu_{D-1}| \equiv |m|$. These mathematical objects have the explicit expression [102, 106–108]

$$\mathcal{Y}_{l,\{\mu\}}(\Omega_{D-1}) = N_{l,\{\mu\}}e^{im\phi}\prod_{j=1}^{D-2}C_{\mu_j-\mu_{j+1}}^{(\alpha_j+\mu_{j+1})}(\cos\theta_j)(\sin\theta_j)^{\mu_{j+1}} \tag{5.92}$$

with the normalization constant

$$N_{l,\{\mu\}}^2 = \frac{1}{2\pi}\prod_{j=1}^{D-2}\frac{(\alpha_j+\mu_j)(\mu_j-\mu_{j+1})!\Gamma^2(\alpha_j+\mu_{j+1})}{\pi 2^{1-2\alpha_j-2\mu_{j+1}}\Gamma(2\alpha_j+\mu_j+\mu_{j+1})} \tag{5.93}$$

where $\alpha_j = (D-j-1)/2$, $\phi \equiv \mu_{D-1}$, $C_n^{\lambda}(t)$ denotes the Gegenbauer or ultraspherical polynomial of degree n and parameter λ, and with the values $0 \leq \theta_j \leq \pi$ $(j = 1, 2, \ldots, D-2)$ and $0 \leq \phi \leq 2\pi$. Moreover, these hyperfunctions satisfy the orthonomalization condition

$$\int_{S_{D-1}} d\Omega_{D-1}\mathcal{Y}_{l',\{\mu'\}}^*(\Omega_{D-1})\mathcal{Y}_{l,\{\mu\}}(\Omega_{D-1}) = \delta_{l,l'}\delta_{\{\mu\},\{\mu'\}}. \tag{5.94}$$

The eigenfunctions $\Psi(\mathbf{r})$ of the problem (5.88)–(5.89) can be separated out as

$$\Psi_{El\{\mu\}}(\mathbf{r}) = R_{nl}(r)\mathcal{Y}_{l,\{\mu\}}(\Omega_{D-1}) \tag{5.95}$$

where the radial eigenfunction $R_{nl}(r)$, according to (5.88), satisfies the following radial Schrödinger equation

$$\left[-\frac{1}{2}\frac{d^2}{dr^2} - \frac{D-1}{2r}\frac{d}{dr} + \frac{l(l+D-2)}{2r^2} + V_D(r)\right] R_{nl}(r) = E_D R_{Dl}(r). \tag{5.96}$$

Then, the probability to find the particle between $\mathbf{r}$ and $\mathbf{r}+d\mathbf{r}$ is given by

$$\rho_{El\{\mu\}}(\mathbf{r})d\mathbf{r} = |\Psi_{El\{\mu\}}(\mathbf{r})|^2 d\mathbf{r} = |R_{nl}(\mathbf{r})|^2 r^{D-1} dr \times |\mathcal{Y}_{l,\{\mu\}}(\Omega_{D-1})|^2 d\Omega_{D-1} \tag{5.97}$$

where the solid angle $d\Omega_{D-1}$ has the expression

$$d\Omega_{D-1} = \left(\prod_{j=1}^{D-2} (\sin\theta_j)^{2\alpha_j} d\theta_j\right) d\phi. \tag{5.98}$$

Taking the Fourier transform of $\Psi_{El\{\mu\}}(\mathbf{r})$ in (5.95), we obtain that the momentum wavefunctions $\hat{\Psi}_{El\{\mu\}}(\mathbf{p})$ can also be written in the form

$$\hat{\Psi}_{El\{\mu\}}(\mathbf{p}) = M_{El}(p)\mathcal{Y}_{l,\{\mu\}}(\hat{\Omega}_{D-1}) \tag{5.99}$$

where $(p, \hat{\Omega}_{D-1})$ denote the spherical polar coordinates in momentum space, $p = |\mathbf{p}|$ and $\hat{\Omega}_{D-1} \equiv (\hat{\theta}_1, \hat{\theta}_2, \ldots, \hat{\theta}_{D-2}, \phi)$. So, the probability that the particle has the momentum value between $\mathbf{p}$ and $\mathbf{p} + d^D p$ is

$$\gamma_{El\{\mu\}}(\mathbf{p})d\mathbf{p} = |M_{El}(\mathbf{p})|^2 p^{D-1} dp \times |\mathcal{Y}_{l,\{\mu\}}(\hat{\Omega}_{D-1})|^2 d\hat{\Omega}_{D-1}. \tag{5.100}$$

The spreading of the probability density $\rho(\mathbf{r})$ all over the D-dimensional space is usually quantified by means of radial expectation values

$$\langle f(r)\rangle = \int_{R_D} f(r)\rho_{El\{\mu\}}(\mathbf{r})d\mathbf{r} = \int_0^\infty f(r)|R_{nl}(r)|^2 r^{D-1} dr \tag{5.101}$$

and, more appropriately, by the use of information-theoretic quantities of global (Shannon, Rényi and Tsallis entropies) and local (Fisher information) types. All these quantities cannot be calculated unless we know the analytical form of the central potential. Nevertheless, we can go farther. The *power moments* $\langle r^\alpha\rangle$ of the particle for central potentials $V_D(r)$ such that $r^2 V_D(r) = 0$ when $r \to 0$, satisfy the hypervirial relation [109] (see also (4.1.3) of [103])

$$\begin{aligned} &2\left(\left\langle r^\alpha \frac{dV_D(r)}{dr}\right\rangle + 2\alpha\langle r^{\alpha-1}V_D\rangle - 2\alpha E_D\langle r^{\alpha-1}\rangle\right) \\ &\quad + \frac{1}{2}(\alpha-1)[(2L+1)^2 - (\alpha-1)^2]\langle r^{\alpha-3}\rangle = (2L+1)^2 C_l^2 \delta_{\alpha,-2L}, \\ &\quad \alpha \geq -2L \end{aligned} \tag{5.102}$$

where we have used the small distance behaviour of the regular wavefunction at the origin:

$$\lim_{r\to 0} r^{-l} R_l(r) = C_l \tag{5.103}$$

and $C_l \equiv N_{l,\{\mu\}}$ denotes the normalization constant of the radial position wavefunction, and $L = l + \frac{D-3}{2}$. Note that C_l is related to the lth derivative to the radial wavefunction at the origin:

$$C_l = \frac{1}{l!}\frac{d^l R_l(0)}{dr^l}. \tag{5.104}$$

In particular, $C_0 = R_0(0)$, the value of the S-wave radial wavefunction at the origin. Moreover, the relevant power moments $(\langle p^2\rangle, \langle r^{-2}\rangle)$ are mutually related [110] as

$$\begin{aligned}\langle p^2\rangle &= J_R(D) + l(l+D-2)\langle r^{-2}\rangle \\ &= J_R(D) + \left[L(L+1) - \frac{1}{4}(D-1)(D-3)\right]\langle r^{-2}\rangle\end{aligned} \tag{5.105}$$

with the radial integral

$$J_R(D) = \int_0^\infty \left[\frac{dR_{nl}(r)}{dr}\right]^2 r^{D-1}dr. \tag{5.106}$$

A similar expression can be written down for the pair $(\langle r^2\rangle, \langle p^{-2}\rangle)$. Let us also highlight that, since the radial integral $J_R(D)$ is non-negative, we have the radial uncertainty-like inequalities

$$\langle p^2\rangle \geq \left[L(L+1) - \frac{1}{4}(D-1)(D-3)\right]\langle r^{-2}\rangle \tag{5.107}$$

and

$$\langle r^2\rangle \geq \left[L(L+1) - \frac{1}{4}(D-1)(D-3)\right]\langle p^{-2}\rangle. \tag{5.108}$$

Note that the inequality (5.107) extends the general inequality given by [110]

$$\langle p^2\rangle \geq \left(\frac{D-2}{2}\right)^2 \langle r^{-2}\rangle \tag{5.109}$$

(see also [73]). The inequalities (5.107)–(5.108) can be improved [95] as

$$\begin{aligned}\langle p^2\rangle &\geq L(L+1)\langle r^{-2}\rangle, \\ \langle r^2\rangle &\geq L(L+1)\langle p^{-2}\rangle.\end{aligned} \tag{5.110}$$

A further improvement [95] has been recently obtained as

$$\begin{aligned}\langle p^2\rangle &\geq \left(L+\frac{1}{2}\right)^2 \langle r^{-2}\rangle, \\ \langle r^2\rangle &\geq \left(L+\frac{1}{2}\right)^2 \langle p^{-2}\rangle.\end{aligned} \tag{5.111}$$

These two inequalities improve for central potentials various similar uncertainty inequalities of general validity; see [95] for a detailed discussion and explanation.

The *Shannon entropy* (5.30) of the particle in the D-dimensional central potential has been shown to be discomposed [108, 111] into two parts:

$$S[\rho] \equiv S_{n,l,\{\mu\}} = S_{n,l}[R;D] + S_{l,\{\mu\}}[\mathcal{Y},D], \tag{5.112}$$

where

$$S_{n,l}[R;D] = -\int r^{D-1} R_{n,l}^2(r) \ln R_{n,l}^2(r) dr \tag{5.113}$$

denotes the radial Shannon entropy, and

$$S_{l,\{\mu\}}[\mathcal{Y};D] = -\int |\mathcal{Y}_{l,\{\mu\}}(\Omega_{D-1}|^2 \ln |\mathcal{Y}_{l,\{\mu\}}(\Omega_{D-1}|^2 d\Omega_{D-1} \tag{5.114}$$

gives the angular or spatial Shannon entropy. Observe that the angular part does not depend on the potential $V_D(r)$, while the radial component is independent from the magnetic quantum numbers $\mu_1, \mu_2, \ldots, \mu_{D-1}$. In momentum space, the corresponding Shannon entropy of the central potential is given by

$$S[\gamma] \equiv S_{n,l,\{\mu\}}[\gamma] = S_{n,l}[M;D] + S_{l,\{\mu\}}[\mathcal{Y};D], \tag{5.115}$$

where the radial momentum Shannon entropy is

$$S_{n,l}[M;D] = -\int_0^\infty p^{D-1} M_{n,l}^2(p) \ln M_{n,l}^2(p) dp. \tag{5.116}$$

The angular contribution to the position and momentum Shannon entropies is given by the entropic integral (or entropy) of hyperspherical harmonics $S_{l,\{\mu\}}[\mathcal{Y};D]$ expressed by (5.114). It has been found [112] to have the following value

$$\begin{aligned} S_{l,\{\mu\}}[\mathcal{Y};D] = \ln(2\pi) + \sum_{j=1}^{D-2} E\left[\tilde{C}_{\mu_j-\mu_{j+1}}^{(\alpha_j+\mu_{j+1})}\right] \\ - 2\sum_{j=1}^{D-2} \mu_{j+1}\Bigg[\psi(2\alpha_j+\mu_j+\mu_{j+1}) - \psi(\alpha_j+\mu_j) \\ - \ln 2 - \frac{1}{2(\alpha_j+\mu_j)}\Bigg] \end{aligned} \tag{5.117}$$

in terms of the quantum numbers $(l, \{\mu\})$ and the dimensionality D, where $E[\rho_n]$ denotes the entropy of the polynomials $p_n(x)$ orthogonal with respect to the weight function $\omega(x)$ on the interval (a,b)

$$E[p_n] := -\int_a^b p_n^2(x) \ln p_n^2(x) \omega(x) dx, \tag{5.118}$$

and $\tilde{C}_n^{(\lambda)}(x)$ denotes the Gegenbauer polynomial orthonormal with respect to the weight function $\omega_\lambda(x) = (1-x^2)^{\lambda-\frac{1}{2}}$ on the interval $[-1,+1]$. The orthonormal Gegenbauer polynomial $\tilde{C}_n^{(\lambda)}(x)$ is related to the orthogonal Gegenbauer polynomial $C_n^{(\lambda)}(x)$ by the relation

$$\tilde{C}_n^{(\alpha)}(x) = \frac{C_n^{(\alpha)}}{h_n}; \quad \text{with } h_n^2 = \frac{2^{1-2\lambda}\pi\Gamma(n+2\lambda)}{[\Gamma(\lambda)]^2(n+\lambda)n!}. \tag{5.119}$$

For further details about the information-theoretic properties of the hyperspherical harmonics, see [112, 113]. Let us underline that the radial part of the Shannon entropy of D-dimensional central potentials in both position and momentum spaces requires the knowledge of the corresponding radial eigenfunctions to go ahead, but the angular part (i.e. the entropy of the hyperspherical harmonics) is under control.

The Fisher information of the D-dimensional central potentials, defined by (5.51), has been analogously shown [110, 114] to have form

$$I[\rho] = I[R; D] + \langle r^{-2} \rangle I[\mathcal{Y}; D] \tag{5.120}$$

in position space, and

$$I[\gamma] = J[M; D] + \langle p^{-2} \rangle I[\mathcal{Y}; D] \tag{5.121}$$

in momentum space. Here, the radial parts are given by

$$I[R; D] = 4 \int_0^\infty \left[R'_{nl}(r)\right]^2 r^{D-1} dr = 4\langle p^2 \rangle \tag{5.122}$$

and

$$J[M; D] = 4 \int_0^\infty \left[M'_{nl}(p)\right]^2 p^{D-1} dp = 4\langle r^2 \rangle \tag{5.123}$$

in position and momentum spaces, respectively. The angular part is given by

$$I[\mathcal{Y}; D] = -2|m|(2l + D - 2) \tag{5.124}$$

in both spaces. Then, we have finally the nice expressions

$$I[\rho] = 4\langle p^2 \rangle - 2|m|(2l + D - 2)\langle r^{-2} \rangle \tag{5.125}$$

and

$$I[\gamma] = 4\langle r^2 \rangle - 2|m|(2l + D - 2)\langle p^{-2} \rangle \tag{5.126}$$

for the position and momentum Fisher informations in terms of the pairs of radial expectation values $(\langle p^2 \rangle, \langle r^{-2} \rangle)$ and $(\langle r^2 \rangle, \langle p^{-2} \rangle)$, respectively.

From here it is straightforward to write down the position-momentum inequalities

$$I[\rho] \geq 4\left(1 - \frac{2|m|}{2L+1}\right)\langle p^2 \rangle \tag{5.127}$$

and

$$I[\gamma] \geq 4\left(1 - \frac{2|m|}{2L+1}\right)\langle r^2 \rangle \tag{5.128}$$

where we have combined the exact expressions (5.125)–(5.126) and the radial uncertainty-like inequalities

$$\langle p^2 \rangle \geq \left(L + \frac{1}{2}\right)^2 \langle r^{-2} \rangle; \qquad \langle r^2 \rangle \geq \left(L + \frac{1}{2}\right)^2 \langle p^{-2} \rangle \tag{5.129}$$

(which improve for central potential the general inequalities (5.63) and (5.65) with $a = 2$).

Let us now state the *uncertainty relations* associated to the second-order power moments ($\langle r^2\rangle$, $\langle p^2\rangle$), or Heisenberg-like relation, and to the Fisher information of a particle in a D-dimensional central potential, which have been recently found [95, 110, 115]. First, the general Heisenberg-like inequality (5.60) transforms [110] into the following expression

$$\langle r^2\rangle\langle p^2\rangle \geq \left(l+\frac{D}{2}\right)^2 = \left(L+\frac{3}{2}\right)^2 \tag{5.130}$$

for central potentials, where $L = l + \frac{D-3}{2}$ is a generalized angular momentum as previously mentioned. This relation, which saturates for the (nodeless) ground-state wavefunction of the isotropic harmonic oscillator, provides a higher (so, better) value for the lower bound of the Heisenberg-like product.

It is worth highlighting that the general logarithmic, entropy, Rényi and Tsallis uncertainty relations given by (5.67)–(5.76) have not yet been improved for central potentials. In contrast, the position and momentum Fisher informations of central potentials satisfy [110] the following relation

$$I[\rho]\times I[\gamma] \geq 16\left[1-\frac{(2l+D-2)|m|}{2l(l+D-2)}\right]^2 \langle r^2\rangle\langle p^2\rangle, \tag{5.131}$$

which illustrates the uncertainty character of the Fisher-information product $I[\rho]\times I[\gamma]$. Moreover, from (5.130) and (5.131), we finally have the Fisher-information-based uncertainty relation [115]

$$I[\rho]\times I[\gamma] \geq 16\left[1-\frac{(2l+D-2)|m|}{2l(l+D-2)}\right]^2 \left(l+\frac{D}{2}\right)^2 \tag{5.132}$$

which extends and improves a similar relation previously obtained in three [116] and D [110] dimensions. Here again the equality is reached for the ground-state oscillator wavefunctions. It is also worth noting that for S states (i.e. when $l=0$), this inequality simplifies as

$$I[\rho]\times I[\gamma] \geq 4D^2. \tag{5.133}$$

In fact, it was proved for general one-dimensional states with even real-valued wavefunctions [94] and, just recently, for general D-dimensional states with general real-valued wavefunctions [93]. Finally for completeness, let us consider the Cramér-Rao inequality (5.52) for general systems, $I[\rho]\times V[\rho] \geq D^2$, which is equivalent to the variance-based Heisenberg uncertainty relation $V[\rho]\times V[\gamma] \geq \frac{D^2}{4}$ at least for real valued wavefunctions. Moreover, the expressions (5.125)–(5.126) together with (5.111) have allowed us to find [45] the following relation between the Cramér-Rao product and the Heisenberg-like product $\langle r^2\rangle\langle p^2\rangle$:

$$\langle r^2\rangle I[\rho] \geq 4\left(1-\frac{2|m|}{2L+1}\right)\langle r^2\rangle\langle p^2\rangle. \tag{5.134}$$

Then, taking into account the D-dimensional Heisenberg relation (5.130) for central potentials we have [45] that

$$\langle r^2\rangle I[\rho] \geq 4\left(1-\frac{2|m|}{2L+1}\right)\left(L+\frac{3}{2}\right)^2 \tag{5.135}$$

in position space, and

$$\langle p^2\rangle I[\gamma] \geq 4\left(1-\frac{2|m|}{2L+1}\right)\left(L+\frac{3}{2}\right)^2 \tag{5.136}$$

in momentum space. These Cramér-Rao relations improve for central potentials the general Cramér-Rao inequality (5.85); hence providing a higher (better) lower bound to the D-dimensional Cramér-Rao complexity $C_{CR}[\rho]$. Note that the lower bound given by (5.135) and (5.136) equals D^2 for S states. Moreover, these inequalities behave as uncertainty relations although in the same space, indicating that the wigglier is the quantum-mechanical wavefunction of the system, the less concentrated around the centroid the associated probability density is, and vice versa.

Furthermore, from (5.127) and the Stam inequality (5.77) we can bound the kinetic energy $T(=\frac{\langle p^2\rangle}{2})$ in both senses as

$$\frac{1}{8}I[\rho] \leq T \leq \frac{1}{8}\frac{2L+1}{2L+1-2|m|}I[\rho], \tag{5.137}$$

in terms of the position Fisher information $I[\rho]$. Similarly, from (5.128) and the Stam inequality (5.77) we find

$$\frac{1}{8}I[\gamma] \leq \langle r^2\rangle \leq \frac{1}{4}\frac{2L+1}{2L+1-2|m|}I[\gamma], \tag{5.138}$$

which allows us to bound numerous physical quantities related to $\langle r^2\rangle$ in terms of the momentum Fisher information $I[\gamma]$. This is the case, for instance, of the Langevin-Pauli diamagnetic susceptibility $\chi = \frac{-\alpha_{FS}\langle r^2\rangle}{6}$, α_{FS} being the fine structure constant.

5.4 Entropy and Complexity Analyses of Hydrogenic Systems

Here we describe the analytic information-theoretic properties of the ground and excited states of a D-dimensional hydrogenic system in both position and momentum spaces. This system [113, 117, 118] is composed of a negatively-charged particle moving around a positively charged core which electromagnetically binds it to its orbital, i.e. moving in the Coulomb potential

$$V_D(r) = -\frac{Z}{r}, \tag{5.139}$$

where Z is the charge of the core, and $r = |\mathbf{r}|$. It includes (models) a large diversity of physical systems and quantum phenomena [113] in quantum cosmology, nanotechnology, quantum computation, quantum field theory, D-dimensional physics and quantum chemistry. The existence of hydrogenic systems has been observed for $D \leq 3$ and suggested for $D > 3$ [23].

The physical solutions of the D-dimensional hydrogenic problem (5.88) with the potential (5.139) are the wavefunctions [113]

$$\Psi_{n,l,\{\mu\}}(\mathbf{r}) = R_{n,l}(r)\mathcal{Y}_{l,\{\mu\}}(\Omega_{D-1}), \tag{5.140}$$

where the hyperspherical harmonics are given by (5.92) and the radial part is given as

$$R_{n,l}(r)=\left(\frac{\lambda^{-D}}{2\eta}\right)^{\frac{1}{2}}\left[\frac{\omega_{2L+1}(\widetilde{r})}{\widetilde{r}^{D-2}}\right]^{\frac{1}{2}}\widetilde{L}_{\eta-L-1}^{(2L+1)}(\widetilde{r}) \tag{5.141}$$

where $\eta=n+\frac{D-3}{2}$, the grand orbital angular momentum quantum number $L=l+\frac{D-3}{2}$, $2L+1=2l+D-2$, the parameter $\lambda=\frac{\eta}{2Z}$, and $\widetilde{r}=\frac{r}{\lambda}$. The symbols $L_m^{(\alpha)}(x)$ and $\widetilde{L}_m^{(\alpha)}(x)$ denotes the usual and orthonormal, respectively, Laguerre polynomials with respect to the weight $\omega_\alpha(x)=x^\alpha e^{-x}$ on the interval $[0,\infty)$, so that

$$\widetilde{L}_m^{(\alpha)}(x)=\left[\frac{m!}{\Gamma(m+\alpha+1)}\right]^{\frac{1}{2}}L_m^{(\alpha)}(x). \tag{5.142}$$

Then, the electronic probability density of the D-dimensional hydrogenic system in position space is

$$\rho(\mathbf{r})=|\Psi_{n,l,\{\mu\}}(\mathbf{r})|^2=R_{n,l}^2(r)|\mathcal{Y}_{l,\{\mu\}}(\Omega_{D-1})|^2. \tag{5.143}$$

Similarly, in momentum space, the wavefunctions of the system are the Fourier transforms of the position wavefunction (5.140), given rise to the expression

$$\widetilde{\Psi}_{n,l,\{\mu\}}(\mathbf{p})=M_{n,l}(p)\mathcal{Y}_{l,\{\mu\}}(\Omega_{D-1}), \tag{5.144}$$

where the radial momentum wavefunction is

$$M_{n,l}(p)=\left(\frac{\eta}{2}\right)^{\frac{D}{2}}(1+y)^{\frac{3}{2}}\left(\frac{1+y}{1-y}\right)^{\frac{D-2}{4}}\sqrt{\omega_{L+1}^*(y)}\widetilde{C}_{\eta-L-1}^{(L+1)}(y), \tag{5.145}$$

with $\eta-L-1=n-l-1$ and $y=\frac{1-\eta^2\tilde{p}^2}{1-\eta^2\tilde{p}^2}$, $\tilde{p}=\frac{p}{Z}$. The symbols $C_m^{(\alpha)}(y)$ and $\widetilde{C}_m^{(\alpha)}(y)$ denote the usual and orthonormal, respectively, Gegenbauer polynomials with respect to the weight function $\omega_\alpha^*(y)=(1-y^2)^{\alpha-\frac{1}{2}}$ on the interval $[-1,+1]$.

Then, the momentum probability density of the D-dimensional hydrogenic system has the expression

$$\gamma(\mathbf{p})=|\widetilde{\Psi}_{n,l,\{\mu\}}(\mathbf{p})|^2=M_{n,l}^2(p)|\mathcal{Y}_{l,\{\mu\}}(\hat{\Omega}_{D-1})|^2. \tag{5.146}$$

The position and momentum D-dimensional wavefunctions (5.140) and (5.144), respectively, reduce to the corresponding three-dimensional wavefunctions (see e.g. [119–122]. Let us now study the spreading of both position and momentum densities given by (5.143) and (5.146), respectively, by means of the power and logarithmic moments, some information-theoretic measures (Shannon entropy, Fisher information), and the LMC shape complexity measures. The associated uncertainty relations are also given.

5.4.1 Power and Logarithmic Moments

The spreading of the position probability density $\rho(\mathbf{r})$, which controls all the macroscopic physical and chemical properties of the hydrogenic system, is conventionally measured by means of the power moments

$$\langle r^\alpha \rangle := \int_{R_D} r^\alpha \rho(\mathbf{r}) d\mathbf{r} = \int_0^\infty r^{\alpha+D-1} R_{n,l}^2(r) dr. \tag{5.147}$$

Taking into account the expression (5.141) for the radial wavefunction $R_{n,l}(r)$, it has been found that they have [113, 123] the values

$$\begin{aligned}\left(\frac{2Z}{\eta}\right)^\alpha \langle r^\alpha \rangle &= \frac{\Gamma(2L+\alpha+3)}{2\eta\Gamma(2L+2)} {}_3F_2\left(\begin{matrix} -\eta+L+1, & -\alpha-1, & \alpha+2 \\ & 2L+2, 1 & \end{matrix} \middle| 1 \right) \\ &= \frac{1}{2n+D-3} \frac{(n-l-1)!}{(n+l+D-3)} \\ &\quad \times \sum_{i=0}^{n-l-1} \binom{\alpha+1}{n-l-i-1}^2 \frac{\Gamma(\alpha+2l+D+i)}{i!},\end{aligned} \tag{5.148}$$

valid for $\alpha > -2l - D$. Moreover, these quantities satisfy [109] the recursion relation

$$\frac{Z}{\eta^2}\langle r^{S-1} \rangle = \frac{2S-1}{S}\langle r^{S-2} \rangle - \frac{1}{Z}\frac{S-1}{4S}\left[(2L+1)^2 - (S-1)^2\right]\langle r^{S-3} \rangle, \tag{5.149}$$

for $S > -2L$. Then, in particular, we have the values $\langle r^\alpha \rangle$, with $\alpha = 1$ and 2,

$$\begin{aligned}\langle r \rangle &= \frac{1}{2Z}[3\eta^2 - L(L+1)], \\ \langle r^2 \rangle &= \frac{\eta^2}{2Z^2}[5\eta^2 - 3L(L+1) + 1].\end{aligned} \tag{5.150}$$

So, the familiar variance of $\rho(\mathbf{r})$ is given by

$$V[\rho] := \langle r^2 \rangle - \langle r \rangle^2 = \frac{\eta^2(\eta^2+2) - L^2(L+1)^2}{4Z^2} \tag{5.151}$$

which extends to D dimensions the known value (see e.g. [122]) of the real hydrogenic atom.

In momentum space we can operate similarly to quantify the spreading of the probability density $\gamma(\mathbf{p})$ by means of the momentum power moments or radial momentum expectation values $\langle p^\alpha \rangle$ given by

$$\langle p^\alpha \rangle := \int_{R_D} p^\alpha \gamma(\mathbf{p}) d\mathbf{p} = \int_0^\infty p^{\alpha+D-1} M_{n,l}^2(p) dp. \tag{5.152}$$

The use of (5.145) for the radial momentum wavefunction $M_{n,l}(r)$ has allowed us to find [124] the values

$$\langle p^\alpha \rangle = \frac{2^{1-2\nu} Z^\alpha \sqrt{\pi}}{k!\eta^{\alpha-1}} \frac{\Gamma(k+2\nu)\Gamma(\nu+\frac{\alpha+1}{2})\Gamma(\nu+\frac{3-\alpha}{2})}{\Gamma^2(\nu+\frac{1}{2})\Gamma(\nu+1)\Gamma(\nu+\frac{3}{z})}$$
$$\times {}_5F_4\left(\begin{matrix} -k, k+2\nu, \nu, \nu+\frac{\alpha+1}{2}, \nu+\frac{3-\alpha}{2} \\ 2\nu, \nu+\frac{1}{2}, \nu+1, \nu+\frac{3}{2} \end{matrix} \middle| 1 \right) \quad (5.153)$$

valid for $-2l - D \leq \alpha \leq 2l + D + 2$, where $\nu \equiv L + 1 = l + \frac{D-1}{2}$ and $k = n - l - 1$. Let us remark that the generalized hypergeometric function ${}_5F_4(1)$, as ${}_3F_2(1)$, is a single sum since it involves a terminating and Saalschutzian (balanced) hypergeometric function. See also [121]. For $\alpha = 0$ and 2, we have the expectation values

$$\langle p^0 \rangle = 1 \quad \text{and} \quad \langle p^2 \rangle = \frac{Z^2}{\eta^2}. \quad (5.154)$$

Moreover, they satisfy the reflection formula [113]

$$\left(\frac{\eta}{Z}\right)^{2-\alpha} \langle p^{2-\alpha} \rangle = \left(\frac{\eta}{Z}\right)^{\alpha} \langle p^\alpha \rangle \quad (5.155)$$

which is not trivial for $\alpha \neq 1$. The momentum expectation values with odd α are not so simple; in particular, for $\alpha = -1$ see also [125].

Alternative measures of spreading of the position and momentum electron densities are provided by the logarithmic moments

$$\langle \ln r \rangle = \int (\ln r)\rho(\mathbf{r})d\mathbf{r} \quad (5.156)$$

and

$$\langle \ln p \rangle = \int (\ln p)\gamma(\mathbf{p})d\mathbf{p}, \quad (5.157)$$

respectively. We have found [124] the values

$$\langle \ln r \rangle = \ln \eta + \frac{2\eta - 2L - 1}{2\eta} + \psi(\eta + L + 1) - \ln 2 - \ln Z \quad (5.158)$$

and

$$\langle \ln p \rangle = -\ln \eta + \frac{2\eta(2L+1)}{4\eta^2 - 1} - 1 + \ln Z, \quad (5.159)$$

also respectively.

5.4.2 Shannon Entropy and Fisher Information

Here, we will show the best measures (according to certain criteria) of the global or bulk extent (Shannon entropy) and the local concentration or gradient content (Fisher information) of the position and momentum electron densities of the D-dimensional hydrogenic systems.

According to (5.30) the *position Shannon entropy* $S[\rho] \equiv S_{n,l,\{\mu\}}[\rho]$ has a radial ($S_{nl}[R; D]$) and an angular ($S_{l,\{\mu\}}[\mathcal{Y}; D]$) part. The latter one is given by

(5.117) in terms of the entropic integral of the orthonormal Gegenbauer polynomials $E[\widetilde{C}_{\mu_j-\mu_{j+1}}^{(\alpha_j+\mu_{j+1})}]$. The radial Shannon entropy has been shown to have the following value [108, 111]

$$S_{nl}[R;D] = A(n,l,D) + \frac{1}{2\eta} E_1\left(\tilde{L}_{\eta-L-1}^{(2L+1)}\right) - D\ln Z \tag{5.160}$$

where

$$A(n,l,D) = \frac{3\eta^2 - L(L+1)}{\eta} - 2l\left[\frac{2\eta-2L-1}{2\eta} + \psi(\eta+L+1)\right] \\ + (D+1)\ln\eta - (D-1)\ln 2. \tag{5.161}$$

The symbols $\widetilde{L}_m^{(\alpha)}$ denote the orthonormal Laguerre polynomials, and $E_1(\widetilde{p}_n)$ denotes [126] the entropic integral of the polynomial orthonormal $\widetilde{p}_n(x)$ with respect to the weight function $\omega_\lambda^*(x) = x^\lambda e^{-x}$ on the interval $[0,\infty)$:

$$E_1(\widetilde{p}_n) = -\int_0^\infty x\omega_\lambda^*(x)\widetilde{p}_n^2(x)\ln\widetilde{p}_n^2(x)dx. \tag{5.162}$$

Then, the combination of (5.30), (5.117) and (5.160) has led [113] us to the following expression for the total Shannon entropy of the D-dimensional hydrogenic system in terms of the hyperquantum numbers $(n,l,\{\mu\})$ characterizing the state under consideration:

$$S[\rho] = A(n,l,D) + B(l,\{\mu\},D) + \frac{1}{2\eta}E_1\left(\widetilde{L}_{\eta-L-1}^{2L+1}\right) \\ + \sum_{j=1}^{D-2} E\left(\widetilde{C}_{\mu_j-\mu_{j+1}}^{\alpha_j+\mu_{j+1}}\right) - D\ln Z \tag{5.163}$$

where A is given by (5.161) and B has the value

$$B(l,\{\mu\},D) = \ln 2\pi - 2\sum_{j=1}^{D-2}\mu_{j+1}\Bigg[\psi(2\alpha_j+\mu_j+\mu_{j+1}) \\ - \psi(\alpha_j+\mu_j) - \ln 2 - \frac{1}{2(\alpha_j+\mu_j)}\Bigg]. \tag{5.164}$$

The entropic integrals $E(\widetilde{C})$ and $E_1(\widetilde{L})$ of the orthonormal Gegenbauer and Laguerre polynomials are given by (5.118) and (5.162) respectively. They have not yet been calculated in an analytical way, except for very special cases (e.g. the ground and circular states). For their numerical evaluation, a recent algorithm [127] has been developed which computes them very accurately. Let us only point out here the exact value of the position Shannon entropy

$$S[\rho_{g.s.}] = \ln\left(\frac{(D-1)^D}{2^D}\pi^{\frac{D-1}{2}}\Gamma\left(\frac{D+1}{2}\right)\right) + D - D\ln Z \tag{5.165}$$

for the hydrogenic ground state (g.s.).

Operating similarly in momentum space, we have obtained [113] the following value for the total *momentum Shannon entropy*

$$S[\gamma] = F(n,l,D) + B(l,\{\mu\},D) + E\left(\tilde{C}^{(L+1)}_{\eta-L-1}\right) + \sum_{j=1}^{D-2} E\left(\tilde{C}^{(\alpha_j+\mu_{j+1})}_{\mu_j-\mu_{j+1}}\right) + D\ln Z \tag{5.166}$$

where

$$F(n,l,D) = -\ln\frac{\eta^D}{2^{2L+4}} - (2L+4)[\psi(\eta+L+1)-\psi(\eta)] + \frac{L+2}{\eta} - (D+1)\left[1-\frac{2\eta(2L+1)}{4\eta^2-1}\right]. \tag{5.167}$$

It is worth noting that this quantity only depends on the entropic integral of the orthonormal Gegenbauer polynomials. This Gegenbauer functional can be numerically computed by means of the highly efficient algorithm of Buyarov et al. [127]; however, its analytical calculation is a formidable task, not yet done save for a few cases (e.g. for ground and circular states). In particular, we found [113] the value

$$S[\gamma_{g.s.}] = \ln\frac{2^{2D+1}\pi^{\frac{D+1}{2}}}{(D-1)^D\Gamma(\frac{D+1}{2})} + D\ln Z + \frac{D+1}{D(D-1)} - (D+1)\left[\psi(D-1)-\psi\left(\frac{D-1}{2}\right)\right] \tag{5.168}$$

for the momentum Shannon entropy of the D-dimensional hydrogenic ground state.

Let us now consider the *position Fisher information* $I[\rho]$ given by (5.51), which measures the gradient content of the electron density $\rho(\mathbf{r})$ of the D-dimensional hydrogenic system in a quantum mechanical state characterized by the hyperquantum numbers $(n,l,\{\mu\})$. We have previously shown that this quantity can be expressed in the form (5.125) in terms of the expectation values $\langle p^2\rangle$ and $\langle r^{-2}\rangle$. Since

$$\langle p^2\rangle = \frac{Z^2}{\eta^2}, \quad \text{and} \quad \langle r^{-2}\rangle = \frac{2Z^2}{\eta^3}\frac{1}{2L+1} \tag{5.169}$$

we have the following value [113, 114]

$$I[\rho] = \frac{4Z^2}{\eta^3}[\eta - |m|], \quad D \geq 2, \tag{5.170}$$

for the total position Fisher information of the D-dimensional hydrogenic system.

The *momentum Fisher information* $I[\gamma]$ of our system can be similarly calculated by means of (5.126) in terms of the expectation values $(\langle r^2\rangle, \langle p^{-2}\rangle)$ together with the expressions (5.150) for $\langle r^2\rangle$ and the value

$$\langle p^{-2}\rangle = \frac{\eta^2}{Z^2}\frac{8\eta-3(2L+1)}{2L+1}. \tag{5.171}$$

We have found consequently the value

$$I[\gamma] = \frac{2\eta^2}{Z^2}[5\eta^2 - 3L(L+1) - (8\eta - 6L - 3)|m| + 1]; \quad D \geq 2, \tag{5.172}$$

for the total momentum Fisher information of the D-dimensional hydrogenic state $(\eta, l, \{\mu\})$.

5.4.3 Uncertainty Relations

Here we shall give the uncertainty relations associated with the following spreading/uncertainty measures of the D-dimensional hydrogenic system: the power moments ($\langle r^2\rangle$, $\langle p^2\rangle$), the logarithmic moments ($\langle \ln r\rangle$, $\langle \ln p\rangle$), the Shannon entropies ($S[\rho]$, $S[\gamma]$) and the Fisher information ($I[\rho]$, $I[\gamma]$). In addition, we provide the Cramér-Rao products in position ($\langle r^2\rangle I[\rho]$) and momentum ($\langle p^2\rangle I[\gamma]$) spaces. Let us begin with the *Heisenberg uncertainty relation*. To find it for the appropriate canonically conjugate radial coordinates, we have to consider the pair (r, p_r) where p_r is the so-called radial momentum operator [128]

$$p_r = -i\hbar \frac{1}{r^{\frac{D-1}{2}}} \frac{\partial}{\partial r} r^{\frac{D-1}{2}} = -i\hbar\left(\frac{\partial}{\partial r} + \frac{D-1}{2r}\right) = \mathbf{p}^2 - \frac{L(L+1)}{r^2} \tag{5.173}$$

which is manifestly hermitian. So, p_r has the expectation $\langle p_r\rangle = 0$ and the second order moment

$$\langle p_r^2\rangle = \langle \mathbf{p}^2\rangle - L(L+1)\langle r^{-2}\rangle = \frac{Z^2}{\eta^2}\left[1 - \frac{2}{\eta}\frac{L(L+1)}{2L+1}\right]. \tag{5.174}$$

Then, the radial momentum standard deviation Δp_r becomes

$$\Delta p_r = \sqrt{\langle p_r^2\rangle - \langle p_r\rangle^2} = \frac{Z}{\eta}\left[1 - \frac{2}{\eta}\frac{L(L+1)}{2L+1}\right] \tag{5.175}$$

and, from (5.150), the standard deviation of the radial position is

$$\Delta r = \sqrt{\langle r^2\rangle - \langle r\rangle^2} = \frac{1}{2Z}\left[\eta^2(\eta^2+2) - L^2(L+1)^2\right]^{\frac{1}{2}} \tag{5.176}$$

so that we have, finally, the Heisenberg uncertainty product as

$$\Delta r \Delta p_r = \frac{1}{2\eta}\left\{\frac{1}{\eta(2L+1)}\left[\eta^2(\eta^2+2) - L^2(L+1)^2\right] \times \left[\eta(2L+1) - 2L(L+1)\right]\right\}^{1/2} \tag{5.177}$$

which does not depend on the nuclear charge Z.

On the other hand, from (5.158) and (5.159) we have the following *logarithmic uncertainty relation*:

$$\langle \ln r\rangle + \langle \ln p\rangle = \frac{2n-2l-1}{2n+D-3} + \frac{(2n+D-3)(2l+D-2)}{(2n+D-3)^2-1} - \ln 2 - 1 + \psi(n+l+D-2). \quad (5.178)$$

Note that it does not depend on Z, and for the ground state (i.e. when $n=1$, $l=0$) this relation simplifies as

$$(\langle \ln r\rangle + \langle \ln p\rangle)(\text{g.s.}) = -\frac{1}{D} - \ln 2 + \psi(D), \quad (5.179)$$

which fulfills the general logarithmic uncertainty relation (5.70).

Now, let us see the Shannon-entropy-based or *entropic uncertainty relation*. From (5.163) and (5.166) we have that the Shannon entropy sum is given by

$$S[\rho] + S[\gamma] = A(n,l,D) + F(n,l,D) + 2B(l,\{\mu\},D) + \frac{1}{2\eta}E_1\left(\widetilde{L}_{\eta-L-1}^{(2L+1)}\right) \times E(\widetilde{C}_{\eta-L-1}^{(L+1)}) + 2\sum_{j=1}^{D-2} E\left(\widetilde{C}_{\mu_j-\mu_{j+1}}^{(\alpha_j+\mu_{j+1})}\right) \quad (5.180)$$

where the terms A, B and F are explicitly given by (5.161), (5.164) and (5.167), respectively. The entropic integrals $E_i(\widetilde{p}_n)$, $i=0$ and 1, are given by (5.118) and (5.162), respectively. In particular, the uncertainty entropy sum for the ground state has the value

$$S[\rho_{g.s.}] + S[\gamma_{g.s.}] = \ln\left(\frac{4^{D+1}\pi^D}{D}\right) + \frac{2}{D-1} + \frac{D^2-1}{D} - (D-1)\left[\psi(D-1) - \psi\left(\frac{D-1}{2}\right)\right] \quad (5.181)$$

which certainly fulfills the entropic uncertainty relation (5.72) valid for general systems.

The position-momentum *Fisher-information-based uncertainty relation* of the hydrogenic state $(n,l,\{\mu\})$ can be obtained from (5.170) and (5.172), yielding the value

$$I[\rho]\times I[\gamma] = \frac{8}{\eta}(\eta-|m|)[5\eta^2 - 3L(L+1) - |m|(8\eta-6L-3)+1]; \quad D\geq 2. \quad (5.182)$$

Here again, this uncertainty product does not depend on the potential strength (nuclear charge Z), and for the ground state (g.s.) it boils down to

$$I[\rho_{g.s.}]\times I[\gamma_{g.s.}] = 4D(D+1) \quad (5.183)$$

which clearly satisfies not only the Fisher-information-based uncertainty relation (5.78) valid for general systems but also the corresponding expression (5.132) valid for systems moving in arbitrary central potentials. The uncertainty-like relations of the D-dimensional hydrogen atom seem to have been recently found [129].

Finally, from (5.150) for $\langle r^2\rangle$ and (5.170) for $I[\rho]$ one has [113] that the position *Cramér-Rao product* of an arbitrary hydrogenic state $(n,l,\{\mu\})$ is given by

$$\langle r^2\rangle I[\rho] = \frac{2}{\eta}[5\eta^2 - 3L(L+1)+1](\eta - |m|). \tag{5.184}$$

And from the values (5.154) for $\langle p^2\rangle$ and (5.172) for $I[\gamma]$, one has

$$\langle p^2\rangle I[\gamma] = 2[5\eta^2 - 3L(L+1) - |m|(8\eta - 6L - 3)+1] \tag{5.185}$$

for any ground and excited states of D-dimensional hydrogenic system. It is worth noting that both Cramér-Rao products do not depend on the nuclear charge Z. Moreover, they fulfill not only the general Cramér-Rao relations (5.85) but also the corresponding relations (5.135) and (5.136) valid for arbitrary central potentials.

5.4.4 Complexity Measures

Here, the LMC shape complexity of the ground and excited states of the D-dimensional hydrogenic system is shown [130]. According to its definition given by (5.79), this quantity has two ingredients: the disequilibrium $D = \langle\rho\rangle$ and the entropic power $N[\rho] = \exp(S[\rho])$. The former quantity can be calculated by use of (5.141) and (5.143) obtaining the expression

$$\langle\rho\rangle := \int_{R_D} \rho^2(\mathbf{r})d\mathbf{r} = \frac{2^{D-2}}{\eta^{D+2}} Z^D K_1(D,\eta,L) \times K_2(l,\{\mu\}) \tag{5.186}$$

where

$$K_1(D,\eta,L) = \int_0^\infty x^{-D-5}\left\{\omega_{2L+1}(x)\left[\widetilde{L}^{(2L+1)}_{\eta-L-1}(x)\right]\right\}^2 dx \tag{5.187}$$

and

$$K_2(l,\{\mu\}) = \int_{S_{D-1}} |\mathcal{Y}_{l,\{\mu\}}(\Omega_{D-1})|^4 d\Omega_{D-1}. \tag{5.188}$$

The entropic power $N[\rho]$ can be straightforwardly written down from the expression (5.163) for the position Shannon entropy $S[\rho]$, or alternatively from the expression (5.160) for the radial position Shannon entropy $S_{nl}[R;D]$ and the expression (5.117) for the angular component $S_{l,\{\mu\}}[\mathcal{Y};D]$. In turn, we have the value

$$\begin{aligned} C_{LMC}[\rho] &:= \langle\rho\rangle \exp(S[\rho]) \\ &= \frac{2^{D-2}}{\eta^{D+2}} K_1(D,\eta,L) K_2(l,\{\mu\}) \\ &\quad \times \exp\left\{A(n,l,D) + \frac{1}{2\eta}E_1\left[\widetilde{L}^{(2L+1)}_{\eta-L-1}\right] + S_{l,\{\mu\}}[\mathcal{Y};D]\right\} \end{aligned} \tag{5.189}$$

for the LMC shape complexity of hydrogenic state $(n,l,\{\mu\})$ in position space, where A and $E_1(\widetilde{p}_n)$ are given by (5.161) and (5.162), respectively. Let us highlight that this complexity measure does not depend on the nuclear charge Z.

The explicit expression for this quantity is not known except for some special cases, such as e.g. the ground state (g.s.); in this case we can easily shown [130] that

$$C_{LMC}[\rho_{g.s}] = \left(\frac{e}{2}\right)^D \tag{5.190}$$

for the LMC shape complexity of the D-dimensional hydrogenic ground state, as previously found [71, 131]. See [130] for further details and to find the explicit values of this quantity in other quantum states.

Finally, let us also mention here that the LMC shape complexity and the Fisher-Shannon and Cramér-Rao complexities of real ($D = 3$) hydrogenic systems in both ground and excited states have been explicitly discussed [118] in terms of their quantum numbers (n, l, m). Let us just point out that the Fisher-Shannon complexity is shown to quadratically depend on the principal quantum number n.

5.5 Conclusion and Open Problems

This work has surveyed the present status of the analytic D-dimensional information theory of the general quantum systems, the single systems in arbitrary central potentials and the hydrogenic systems. We have shown the present results, to the best of our knowledge, about not only the power, logarithmic and entropic moments but also the indirect one-ingredient (Rényi, Shannon and Tsallis entropies, and the Fisher information) and two-ingredient (Cramér-Rao, Fisher-Shannon and LMC shape complexities) information-theoretic measures of these systems, together with their associated uncertainty relations.

We have identified a number of open problems. First, the conditions that a set of real numbers must satisfy so that a density exists having them as entropic moments are not yet known. Second, the behaviour of the D-dimensional systems in external fields has not been studied yet, although some results in presence of electric fields have been published [15, 132]. Third, it would be very interesting to calculate the direct spreading measures of the D-dimensional systems so as to be able to mutually compare them in a proper way; in particular, by calculating the ensemble spreading lengths in the sense of Hall [133]. Fourth, to include the relativistic effects because the conceptual importance of information is greatest in the interplay between dimensionality and the relativistic effects that is laid on atomic wavefunctions. It is in this interplay that one finds the origin of the physical phenomena. Some authors [10, 134–137] have begun to explore this fairly wild territory in both Klein-Gordon and Dirac cases. Fifth, to study the information-theoretic measures of many-electron systems moving on a D-dimensional hypershere [138]. Finally, it would be physically interesting to improve the Rényi, Shannon and Tsallis uncertainty relations for central potentials.

Acknowledgements This work was partially supported by the projects FQM-2445 and FQM-4643 of the Junta de Andalucia (Spain, EU), and the grant FIS2008-2380 of the Ministerio de Innovación y Ciencia. We belong to the Andalusian research group FQM-207. We are very grateful to C. Vignat and S. Zozor for their critical reading of the manuscript.

References

1. Witten E (1980) Quarks, atoms and the $1/N$ expansion. Phys Today 38:33, July
2. Weinberg S, Piran T (eds) (1986) Physics in higher dimensions. World Scientific, Singapore
3. Kunstatter G (2003) D-dimensional black hole entropy spectrum from quasinormal modes. Phys Rev Lett 90:161301
4. Adhav KS, Nimkar AS, Dawande MV (2007) Astrophys Space Sci 310:321
5. Avelino-Camelia G, Kowalski-Glikman J (eds) (2005) Planck scale effects in astrophysics and cosmology. Springer, Berlin
6. Harrison P (2005) Quantum wells, wires and dots: theoretical and computational physics of semiconductors nanostructure, 2nd edn. Wiley-Interscience, New York
7. Li SS, Xia JB (2007) Electronic states of a hydrogenic donor impurity in semiconductor nano-structures. Phys Lett A 366:120
8. McKinney BA, Watson DK (2000) Semiclassical perturbation theory for two electrons in a d-dimensional quantum dot. Phys Rev B 61:4958
9. Dykman MI, Platzman PM, Seddgard P (2003) Qubits with electrons on liquid helium. Phys Rev B 67:155402
10. Nieto MN (2000) Electrons above a helium surface and the one-dimensional Rydberg atom. Phys Rev A 61:034901
11. Maia A, Lima JA (1998) D-dimensional radiative plasma: a kinetic approach. Class Quantum Gravity 15:2271
12. Herschbach DR, Avery J, Goscinski O (eds) (1993) Dimensional scaling in chemical physics. Kluwer, Dordrecht
13. Tsipis CA, Herschbach DR, Avery J (eds) (1996) NATO conference book, vol 8. Kluwer, Dordrecht
14. Aquilanti V, Cavalli S, Colleti C (1997) The d-dimensional hydrogen atom: hyperspherical harmonics as momentum space orbitals and alternative Sturmian basis sets. Chem Phys 214:1–13
15. Sälen L, Nepstad R, Hansen JR, Madsen LB (2007) The D-dimensional Coulomb problem: Stark effect in hyperparabolic and hyperspherical coordinates. J Phys A, Math Gen 40:1097
16. Mlodinow LD, Papanicolaou N (1980) SO(2, 1) algebra and the large n expansion in quantum mechanics. Ann Phys 128:314–334
17. Jakubith S, Rotermund HH, Engel W, von Oertzen A, Ertl G (1985) Spatio-temporal concentration patterns in a surface reaction: Propagating and standing waves, rotating spirals, and turbulence. Phys Rev Lett 65:3013–3016
18. Beta C, Moula MG, Mikhailov AS, Rotermund HH, Ertl G (2004) Excitable CO oxidation on Pt(110) under nonuniform coupling. Phys Rev Lett 93:188302
19. Cox MP (1985) Spatial self-organization of surface structure during an oscillating catalytic reaction. Phys Rev Lett 54:1725
20. Gleick J (1987) Chaos making a new science. Viking Penguin, New York
21. Mandelbrot BB (1983) The fractal geometry of nature. Freeman, San Francisco
22. Aquilanti V, Cavalli S, Coletti C, di Domenico D, Grossi G (2001) Hyperspherical harmonics as Sturmian orbitals in momentum space: A systematic approach to the few-body Coulomb problem. Int Rev Phys Chem 20:673
23. Burgbacher F, Lämmerzahl C, Macias A (1999) Is there a stable hydrogen atom in higher dimensions? J Math Phys 40:625
24. Gurevich L, Mostepanenko V (1971) On the existence of atoms in n dimensional space. Phys Lett A 35:201
25. Kleppner D (1999) The yin and yang of hydrogen. Phys Today 52:11, April
26. Parr RG, Yang W (1989) Density-functional theory of atoms and molecules. Oxford University Press, New York
27. Angulo JC, Dehesa JS (1992) Tight rigorous bounds to atomic information entropies. J Chem Phys 97:6485. Erratum 1 (1993)
28. Koga T, Angulo JC, Dehesa JS (1994) Electron-electron coalescence and interelectronic log-moments in atomic and molecular moments. Proc Indian Acad Sci, Chem Sci 106(2):123

29. Lenz F, Rosenfelder R (1971) Nuclear radii in the high-energy limit of elastic electron scattering. Nucl Phys A 176:571
30. Liu S, Parr RG (1996) Expansions of the correlation-energy density functional and its kinetic-energy component in terms of homogeneous functionals. Phys Rev A 53:2211
31. Liu S, Parr RG (1997) Expansions of density functionals: Justification and nonlocal representation of the kinetic energy, exchange energy, and classical Coulomb repulsion energy for atoms. Physica A 55:1792
32. Nagy A, Liu S, Parr RG (1999) Density-functional formulas for atomic electronic energy components in terms of moments of the electron density. Phys Rev A 59:3349
33. Angulo JC, Romera E, Dehesa JS (2000) Inverse atomic densities and inequalities among density functionals. J Math Phys 41:7906
34. Pintarelli MB, Vericat F (2003) Generalized Hausdorff inverse moment problem. Physica A 324(3–4):568–588
35. Romera E, Angulo JC, Dehesa JS (2001) The Hausdorff entropic moment problem. J Math Phys 42:2309. Erratum 44:1 (2003)
36. Leonenko N, Pronzato L, Savani V (2008) A class of Rényi information estimator for multidimensional densities. Ann Stat 40(4):2153–2182
37. Dehesa JS, Galvez FJ, Porras I (1989) Bounds to density-dependent quantities of D-dimensional many-particle systems in position and momentum spaces: Applications to atomic systems. Phys Rev A 40:35
38. Dehesa JS, Galvez FJ (1988) Rigorous bounds to density-dependent quantities of D-dimensional many-fermion systems. Phys Rev A 37:3634
39. Galvez FJ, Porras I (1991) Improved lower bounds to the total atomic kinetic energy and other density-dependent quantities. J Phys B, At Mol Opt Phys 24:3343
40. Lieb EH, Oxford S (2004) Improved lower bound on the indirect Coulomb energy. Int J Quant Chem 19:427
41. Odashima MM, Capelle K (2008) Empirical analysis of the Lieb-Oxford in ions and molecules. Int J Quant Chem 108:2428
42. Lieb EH (1976) The stability of matter. Rev Mod Phys 48:553
43. Lieb EH (2000) Kluwer encyclopedia of mathematics supplement, vol II. Kluwer, Dordrecht
44. Lieb EH, Seiringer R (2010) The stability of matter in quantum mechanics. Cambridge University Press, Cambridge
45. Dehesa JS, González-Férez R, Sánchez-Moreno P, Yáñez RJ (2007) Kinetic energy bounds for particles confined in spherically-symmetric traps with nonstandard dimensions. New J Phys 9:131
46. Beckner W (1995) Pitt's inequality and the uncertainty principle. Proc Am Math Soc 123:159
47. Daubechies I (1983) An uncertainty principle for fermions with generalized kinetic energy. Commun Math Phys 90:511
48. Rényi A (1970) Probability theory. Academy Kiado, Budapest
49. Tsallis C (1988) Possible generalization of Boltzmann-Gibbs statistics. J Stat Phys 52:479
50. Havrda JH, Charvát F (1967) Quantification methods of classification processes: Concept of α-entropy. Kybernetica (Prague) 3:95–100
51. Shannon CE (1948) A mathematical theory of communication. Bell Syst Tech J 27:379
52. Gyftopoulos EP, Cubukcu E (1997) Entropy: Thermodynamic definition and quantum expression. Phys Rev E 55:3851
53. Bialynicki-Birula I, Mycielski J (1975) Uncertainty relations for information entropy in wave mechanics. Commun Math Phys 44:129
54. Brukner C, Zeilinger A (1999) Operationally invariant information in quantum measurements. Phys Rev Lett 83:3354
55. Zurek WH, Habib S, Paz JP (1993) Coherent states via decoherence. Phys Rev Lett 70:1187
56. Angulo JC (1994) Information entropy and uncertainty in D-dimensional many-body systems. Phys Rev A 50:311
57. López-Rosa S, Angulo JC, Dehesa JS, Yáñez RJ (2008) Existence conditions and spreading properties of extreme entropy D-dimensional distributions. Physica A 387:2243–2255. Erratum, ibid 387:4729–4730 (2008)

58. López-Rosa S, Angulo JC, Dehesa JS (2009) Spreading measures of information-extremizer distributions: applications to atomic electron densities in position and momentum spaces. Eur J Phys D 51:321–329
59. Brody DC, Buckley IRC, Constantinou IC (2007) Option price calibration from Rényi entropy. Phys Lett A 366:298–307
60. Bashkirov AG (2004) Maximum Rényi entropy principle for systems with power-law Hamiltonians. Phys Rev Lett 93:130601
61. Dehesa JS, Galvez FJ (1985) A lower bound for the nuclear kinetic energy. Phys Lett B 156:287
62. Fisher RA (1925) Theory of statistical estimation. Proc Camb Philos Soc 22:700. Reprinted in Collected papers of RA Fisher, edited by JH Bennet, University of Adelaide Press, Australia, 1972, pp 15–40
63. Frieden BR (2004) Science from Fisher information. Cambridge University Press, Cambridge
64. Sears SB, Parr RG (1980) On the quantum-mechanical kinetic energy as a measure of the information in a distribution. Isr J Chem 19:165–173
65. March NH, Kais S (1998) Kinetic energy functional derivative for the Thomas-Fermi atom in D dimensions. Int J Quant Chem 65:411
66. Massen SE, Panos CP (2001) A link of information entropy and kinetic energy for quantum many-body systems. Phys Lett A 280:65
67. Luo S (2002) Fisher information, kinetic energy and uncertainty relation inequalities. J Phys A, Math Gen 35:5181
68. Romera E, Dehesa JS (1994) Weiszäcker energy of many electron systems. Phys Rev A 50:256
69. Dembo A, Cover TM, Thomas JA (1991) Information theoretic inequalities. IEEE Trans Inf Theory 37:1501
70. Cover TM, Thomas JA (1991) Elements of information theory. Wiley-Interscience, New York
71. López-Rosa S, Angulo JC, Antolin J (2009) Rigorous properties and uncertainty-like relationships on product-complexity measures: Applications to atomic systems. Physica A 388:2081–2091
72. González-Férez R, Dehesa JS (2005) Characterization of atomic avoided crossing by means of Fisher's Information. Eur J Phys D 32:39–43
73. Romera E (2002) Stam's principle, D-dimensional uncertainty-like relationships and some atomic properties. Mol Phys 100:3325
74. López-Rosa S, Esquivel RO, Angulo JC, Antolín J, Dehesa JS, Flores-Gallegos N (2010) Fisher information study in position and momentum spaces for elementary chemical reactions. J Chem Theory Comput 6:145–154
75. Plastino A, Plastino AR (2006) Fisher info and thermodynamics first law. Physica A 369:432
76. Romera E, Angulo JC, Dehesa JS (1999) Fisher entropy and uncertainty-like relationships in many-body systems. Phys Rev A 59:4064
77. Angulo JC (1993) Uncertainty relationships in many-body systems. J Phys A, Math Gen 26:6493
78. Folland GB, Sitaram A (1997) The uncertainty principle: A mathematical survey. J Fourier Anal Appl 3:207
79. Wang YA, Carter EA (1999) Improved lower bounds for uncertainty like relationships in many-body systems. Phys Rev A 60:4153
80. Beckner W (1975) Inequalities in Fourier analysis. Ann Math 102:159
81. Rajagopal AK (1995) The Sobolev inequality and the Tsallis entropic uncertainty relation. Phys Lett A 205:32
82. Maassen H, Uffink JBM (1988) Generalized entropic uncertainty relations. Phys Rev Lett 60:1103
83. Dodonov DD, Man'ko VI (1989) Invariants and the evolution of non-stationary quantum states. Nova Publ, New York
84. Zakai M (1960) A class of definitions of duration (or uncertainty) and the associated uncertainty relations. Inf Control 3:101

85. Hirschman II (1957) New bounds for the uncertainly principle. Am J Math 79:152
86. Guevara NL, Sagar RP, Esquivel RO (2003) Shannon-information entropy sum as a correlation measure in atomic systems. Phys Rev A 67:012507
87. Gadre SR (2003) In: Reviews of modern quantum chemistry: a celebration in the contributions of Robert G Parr, vol 1. World Scientific, Singapore
88. Gadre SR, Sears SB, Chakravorty SJ, Bendale RD (1985) Some novel characteristics of atomic information entropies. Phys Rev A 32:2602
89. Bialynicki-Birula I (2006) Formulation of the uncertainty relations in terms of the Rényi entropies. Phys Rev A 74:052101
90. Zozor S, Vignat C (2007) On classes of non-Gaussian asymptotic minimizers in entropic uncertainty principles. Physica A 375:499
91. Zozor S, Portesi M, Vignat C (2008) Some extensions of the uncertainty principle. Physica A 387:4800–4808
92. Stam AJ (1959) Some inequalities satisfied by the quantities of information. Inf Control 2:101
93. Dehesa JS, Plastino AR, Sánchez-Moreno P (2011) A quantum uncertainty relation based on Fisher's information. J Phys A 44:065301
94. Dehesa JS, Martínez-Finkelshtein A, Sorokin V (2006) Information-theoretic measures for Morse and Pöschl-Teller potentials. Mol Phys 104:613
95. Dehesa JS, González-Férez R, Sánchez-Moreno P (2007) The Fisher-information-based uncertainty relation, Cramér-Rao inequality and kinetic energy for the D-dimensional central problem. J Phys A, Math Gen 40:1845–1856
96. Angulo JC, Antolin J, Esquivel RO (2010) Atomic and molecular complexities: their physical and chemical interpretations. Springer, Berlin. See Chap 6 of this book
97. López-Ruiz R, Mancini HL, Calvet X (1995) A statistical measure of complexity. Phys Lett A 209:321
98. Catalán RG, Garay J, López-Ruiz R (2002) Features of the extension of a statistical measure of complexity to continue systems. Phys Rev E 66:011102
99. Anteneodo C, Plastino AR (1996) Some features of the López-Ruiz-Mancini-Calbet (LMC) statistical measure of complexity. Phys Lett A 223:348–354
100. López-Rosa S, Montero J, Sánchez-Moreno P, Venegas J, Dehesa JS (2010) Position and momentum information-theoretic measures of a D-dimensional particle-in-a-box. J Math Chem 49:971
101. López-Ruiz R, Nagy A, Romera E, Sañudo IJ (2009) A generalize statistical complexity measure: Applications to quantum systems. J Math Phys 50:123528
102. Avery J (2000) Hyperspherical harmonics and generalized Sturmians. Kluwer, Dordrecht
103. Chatterjee A (1990) Large-N expansions in quantum mechanics, atomic physics and some $O(N)$ invariant systems. Phys Rep 186:249
104. Avery JS (2010) Harmonic polynomials, hyperspherical harmonics and atomic spectra. J Comput Appl Math 233:1366
105. Avery J (1998) A formula for angular and hyperangular integration. J Math Chem 24:169
106. Louck JD (1960) Generalized orbital angular momentum and the N-fold degenerated quantum-mechanical oscillator. J Mol Spectrosc 4:334
107. Nikiforov AF, Suslov SK, Uvarov VB (1991) Classical orthogonal polynomials of a discrete variable. Springer, Berlin
108. Yáñez RJ, Van Assche W, Dehesa JS (1994) Position and momentum information entropies of the d-dimensional harmonic oscillator and hydrogen atom. Phys Rev A 50(4):3065–3079
109. Ray A, Kalyaneswari M, Ray PP (1988) Moments of probability distributions, wavefunctions, and their derivatives at the origin of N-dimensional central potentials. Am J Phys 56:462
110. Romera E, Sánchez-Moreno P, Dehesa JS (2006) Uncertainty relation for Fisher information of D-dimensional single-particle systems with central potentials. J Math Phys 47:103504
111. Dehesa JS, Martínez-Finkelshtdein A, Sánchez-Ruiz J (2001) Quantum information entropies and orthogonal polynomials. J Comput Appl Math 133:23–46

112. Yáñez RJ, Van Assche W, Gonzalez-Ferez R, Dehesa JS (1999) Entropic integrals of hyperspherical harmonics and spatial entropy of D-dimensional central potentials. J Math Phys 40:5675
113. Dehesa JS, López-Rosa S, Martínez-Finkelshtein A, Yáñez RJ (2010) Information theory of D-dimensional hydrogenic systems. Application to circular and Rydberg states. Int J Quant Chem 110:1529
114. Dehesa JS, López-Rosa S, Olmos B, Yáñez RJ (2006) Fisher information of D-dimensional hydrogenic systems in position and momentum spaces. J Math Phys 47:052104–1–13
115. Sánchez-Moreno P, González-Férez R, Dehesa JS (2006) Improvement of the Heisenberg and Fisher-information-based uncertainty relations for D-dimensional central potentials. New J Phys 8:330
116. Romera E, Sánchez-Moreno P, Dehesa JS (2005) The Fisher information of single-particle systems with a central potential. Chem Phys Lett 414:468–472
117. Andrew K, Supplee K (1990) A hydrogenic atom in D-dimensions. Am J Phys 58:1177
118. Dehesa JS, López-Rosa S, Manzano D (2009) Configuration complexities of hydrogenic atoms. Eur J Phys D 55:539–548
119. Fock V (1935) Zur theorie des wasserstoffatoms. Z Phys 98:145
120. Lombardi JR (1980) Hydrogen atom in the momentum representation. Phys Rev A 22:797
121. Hey JD (1993) On the momentum representation of hydrogenic wave functions: Some properties and applications. Am J Phys 61:28
122. Hey JD (1993) Further properties of hydrogenic wave functions. Am J Phys 61:741
123. Tarasov VF (2004) Exact numerical values of diagonal matrix elements $\langle rk\rangle_{nl}$, as $n \leq 8$ and $-7 \leq k \leq 4$, and the symmetry of Appell's function $F2(1,1)$. Int J Mod Phys B 18:3177–3184
124. Van Assche W, Yáñez RJ, González-Férez R, Dehesa JS (2000) Functionals of Gegenbauer polynomials and D-dimensional hydrogenic momentum expectation values. J Math Phys 41:6600
125. Delburgo R, Elliott D (2009) Inverse momentum expectation value for hydrogenic systems. J Math Phys 50:062107
126. Dehesa JS, Yáñez RJ, Aptekarev AA, Buyarov V (1998) Strong asymptotics of Laguerre polynomials and information entropies of 2D harmonic oscillator and 1D Coulomb potentials. J Math Phys 39:3050
127. Buyarov V, Dehesa JS, Martínez-Finkelshtein A, Sánchez-Lara J (2004) Computation of the entropy of polynomials orthogonal on an interval. SIAM J Sci Comput 26:488
128. Schleich VP, Dahl JP (2002) Dimensional enhancement of kinetic energies. Phys Rev A 65:052109
129. Sen KD et al (2009) Private communication
130. López-Rosa S, Manzano D, Dehesa JS (2009) Complexity of D-dimensional hydrogenic systems in position and momentum spaces. Physica A 388:3273–3281
131. Sañudo J, López-Ruiz R (2009) Generalized statistical complexity and Fisher-Rényi entropy product in the H-atom. IREPHY 3:207
132. Cizek J, Vinette F (1987) N-dimensional hydrogen atom in an external spherically symmetric-field. Theor Chim Acta 72:497
133. Hall MJW (1999) Universal geometric approach to uncertainty, entropy and information. Phys Rev A 59:2602
134. Dong SH, Gu XY, Ma ZQ (2003) The Klein-Gordon equation with Coulomb potential in D dimensions. Int J Mod Phys E 12:555
135. Howard IA, March NH (2004) Relativistic effects when many independent fermions are confined in D dimensions. J Phys A, Math Gen 37:965
136. Goodson DZ, Morgan JD III, Herschbach DR (1991) Dimensional singularity analysis of relativistic equations. Phys Rev A 43:4617
137. Bencheikh K, Nieto LM (2007) On the density profile in Fourier space of a harmonically trapped ideal Fermi gas in d dimensions. J Phys A, Math Gen 40:13503
138. Loos PF, Gill PMW (2009) Two electrons on a hypersphere: a quasiexactly solvable model. Phys Rev Lett 103:123008

Chapter 6
Atomic and Molecular Complexities: Their Physical and Chemical Interpretations

J.C. Angulo, J. Antolín, and R.O. Esquivel

Abstract Within the present work on the meaning, interpretation and applications of the complexity measures, different order-uncertainty planes embodying relevant information-theoretical magnitudes are studied in order to analyse the information content of the position and momentum electron densities of several atomic (neutrals, singly-charged ions, isoelectronic series) and molecular (closed shells, radicals, isomers) systems. The quantities substaining those planes are the exponential and the power Shannon entropies, the disequilibrium, the Fisher information and the variance. Each plane gives rise to a measure of complexity, determined by the product of its components. In the present work, the values of the so-called López-Ruiz, Mancini and Calbet (LMC), Fisher-Shannon (FS) and Cramér-Rao (CR) complexities will be provided in both conjugated spaces and interpreted from physical and chemical points of view. Computations for atoms were carried out within a Hartree-Fock framework, while for molecules by means of CISD(T)/6-311++G(3df, 2p) wave functions. In order to have a complete information-theoretical description of these systems, it appears relevant to consider simultaneously the results in both spaces.

6.1 Introduction

There has been a tremendous interest in the literature to apply information theory to the electronic structure theory of atoms and molecules [1, 2]. The concepts of uncertainty, randomness, disorder or delocalization, are basic ingredients in the study, within an information theoretical framework, of relevant structural properties for many different probability distributions appearing as descriptors of several chemical and physical systems and/or processes.

Following the usual procedures carried out within the Information Theory for quantifying the aforementioned magnitudes concerning individual distributions,

J.C. Angulo (✉)
Departamento de Física Atómica, Molecular y Nuclear, Universidad de Granada, 18071 Granada, Spain
e-mail: angulo@ugr.es

K.D. Sen (ed.), *Statistical Complexity*,
DOI 10.1007/978-90-481-3890-6_6, © Springer Science+Business Media B.V. 2011

some other extensions have been done in order to introduce and to apply the concepts of 'similarity' [3–5] or 'divergence' [6–9] between two distributions, as comparative measures. Quantum similarity theory was originally developed in order to establish quantitative comparisons between molecular systems by means of their fundamental structure magnitudes: electron density functions. Applications of this important theory have been one of the cornerstones of recent chemical research in molecules [10–12].

Some pioneering efforts relating Information Theory to electronic structure and properties of molecules can be already found in the seminal papers by Daudel in the framework of loge theory [13, 14], subsequently followed by Mezey [15] and reexamined later by Nalewajski [16]. The studies of Mezey [17] and Avnir [18] on symmetry and chirality-related problems in molecules, and in other very diverse fields (e.g. image and texture analysis), are also examples of applications of informational measures on specific aspects of shape, disorder and complexity.

This kind of measures and techniques, which in fact characterize most of the information theory aims and tools, have been widely employed in recent years also within the atomic and molecular physics framework. The present work includes a survey of some of those applications for obtaining relevant information on different properties of atomic and molecular systems, including structural and experimental ones.

The role played by the two conjugated variables, namely position and momentum, appears fundamental for a complete description of the atomic and molecular information features. For example, it is shown that, in spite of their simplicity among the many-body systems, the atomic ones posses a highly enough level of organization and hierarchy so as to be considered as an appropriate benchmark for the suggested complexity study. As should be expected, the same is true also for much more complex systems such as molecules.

The relevancy of the above concepts motivates the search for an appropriate quantification, giving rise to a variety of density functionals, each one with its own characteristics and properties which make them more or less useful attending to the specific problem we are dealing with.

Diverse information measures for probability distributions of arbitrary dimensionality have been widely applied with the aim of describing a great variety of systems or processes in many different scientific fields. One of the pioneering and most well-known of such measures is the variance [19], but later on many others have been also considered for these kind of applications. Among them, it should be emphasized the role played by the Shannon entropy S [20]

$$S(\rho) \equiv -\int \rho(\mathbf{r}) \ln \rho(\mathbf{r}) d\mathbf{r} \tag{6.1}$$

and the Fisher information I [21, 22]

$$I(\rho) \equiv \int \rho(\mathbf{r}) |\nabla \ln \rho(\mathbf{r})|^2 d\mathbf{r} \tag{6.2}$$

of a distribution $\rho(\mathbf{r})$. In fact, S is a basic quantity in statistical thermodynamics [23] and it is the essential tool on the application of the 'Maximum Entropy' technique based on Jaynes' principle. More recently, Fisher information appeared as a fundamental magnitude for deriving important laws of e.g. density functional theory [24, 25] or quantum mechanics [26–28] by means of the extremization Frieden principle [22]. The numerous applications of tools based on both S and I suggest the relevancy of using them in a complementary way, attending to their main characteristics and properties as will be described later.

6.1.1 Complexity: Meaning and Definitions

Another relevant concept within information theory, in some cases strongly related to the aforementioned measures, is the so-called 'complexity' of a given system or process. The study of complexity in physical, chemical, biological and social systems or processes is a topic of great contemporary research interest. A quantitative measure of complexity is useful to estimate the ability of systems for organization and it is also proposed as a general indicator of structure or correlation.

Fundamental concepts such as entropy or information are frequently present in the proposals for characterizing complexity, but it is known that other ingredients capturing not only randomness are also necessary. In fact one would wish also to detect, for instance, clustering or pattern.

There is not a unique and universal definition of complexity for arbitrary distributions, but it could be roughly understood as an indicator of pattern, structure and correlation associated to the system that the distribution describes. Nevertheless. many different mathematical quantifications exist under such an intuitive description. This the case of the algorithmic [29–31], Lempel-Ziv [32] and Grassberger [33] complexities, as well as the logical and thermodynamical depths by Bennett [34] and Lloyd and Pagels [35], respectively, all of them with diverse scientific applications, as long as other complexity definitions [36]. Some of them share rigorous connections with others as well as with Bayes and information theory [37].

Complexity is used in very different fields (dynamical systems, time series, quantum wave functions in disordered systems, spatial patterns, language, analysis of multi-electronic systems, cellular automata, neuronal networks, self-organization, molecular or DNA analyses, social sciences, etc.) [38–40]. Although there is no general agreement about the definition of what complexity is, its quantitative characterization is a very important subject of research and has received considerable attention over the past years [41, 42].

The characterization of complexity cannot be univocal and must be adequate for the type of structure or process we study, the nature and the goal of the description we want and for the level or scale of the observation that we use. Thus it is interesting to combine the properties of the new proposals to characterize complexity and test them on diverse and known physical systems or processes. Fundamental concepts such as information or entropy are frequently present in the proposals for characterizing complexity, but some other ingredients capturing not only uncertainty or randomness can also be searched. One wishes also to capture some other properties such as clustering, order or organization of the systems or process. Some of the definitions and relations between the above concepts are not clear; and even less understood is how disorder or randomness takes part in the aforementioned properties of the system and vice versa.

The initial form of complexity is designed in such a way that it vanishes for the two extreme probability distributions (lesser complex ones), corresponding to perfect order (represented by a Dirac-delta) and maximum disorder (associated with a highly flat distribution). Most of those definitions take into account elements of Bayesian and information theories. Some of the more recent ones consist of the product of two factors, measuring, respectively, order and disorder on the given systems or, equivalently, localization and delocalization [43, 44]. They will be referred to as product-complexities.

These product complexity measures have been criticized and consequently modified, leading to powerful estimators successfully applied in a wide variety of fields [45–50]. Fundamental concepts such as entropy or information are frequently present in the proposals for characterizing complexity, but it is known that other ingredients capturing not only randomness are also necessary. In fact one would wish also to detect, for instance, clustering or pattern.

Even restricting ourselves to the aforementioned product complexity measures, there is no unique definition for complexity. The reason is that there exist different candidates for being one of the coupled factors which give rise to complexity. The most popular ones are well-known to play a relevant role in an information-theoretical framework. Among them, let us mention the Shannon entropy S, the disequilibrium D, the Fisher information I and the variance V, which will be defined below. Much work has been done using these quantities as basic measures, not only for quantifying the level of spreading of distributions but also for many other applications, such as, for instance, maximum-entropy estimation and reconstruction of an unknown distribution from very limited information on it.

Other authors have recently dealt with some particular factors of the complexity measures. In particular, Shannon entropy has been extensively used in the study of many important properties of multielectronic systems, such as, for instance, rigorous bounds [51], electronic correlation [52], effective potentials [53], similarity [54] and maximum entropy [55, 56] and minimum cross entropy approximations [57].

More recently, Fisher information has been studied as an intrinsic accuracy measure for specific atomic models and densities [58, 59] and also for quantum mechanics central potentials [27]. Also, the concept of phase space Fisher information, where position and momentum variables are included, was analyzed for hydrogen-like atoms and the isotropic harmonic oscillator [60]. The net Fisher information measure is found to correlate well with the inverse of the ionization potential and dipole polarizability [59].

Quantum similarities and self-similarities D for neutral atoms were computed for nuclear charges $Z = 1$–54 only in the position space [61, 62], but afterwards a more complete analysis including $Z = 1$–103 neutral systems and singly charged ions has been done in position and momentum spaces [5].

Some studies on similarity, using magnitudes closely related to D or to relative Shannon entropies, have been also reported [4, 63]. Very recently a comparative analysis of I and D shows that they both vary similarly with the nuclear charge Z within the neutral atoms, exhibiting the same maxima and minima, but Fisher information presents a significantly enhanced sensitivity in the position and momentum spaces in all systems considered [64].

6.1.1.1 LMC Complexity

Among the more recent and successful definitions of complexity, especially remarkable is the one provided by López-Ruiz, Mancini and Calbet [43], to be denoted by $C(LMC)$ due to its pioneering authors, which satisfies as others do the condition of reaching minimal values for both extremely ordered and disordered limits. Additional relevant properties are the invariance under scaling, translation and replication.

The initial definition of the LMC complexity (also known as 'shape complexity') has been criticized [41] and modified [48] in order to the aforementioned properties to be satisfied, giving rise to the expression

$$C(LMC) \equiv D \cdot e^S = D \cdot L, \tag{6.3}$$

of a distribution $\rho(\mathbf{r})$. It is built up as the product of two relevant quantities within an information-theoretical framework: the 'disequilibrium' D [3, 65, 66],

$$D(\rho) \equiv \int \rho^2(\mathbf{r})d\mathbf{r} \tag{6.4}$$

which quantifies the departure of $\rho(\mathbf{r})$ from equiprobability, and the aforementioned Shannon entropy S as measure of randomness or uncertainty on the distribution. The usefulness of $C(LMC)$ has been shown in different fields, allowing detection of periodic, quasiperiodic, linear stochastic and chaotic dynamics [43, 49, 50].

6.1.1.2 Fisher-Shannon Complexity

It appears also interesting to look for statistical complexities involving also a local information measure. This can be achieved by replacing one of the LMC global factors by a 'local' measure of intrinsic accuracy. In this sense, the main properties of Fisher information I make this quantity to be an appropriate candidate with the aim of defining a complexity measure in terms of complementary global and local factors. Very recently, the Fisher-Shannon complexity $C(FS)$ has been defined [64, 67] in terms of both Fisher information and Shannon entropy and, consequently, providing a measure which combines the global and local characters, and also preserving the desirable properties for any complexity measure as previously described. The Fisher information I itself plays a fundamental role in different physical problems, such as the derivation of non-relativistic quantum-mechanical equations by means of the minimum I principle, as it also does for the time-independent Kohn-Sham equations and the time-dependent Euler equation [25, 68].

The FS complexity in the n-dimensional space is defined in terms of the power Shannon entropy $J \equiv \frac{1}{2\pi e} e^{2S/n}$ and the Fisher information I as

$$C(FS) \equiv I \cdot J \tag{6.5}$$

where definition of J is chosen in order to preserve general complexity properties, such as the scaling invariance and the minimum value n because of the dimension of the space. In contrast with the LMC complexity, and apart from the explicit dependence on Shannon entropy, $C(FS)$ replaces the disequilibrium global factor by the Fisher local one. The $C(FS)$ expression arises from the isoperimetric n-dimensional inequality $I \cdot J \geq n$ [6, 69, 70] providing a universal lower bound to FS complexity. Among the main applications carried out, it should be remarked those concerning with atomic distributions in position and momentum spaces where FS complexity is shown to provide relevant information on atomic shell structure and ionization processes [64, 67, 71, 72] as well as in molecular systems [73].

6.1.1.3 Cramér-Rao Complexity

Aside of the $C(LMC)$ and $C(FS)$, in the present work we will also analyze the 'Cramér-Rao' complexity $C(CR)$, given also as the product of a local and a global measure, keeping the first one as the Fisher information I, and replacing the Shannon entropy exponential by the variance V, giving rise to

$$C(CR) \equiv I \cdot V, \tag{6.6}$$

product which has been considered in different contexts [71, 72, 74]. Especially remarkable is the existence of a lower bound, in spite of the factors being of very different origin as well as their definition in terms of the distribution, emphasizing again the strong connection between both the local and global level of uncertainty.

6.1.1.4 Generalized Rényi-like Complexities

Let us study a generalization of the LMC and FS complexities by replacing the Shannon entropy functional by a more general and powerful magnitude as the Rényi entropy. Hence we deal with a one-parameter (to be denoted by α) generalized complexity which weights different regions of the position or momentum spaces according to the value of α. The LMC and FS complexities are particular cases of these Rényi complexities.

In particular, the so-called 'shape Rényi complexity' (SR) is characterized as a difference between the α-order Rényi entropy and the second order one (expressed in terms of the disequilibrium, D), and it has been extended to continuous systems [75], theoretically studied and tested for the binary symmetric channel (BSC) and the logistic map [50]. A more extended family of generalized complexity measures has been proposed, and rigorous bounds, geometrical properties and several applications have been also studied [76]. Moreover, the 'Fisher Rényi complexity' (FR) is defined by simply replacing the Shannon entropy with the Rényi entropy in the expression of the Fisher-Shannon complexity. Some rigorous properties for this entropic product, also called Fisher-Rényi product, and similar ones have been recently obtained [77]. Previous applications of both the SR and FR complexities are very scarce in the literature. To the best of our knowledge, the first time in which the SR and FR complexities were considered and analysed for atomic systems was in [78], apart from recently derived results [77], mainly concerning with uncertainty-like relationships but not with the atomic structure and the shell-filling process.

In this chapter we consider both generalized Rényi complexities, SR and FR, in order to study their behavior for atomic systems, in particular the neutral atoms throughout the whole Periodic Table of elements. In spite of their simplicity as compared to others quantum-mechanical systems, they display a strongly organized and hierarchical structure.

The concept of Rényi complexity arises from the Rényi entropy, widely used in the literature when affording different problems within an information-theoretical framework. The Rényi entropy [79] plays a similar role to those of other density functionals as descriptors of the uncertainty on a distribution, including very well-known ones such as the Shannon [20] and the Tsallis [80] entropies.

The Rényi entropy of order α for the distribution $\rho(\mathbf{r})$ is defined as

$$R^{(\alpha)} \equiv \frac{1}{1-\alpha} \ln \omega_\alpha \tag{6.7}$$

where the quantity ω_α is the so-called 'α-order frequency moment' of $\rho(\mathbf{r})$ [81], given by

$$\omega_\alpha \equiv \int \rho^\alpha(\mathbf{r}) d\mathbf{r}, \tag{6.8}$$

which has also been employed in diverse fields, being especially remarkable in Density Functional Theory for some specific α values [82] (e.g. Thomas-Fermi kinetic and exchange energies), as well as the own disequilibrium [43, 83]. The normalization to unity of the distribution can be expressed as $\omega_1 = 1$.

The allowed range of values for the characteristic parameter α of the Rényi entropy is determined by the convergence conditions on the integral in (6.8), being imposed by the short- and long-range behaviors of the distribution $\rho(\mathbf{r})$. Apart from the necessary (but not sufficient) condition $\alpha > 0$ for the finiteness of $R^{(\alpha)}$, the particular value $\alpha = 1$ appears as a limiting case, because both the numerator and the denominator in (6.7) vanish, the limit giving rise to

$$R^{(1)} = S = -\int \rho(\mathbf{r}) \ln \rho(\mathbf{r}) d\mathbf{r}, \tag{6.9}$$

that is, the Rényi entropy of order 1 is the Shannon entropy S or, in other words, the Rényi entropy $R^{(\alpha)}$ represents an extension or generalization of the Shannon entropy.

The power α of the distribution in (6.8), where ω_α is defined, allows to enhance or diminish, by increasing or decreasing its value, the contribution of the integrand over different regions to the whole integral and, consequently, to the frequency moments and the Rényi entropy $R^{(\alpha)}$. Higher values of α make the function $\rho^\alpha(\mathbf{r})$ to concentrate around the local maxima of the distribution, while the lower values have the effect of smoothing that function over its whole domain. It is in that sense that the parameter α provides with a powerful tool to get information on the structure of the distribution by means of the Rényi entropy.

Another relevant particular case of the Rényi entropy and the frequency moments corresponds to $\alpha = 2$, giving rise to the disequilibrium D as the second-order frequency moment ω_2, namely

$$D = \int \rho^2(\mathbf{r})d\mathbf{r}, \tag{6.10}$$

which measures the 'level of departure from uniformity' of the distribution [43, 83]. According to its definition and that of $R^{(\alpha)}$ it is immediate to observe that $R^{(2)} = -\ln D$, establishing a link between the Rényi entropy and the disequilibrium.

For the LMC and FS complexities, the Shannon entropy S is employed as a measure of information in one factor, the other factor (measuring order) being the disequilibrium D and the Fisher information I for the LMC and FS complexities, respectively. A 'generalized' version is obtained when the Shannon entropy contribution is replaced by the Rényi entropy $R^{(\alpha)}$, giving rise to generalized complexity measures which will be referred as 'Shape Rényi complexity' $SR^{(\alpha)}$ and 'Fisher-Rényi complexity' $FR^{(\alpha)}$, defined as

$$SR^{(\alpha)} \equiv D \cdot \exp\{R^{(\alpha)}\}, \tag{6.11}$$

with the exponential Rényi entropy being also denoted as $L^{(\alpha)} \equiv \exp\{R^{(\alpha)}\}$, and

$$FR^{(\alpha)} \equiv I \cdot J^{(\alpha)}, \tag{6.12}$$

where

$$J^{(\alpha)} = \frac{1}{2\pi e} \exp\left\{\frac{2}{n} R^{(\alpha)}\right\} \tag{6.13}$$

is the 'α-order power entropy' of the n-dimensional distribution.

Some comments are in order: (i) the particular cases $SR^{(1)}$ and $FR^{(1)}$ corresponding to $\alpha = 1$ provide, respectively, the expressions of the LMC and FS complexities, (ii) all relevant invariance properties of LMC and FS also hold for arbitrary $\alpha > 0$, (iii) the weighting effect of the parameter α over specific regions, as previously mentioned for the Rényi entropy, now translates into the associated complexities, and (iv) attending to its definition, the composing factors of the second order shape Rényi complexity are the inverse of each other, and consequently $SR^{(2)} = 1$.

Other Rényi products have been also considered in the literature, for which different properties such as, e.g., bounds and uncertainty-like relationships are known for very specific α ranges [77]. The analysis of those properties is beyond the scope of the present work, wherein a much wider interval for the α parameter is considered.

6.1.2 Selected Relationships and Rigorous Properties on Complexities

Most of the research on complexities and its corresponding conclusions have been obtained, so far, by numerically quantifying their values, and much less attention has been paid to their theoretical properties and exact meaning within an statistical framework, valid for any arbitrary n-dimensional distribution [84]. In the present section, different product-complexities are investigated, obtaining results such as rigorous bounds, uncertainty-like inequalities, relationships among different complexities. Additionally statistical interpretations will be provided. For the sake of completeness, some of these analytical results on product-complexities will be numerically analyzed for the one-particle densities of atomic systems in both conjugated spaces.

Let us consider an arbitrary n-dimensional distribution $\rho(\mathbf{r})$, whose normalization is given by

$$\int \rho(\mathbf{r})d\mathbf{r} = 1 \qquad (6.14)$$

and the integration is performed over the n-dimensional space R^n. Consequently, the vector $\mathbf{r}$ consists of n components, which can be expressed equivalently in Cartesian or spherical coordinates, namely $\mathbf{r} = (x_1, \ldots, x_n) = (r, \theta_1, \ldots, \theta_{n-2}, \phi)$, where $r = |\mathbf{r}|$ is the modulus of the spatial vector.

It is usual to deal additionally with the corresponding distribution $\gamma(\mathbf{p})$ in the conjugated space. Many properties and characteristics of both densities $\rho(\mathbf{r})$ and $\gamma(\mathbf{p})$ are well-known to be strongly related. Such is the case, for instance, of the one-particle densities of many-particle systems (e.g. atoms, molecules), in which $\rho(\mathbf{r})$ quantify the mass density around location $\mathbf{r}$ and $\gamma(\mathbf{p})$ the linear momentum distribution around the momentum vector $\mathbf{p}$. Different relationships involving quantities associated with both complementary densities are of capital importance through the concept of *uncertainty* of the system.

For illustration, an analysis of different relationships among information-theoretical quantities will be carried out later for atomic systems. Consequently we will deal with densities whose domain is the three-dimensional ($n = 3$) space. Studies on complexity measures have been also done in previous works, mainly by computing their numerical values [40]. However, studies on rigorous relationships among different complexities and/or other information magnitudes, valid for arbitrary systems and dimensionalities, are very scarce.

In what follows, atomic units (a.u.) will be considered for variables, densities, functionals and complexities (i.e. $\hbar = |m| = e = 1$ and, consequently, also the Bohr radius $a_0 = 1$) when carrying out the numerical analysis for atomic and molecular systems. Fixing the system of units is essential for a proper description of different quantities, according to their definition.

It is also worthy to mention that (i) the product distribution $f(\mathbf{r}, \mathbf{p}) \equiv \rho(\mathbf{r})\gamma(\mathbf{p})$ will be also considered in order to have a more complete informational description of the system, and (ii) in some cases (e.g. atomic systems), it will be sufficient to deal with the spherically averaged densities $\rho(r)$ and $\gamma(p)$, for which the independent variable range is the non-negative real line $[0, \infty)$.

Let us consider the frequency or entropic moments, denoted as functions of the characteristic parameter 'q' as $\omega_r(q) \equiv \int \rho^q(\mathbf{r})d\mathbf{r}$, strongly related to the Rényi [79] and Tsallis [80] entropies, denoted as $R_r(q)$ and $T_r(q)$ respectively:

$$R_r(q) = \frac{\ln \omega_r(q)}{1-q}, \tag{6.15}$$

$$T_r(q) = \frac{1-\omega_r(q)}{q-1}, \tag{6.16}$$

and similarly for the corresponding quantities in momentum and product spaces (denoted with the subscripts 'p' or 'rp' instead of 'r') by only replacing $\rho(\mathbf{r})$ by $\gamma(\mathbf{p})$ or $f(\mathbf{r}, \mathbf{p})$. At times, subscripts will be omitted in equations and definitions, understanding their validity for any arbitrary space.

The LMC, FS and CR definitions allow one to observe the small complexity values for extreme distributions, because the Shannon entropy approaches $-\infty$ for highly concentrated distributions (and consequently $L \to 0$), while the disequilibrium and the Fisher information go to zero as the distribution spreads uniformly over its domain.

Contrary to the case of the isolated factors which define $C(LMC)$, namely, the disequilibrium and the exponential Shannon entropy, not many rigorous properties and/or relationships are known on complexity [84]. Concerning the aforementioned factors, it is worth noting, among others, the variational upper and/or lower bounds on each factor in terms of radial expectation values of the density [85, 86].

The Fisher information and the power Shannon entropy determine the so-called *Fisher-Shannon plane*, where the definition of the power entropy is based on well-known inequalities on such a product [69].

Cramér-Rao complexity emerges when, as done in [72], the product of a local and a global measure is considered, involving, on one hand, the Fisher information I as the local one and, on the other, the variance V of the distribution for measuring the degree of deviation from the mean value. Such a complexity corresponds essentially to the 'Cramér-Rao product', giving rise to [72]

$$C(CR) \equiv I \cdot V \tag{6.17}$$

for which the inequality $C(CR) \geq n^2$ is also known [6, 69].

Now, additional rigorous properties on the aforementioned complexities will be shown, being valid for arbitrary n-dimensional distributions. In some cases (as is well known for the Fisher-Shannon and Cramér-Rao complexities), there exist lower bounds given as universal constant values (not necessarily dependent on the dimensionality). In the LMC case, however, such a value has been only shown to exist for the one-dimensional case [46] (i.e. for densities having as domain the real line), what is generalized for arbitrary dimensionality in the present section. Other results are here expressed as bounds in terms of expectation values and/or density functionals. For the sake of completeness, a numerical analysis will be also carried out for the one-particle densities of atomic systems within a Hartree-Fock framework.

6.1.2.1 Lower Bound on the LMC Complexity

Starting with López-Ruiz, Mancini and Calbet complexity $C(LMC) = D \cdot e^S$, let us first observe that it can be also expressed in terms of frequency moments $\omega(q)$. In doing so, it is convenient to define the function

$$f(q) \equiv \ln \omega(q) \tag{6.18}$$

which, due to the normalization constraint, takes the particular value $f(1) = 0$. Additionally, let us write the LMC complexity as $C(LMC) = \exp\{\ln D + S\}$.

Attending to the definition of S and D, it is easy to check that $S = -f'(1)$, i.e. minus the slope of the function $f(q)$ at $q = 1$, and

$$\ln D = \ln \omega(2) = f(2) = \frac{f(2) - f(1)}{2 - 1} \tag{6.19}$$

where the normalization constraint has been taken into account. Last equality is written in order to point out that $\ln D$ represents the slope of the straight line connecting points of the function $f(q)$ at $q = 1$ and $q = 2$. Finally, and having in mind the convexity of $f(q)$ (or equivalently, the log-convexity of frequency moments $\omega(q)$ as can be easily shown by using Hölder's inequality), it is concluded that the single exponent on $C(LMC)$ written as above is non-negative and, consequently, that $C(LMC) \geq 1$.

Moreover, from this proof it is immediately concluded that equality $C(LMC) = 1$ is only reached for uniform distributions with a finite volume support. In doing so, it is enough to observe that equality is only possible for a linear $f(q)$ over the range $1 \leq q \leq 2$. Such a linearity translates on frequency moments as $\omega(q) = D^{q-1}$ (where the values $\omega(1) = 1$ and $\omega(2) = D$ have been considered). This means that

$$\int \left(\frac{\rho(\mathbf{r})}{D}\right)^q d\mathbf{r} = \frac{1}{D}. \tag{6.20}$$

The non-dependence on 'q' of the right-hand-side requires the fraction on the integral to take only the values 0 or 1. Then, the density has the constant value $\rho(\mathbf{r}) = D$ on its whole support Λ (apart from, at most, a zero-measure set of points), being the volume of the support $1/D$ in order to keep the normalization condition.

It is remarkable that lowest LMC complexity corresponds to step distributions over a finite set Λ, which are precisely the maximum-entropy ones among those with domain Λ. Then, they necessarily minimize the disequilibrium also, as it is well-known when dealing with finite-size domains.

In summary, uniform distributions simultaneously minimize the disequilibrium D (i.e. localization) and maximize the Shannon entropy, and consequently the exponential entropy L (i.e. delocalization). But the joint effect of the two opposite ones on each factor of the $C(LMC)$ complexity is dominated by the minimizer one (i.e. the disequilibrium), giving rise to the minimum LMC complexity for uniform densities.

6.1.2.2 Fisher-Shannon Complexity as Perturbed Power Entropy

Let us now concentrate on the Fisher-Shannon complexity $C(FS) = I \cdot J$, in which the localization factor, namely the Fisher information I, has a *local* character, in the sense that it constitutes a sensitive measure of the variation of the gradient along the domain of the distribution. Such a measure is of different character than the other component, defined in terms of a global measure as the Shannon entropy.

The inequality $I \cdot J \geq n$ [87], which provides a lower bound to the product of both quantities, is consequently written in terms of the Fisher-Shannon complexity, as

$$C_r(FS) \geq n, \tag{6.21}$$

$$C_p(FS) \geq n, \tag{6.22}$$

$$C_{rp}(FS) \geq 2\pi e n^2, \tag{6.23}$$

where 'n' is the dimension of the space, and last inequality contains an additional factor apart from the product of complexities in conjugated spaces because of the definition of the power entropy J_{rp} in the product space due to its Shannon entropy $S_{rp} = S_r + S_p$. It is worthy to mention that the above inequalities are valid for arbitrary distributions on n-dimensional spaces, in the same line as the lower bound $C(LMC) \geq 1$ previously obtained for LMC complexities independently of the space we are dealing with.

In spite of the different characteristics of the two main components of the Fisher-Shannon complexity $C(FS) = I \cdot J$, it is known a result which provides a connection between both information measures $I(\rho)$ and $J(\rho)$, which as we are going to show can be also expressed and interpreted in terms of complexities. The above mentioned connection arises from the so-called effect of *Gaussian perturbation*, and it provides information on the variation suffered by the information content of a distribution ρ when adding a very small Gaussian one. Concerning the Fisher and Shannon measures, it is known that

$$\left.\frac{d}{d\varepsilon} S(\rho + \sqrt{\varepsilon}\rho_G)\right|_{\varepsilon=0} = \frac{1}{2} I(\rho), \tag{6.24}$$

as shown by de Bruijn [6], where ρ_G denotes the standard Gaussian distribution with mean 0 and variance 1. In this sense, the Fisher information could be understood as a measure of the variation of Shannon entropy of the starting density under a Gaussian perturbation.

Keeping in mind this result, let us consider the power entropy of the perturbed distribution $\rho_\varepsilon \equiv \rho + \sqrt{\varepsilon}\rho_G$, namely

$$J(\rho_\varepsilon) = \frac{1}{2\pi e} e^{\frac{2}{n} S(\rho_\varepsilon)}. \tag{6.25}$$

Carrying out the same derivation and limiting operations as in (6.24), but on the power entropy $J(\rho_\varepsilon)$ instead of the Shannon entropy $S(\rho_\varepsilon)$, it is immediate to check that

$$C(FS) = n \left.\frac{d}{d\varepsilon} J(\rho + \sqrt{\varepsilon}\rho_G)\right|_{\varepsilon=0}, \tag{6.26}$$

which gives rise to an additional interpretation of the Fisher-Shannon complexity: it represents the variation of the power entropy J of a given density ρ when perturbed by a Gaussian distribution ρ_G. So, the interpretation of the Fisher information according to the process of Gaussian perturbation, as shown in (6.25), is now extended to the interpretation, for the same process, in terms of FS complexity (a product-complexity involving simultaneously two information measures) as shown in (6.26).

6.1.2.3 Lower Bounds on the Cramér-Rao Complexity

Concerning the Cramér-Rao complexity $C(CR)$, we next show that it can be also bounded from below in terms of radial expectation values of the density. This kind of upper and/or lower bounds are extensively found in the literature for different density functionals, such as the Shannon entropy [86] or frequency moments [85], but this is not the case for complexities, since they are formed through a product of two different factors, what involves a lot the bounding procedure as compared to its application for single density functionals.

Here we are going to take advantage of the non-negativity of the so-called relative Fisher information between two functions. For simplicity, we will restrict ourselves to the case of spherically symmetric densities, which only applies to atoms.

In doing so, let us consider the non-negative integral

$$\int \rho(\mathbf{r}) \left(\frac{d}{dr} \ln \frac{\rho(r)}{f(r)} \right)^2 d\mathbf{r} \geq 0 \tag{6.27}$$

where $f(r)$ is a function (not necessarily normalized to unity), on which some conditions will be imposed below. By only carrying out the processes of derivation and squaring in (6.27), and defining

$$F(r) \equiv \frac{f'(r)}{f(r)} \tag{6.28}$$

it is not difficult to find the relationship

$$\bar{I}_r \geq -\langle F^2(r) \rangle - 2\langle F'(r) \rangle - 2(n-1) \left\langle \frac{F(r)}{r} \right\rangle \tag{6.29}$$

where $\bar{I}_r$ refers to the Fisher information of the spherically averaged density $\rho(r)$, and the function $F(r)$ has to fulfill the condition $r^{n-1}\rho(r)F(r)|_0^\infty = 0$ for the finiteness of the expectation values on $\rho(r)$.

So, for an appropriate choice of $F(r)$ the above expression provides a lower bound on the Fisher information in terms of expectation values of the density. Let us consider a choice of $F(r)$ for which the right-hand-side of (6.29) consists of a rational function in which the denominator is the variance $V_r = \langle r^2 \rangle - \langle r \rangle^2$ and, consequently, the inequality transforms into a lower bound of the Cramér-Rao complexity $C_r(CR)$. Such a $F(r)$ is given by

$$F(r) = -\alpha\beta r^{\alpha-1} - \nu\gamma r^{\gamma-1}. \tag{6.30}$$

First, we optimize the resulting bound on the parameters (β, ν) and then we consider the particular case $(\alpha = 2, \gamma = 1)$, giving rise to

$$C_r(CR) \geq n^2 + (n-1)\langle r^{-1}\rangle\big[(n-1)\langle r^2\rangle\langle r^{-1}\rangle - 2n\langle r\rangle\big]. \tag{6.31}$$

Some comments are in order: (i) a similar bound for the corresponding quantity in conjugated space is obtained by considering the momentum density, and (ii) some radial moments (in both spaces) are specially relevant from a physical point of view. It is well known [88], for instance that, for many-electron systems, $\langle r^{-1}\rangle$ is essentially the electron-nucleus attraction energy, $\langle r^2\rangle$ is related to the diamagnetic susceptibility [88], $\langle p^{-1}\rangle$ is twice the height of the peak of the Comptom profile [89], and $\langle p^2\rangle$ is twice the kinetic energy [89]. So, these physically relevant and/or experimentally accessible quantities provide also information on the Cramér-Rao complexity of the system.

6.1.2.4 Complexities of Hydrogen-like Systems

It is worth noting that, contrary to the multi-electronic systems, none of the three complexities depends on the nuclear charge Z for one-electron systems (hydrogenic atoms), although the individual factors do (e.g. D_r is proportional to Z^3 and I_r to Z^2, and inversely in conjugated space). In this sense, it is interesting to observe that such complexities can be analytically determined, their values being

$$C_r(LMC) = \frac{e^3}{8} = 2.5107, \tag{6.32}$$

$$C_r(FS) = \frac{2e}{\pi^{1/3}} = 3.712, \tag{6.33}$$

$$C_r(CR) = 3, \tag{6.34}$$

$$C_p(LMC) = \frac{66}{e^{10/3}} = 2.3545, \tag{6.35}$$

$$C_p(FS) = \frac{48(2\pi)^{1/3}}{e^{29/9}} = 3.5311, \tag{6.36}$$

$$C_p(CR) = 12\left(1 - \frac{64}{9\pi^2}\right) = 3.354. \tag{6.37}$$

As observed before, the multi-electronic character of the systems makes their complexities increase considerably as compared to those of the corresponding one-electron ions.

6.1.2.5 Uncertainty-like Relationships for Complexities

It is natural to look also for the existence of uncertainty-like relationships (i.e. involving the same density functionals simultaneously in both conjugated spaces) on the complexities considered in the present work. It appears very interesting to find rigorous and universal relationships among *conjugated complexities*, in a similar fashion to those verified, for instance, by the Shannon entropy (Bialynicki-Birula and Mycielski (BBM) lower bound $S_r + S_p \geq n(1 + \ln \pi)$ [90]), the variance (Heisenberg uncertainty principle [91]), the Fisher information [92] and the Rényi [93] and Tsallis [94] entropies. In all cases, the uncertainty inequality provides a constant lower bound (sometimes dependent on dimensionality) on the sum or product of the aforementioned conjugated information factors.

Apart from those inequalities on density functionals, there also exist uncertainty-like relationships between products of expectation values, as shown for instance in [95] for the radial ones.

So, in spite of existing well known uncertainty inequalities on most of the individual factors composing complexities, that is not the case (to the best of our knowledge) of the complexities themselves. The main reason is that, usually, the two inequalities associated with each factor work in opposite directions, making consequently impossible to combine both together in order to obtain a coherent bound on the whole complexity.

In this section, different uncertainty-like complexity inequalities are obtained, all of them of universal validity. This means that they hold for any pair of functions related via Fourier transform in the same way as the one-particle densities in the conjugated spaces do.

Let us also remark that, for the particular case of analyzing products of complexities $C_r \cdot C_p$, such a study is equivalent to consider the product or phase space complexity C_{rp}. However, the simple product operation of complexities is not at all the only way of getting uncertainty-like relationships, appearing also interesting to deal with quotients of complexities or of some of their powers, among others.

In this sense, it is worthy to mention that the ratio between the Fisher-Shannon and Cramér-Rao complexities on a given space can be bounded in terms of the so-called *uncertainty products*, expressed in terms, as mentioned above, of radial expectation values of both conjugated spaces. In doing so, let us consider the ratio $C(FS)/C(CR)$ (in any space) which, attending to the definition of both complexities, turns out to be J/V in the same space. For simplicity, let us consider the position space ratio J_r/V_r, keeping in mind that all results obtained below will be also valid for the conjugated quantities. The BBM inequality between Shannon entropies can be written in terms of the power entropies as $J_r \cdot J_p \geq (\pi e)^{n-1}/2$, giving rise to a lower bound on the numerator J_r in terms of the power entropy J_p. On the other hand, upper bounds on the power entropy J_p in terms of any non-negative order radial expectation value $\langle p^\alpha \rangle$ are also well known [86]. Both inequalities together provide a lower bound on J_r in terms of $\langle p^\alpha \rangle$ with $\alpha > 0$.

Concerning the denominator V_r on the studied ratio, it is straightforward to obtain from its definition that $1/V_r \geq 1/\langle r^2 \rangle$. Finally, combining the results of both lower bounds, the relationship

$$C_r(FS) \geq \frac{e^{1-\frac{2}{\alpha}}}{2} \left(\frac{n}{\alpha}\right)^{2/\alpha} \left(\frac{\alpha \Gamma(n/2)}{2\Gamma(n/\alpha)}\right)^{2/n} \frac{1}{\langle r^2 \rangle \langle p^\alpha \rangle^{2/\alpha}} C_r(CR) \qquad (6.38)$$

is obtained, and similarly in the conjugated space by exchanging the involved variables. Let us finally remark an uncertainty-like inequality on the LMC complexity product $C_r(LMC) \cdot C_p(LMC) = C_{rp}(LMC)$ in terms of uncertainty products of radial expectation values of arbitrary orders. In order to obtain it, let us remember that the factors appearing in the product space complexity consist of (i) the exponential entropy on rp-space, bounded from below by means of BBM inequality, and (ii) the disequilibriums (i.e. second-order frequency moments $D_r = \omega_r(2)$ and $D_p = \omega_p(2)$, as explained in Sect. 6.1.1.4), which are known both to be bounded from below in terms of two arbitrary radial expectation values of the associated density [85]. Choosing the normalization as one radial constraint in both spaces, the resulting bound on the uncertainty LMC complexity product results in

$$C_r(LMC) \cdot C_p(LMC) \geq e^n \Gamma^2(n/2)(n+\alpha)(n+\beta) \left(\frac{n}{n+2\alpha}\right)^{1+\frac{n}{\alpha}} \times \left(\frac{n}{n+2\beta}\right)^{1+\frac{n}{\beta}} \frac{1}{\langle r^\alpha \rangle^{n/\alpha} \langle p^\beta \rangle^{n/\beta}} \tag{6.39}$$

for radial expectation values with orders $\alpha, \beta > -n/2$ in n-dimensional conjugated spaces. This result confirms the existence of a strong relationship between *complexity uncertainty* and uncertainty products in terms of radial expectation values. So, the knowledge of an uncertainty product imposes a constraint on the minimal value the LMC complexity product can reach.

6.1.3 Applications in Many-Electron Systems

The main aim of the present chapter is to analyze the above defined LMC, FS and CR complexities associated to the one-particle densities in both conjugated spaces, namely position $\rho(\mathbf{r})$ and momentum $\gamma(\mathbf{p})$ densities, as well as the product or phase-space distribution $f(\mathbf{r},\mathbf{p}) \equiv \rho(\mathbf{r})\gamma(\mathbf{p})$, for a significant number of atomic and molecular systems.

The analysis of information-theoretical properties of many-electron systems has been a major area of inquiry, studied by means of different procedures and quantities, in particular for atomic and molecular systems in both spaces. It is worthy to remark the pioneering works of Gadre et al. [96, 97] where the Shannon entropy plays a fundamental role, as well as the more recent ones concerning electronic structural complexity [40, 98], the connection between information measures (e.g. disequilibrium, Fisher information) and experimentally accessible quantities such as the ionization potentials or the static dipole polarizabilities [59], interpretation of chemical phenomena from the Shannon entropy in momentum space [99, 100], applications of the LMC complexity [49, 50] and the quantum similarity measure [61] to the study of neutral atoms, and their extension to the FS and CR complexities [64, 72] as well as to ionized systems [52, 67, 71, 101].

The applications on a global set of 370 atomic systems and 92 molecules will be carried out in order to gain insight not only on the information content of those systems, but also to interpret the complexity values in terms of physical and chemical properties. Also the associated informational planes substended by the factors composing each complexity will allow to obtain relevant interpretations on the main physical processes and characteristics of the distributions here studied.

The study on atomic systems is done in Sect. 6.2 for the neutral ones, and in Sect. 6.3 for those involved in ionization processes or belonging to one of the so-called 'isoelectronic series'. The informational analysis for the molecules here considered is contained in Sect. 6.4. In all cases, the appropriate numerical frameworks are taken into account for an accurate description, as well as the more relevant physical and chemical characteristics in order to better interpret the results obtained.

6.2 Atomic Complexity and Shell Structure

In this section, several applications on a global set of 103 neutral atoms, i.e. throughout the whole Periodic Table, are carried out in order to gain insight not only on the information content of those systems, but also on the associated informational planes substended by the factors composing each complexity. Their analysis will allow to obtain relevant interpretations on the main physical processes and characteristics of the distributions here studied.

In doing so, Near-Hartree-Fock wavefunctions [102, 103] will be employed to compute the atomic densities and the associated information measures and planes as well as complexities. The one particle density in position space $\rho(\mathbf{r})$ as well as the total wavefunction is expressed in terms of Slater-type orbitals, from which the Fourier transform provides the corresponding quantities in momentum space, including the one particle density $\gamma(\mathbf{p})$. For atomic systems in the absence of external fields (as is the case of this work) it is sufficient to deal with the spherically averaged densities $\rho(r)$ and $\gamma(p)$.

Previous complexity studies for atoms have been carried out, but most of them are only for nuclear charges $Z = 1$–54 [40, 98]. Recent complexity computations, using relativistic wave functions in the position space, were also done [104]. Some other complexity works simply take the position density, not the momentum one, as basic variable [105]. In this sense, it is worthy to point out the different behaviors displayed by some of these quantities in position and momentum spaces for atomic systems, as recently shown [4, 64].

In particular, it has been shown that it is not sufficient to study the above measures only in the usual position space, but also in the complementary momentum space, in order to have a complete description of the information theoretical internal structure and the behaviour of physical processes suffered by these systems.

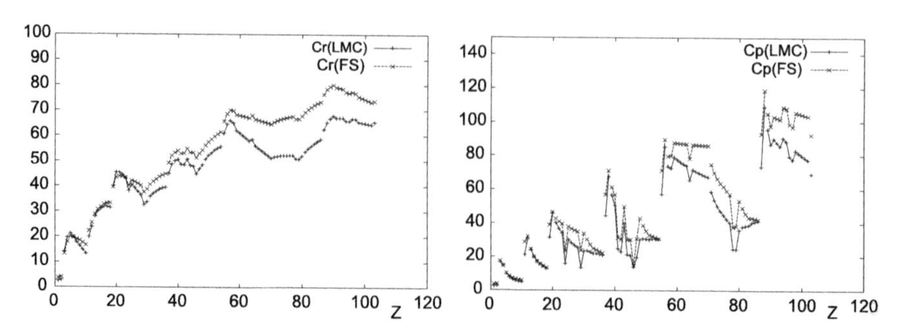

Fig. 6.1 LMC and FS complexities for neutral atoms with nuclear charge Z in position (*left*) and momentum (*right*) spaces. Atomic units (a.u.) are used

6.2.1 Comparison Between Atomic LMC and FS Complexities

First, let us compare the LMC and FS complexities for those systems, as done in Fig. 6.1 for position and momentum spaces ((a) and (b), respectively). It is remarkable, attending to the curves displayed in these figures, the similar structure of LMC and FS complexities in both spaces, in spite of their strongly different definition, mainly due to the information measure accompanying the Shannon factor, namely the 'global' disequilibrium for LMC and the 'local' Fisher information for FS. It is worthy to point out not only the almost identical magnitude orders of both complexities, but also the strong correlation between their structure, characterized by the number and location of extrema, and the shell-filling process as well as the groups the atoms belong to. Last comment is supported by the fact that both complexities in the two conjugated spaces display local minima for noble gases as well as for some atoms involved in the so-called 'anomalous shell-filling' (being specially relevant the systems $Z = 24$, 29, 46). Similar comments can be done concerning maximal values.

Attending to the factors which compose complexities, it is also interesting to analyze the individual contribution of each one to the total complexity. For illustration, the 'disequilibrium-Shannon plane' is shown in Fig. 6.2, drawn in terms of (D, L), as components of the LMC complexity, in position and momentum spaces (Figs. 6.2(a) and 6.2(b), respectively). Both figures again reveal the shell-filling patterns, much clearly in momentum than in position space. In fact, the different pieces of curves in momentum space belong to disjoint exponential entropy (L_p) values. Adding a new subshell makes L_p to increase, the disequilibrium D_p decreasing within each subshell. Opposite behaviors are displayed in position space concerning not only monotonicity, but also location of regions within the planes where heavy atoms concentrate around: high disequilibrium in position space and high disorder (entropy) in the momentum one.

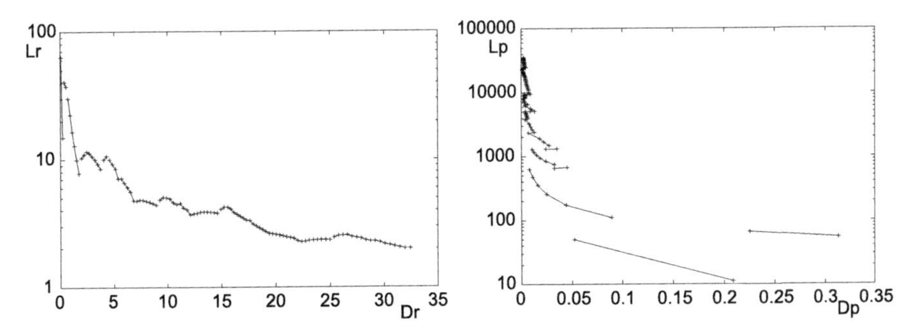

Fig. 6.2 Disequilibrium-Shannon plane (D, L) for neutral atoms with nuclear charge Z in position (*left*) and momentum (*right*) spaces. Atomic units (a.u.) are used

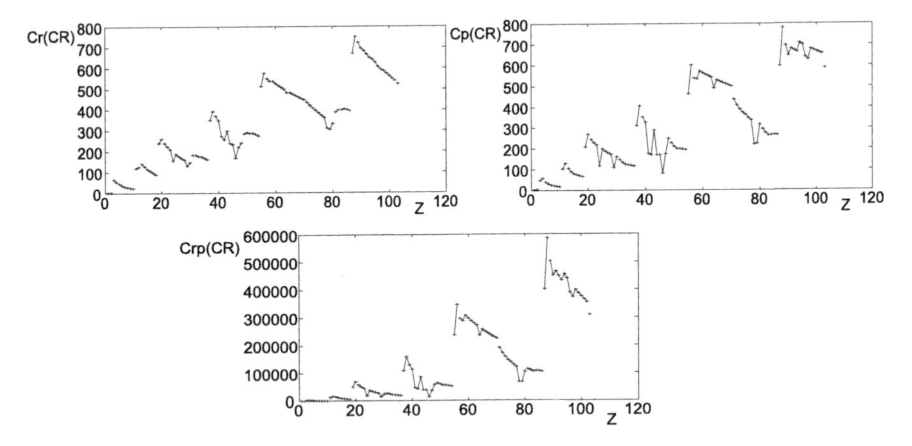

Fig. 6.3 CR complexity for neutral atoms with nuclear charge Z in position (*upper left*), momentum (*upper right*) and product (*lower*) spaces. Atomic units (a.u.) are used

6.2.2 Atomic CR Complexity

Concerning Cramér-Rao complexity $C(CR)$, main numerical results for atomic systems are displayed in Fig. 6.3 for position, momentum and product spaces.

In analyzing their structure as functions of the nuclear charge Z it is interesting to observe that most of the minima of $C_r(CR)$ and all of $C_p(CR)$ are the same of the LMC and FS complexities, previously specified. In fact, shell structure patterns are very similar for the three complexities, in spite of being determined by four quantities (S, D, I and V) of very different character. The same also occurs for some of those isolated factors in all spaces, such as e.g. the exponential entropy L and the variance V, which figures are not shown for the sake of shortness.

The Cramér-Rao (I, V) information plane is shown in Fig. 6.4 for the two conjugated spaces, in order to check to which extent each composing factor is responsible of the shell-filling pattern displayed. In position space (Fig. 6.4(a)), adding a new subshell makes Fisher information I_r to appreciably increase, its values belong-

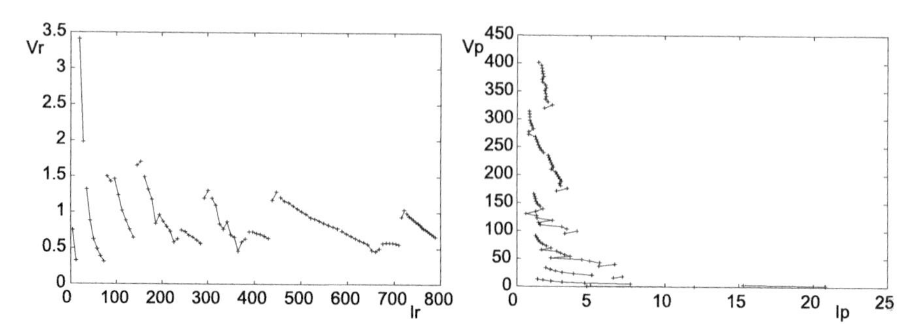

Fig. 6.4 Cramér-Rao plane (I, V) for neutral atoms with nuclear charge Z in position (*left*) and momentum (*right*) spaces. Atomic units (a.u.) are used

ing to disjoint intervals determined by the valence subshell. However, the variance V_r ranges over a unique interval for all systems without distinguishing their shell structure, but displaying a monotonically decreasing behavior (with few exceptions) within each specific subshell. Just the opposite behaviors for the corresponding momentum quantities I_p and V_p are observed in Fig. 6.4(b), in what ranges of values and monotonicity is concerned.

It is worthy to notice how the three complexity measures here considered are able to provide information not only on randomness or disorder, but also on the structure and organization of the atomic systems. The same is not always true for the individual factors, appearing relevant to deal simultaneously with the localization and randomness factors, as well as the complementary conjugated spaces, in order to have a more complete description of the information content of atomic systems.

Summarizing the results of this section, (i) a complete description of the information-theoretic characteristics of atomic systems requires the complementary use of position and momentum spaces, (ii) LMC and FS complexities provide similar results (qualitatively and quantitatively) for all neutral atoms in both spaces, displaying periodicity and shell-filling patterns as also CR complexity does, and (iii) such patterns of the localization-delocalization planes in one space are inverse to those of the conjugated space.

6.2.3 Generalized Atomic Complexities

The next purpose is to analyze numerically the Shape Rényi and Fisher-Rényi complexities of the one-particle densities in position and momentum spaces, $\rho(\mathbf{r})$ and $\gamma(\mathbf{p})$ respectively, for neutral atoms. The Shape Rényi complexity in position and momentum spaces, to be denoted by $SR_r^{(\alpha)}$ and $SR_p^{(\alpha)}$ respectively, are shown for these atomic systems in Figs. 6.5(a) (position) and 6.5(b) (momentum), for diverse values of the parameter α within the range $0.4 \leq \alpha \leq 3.6$, corresponding to the

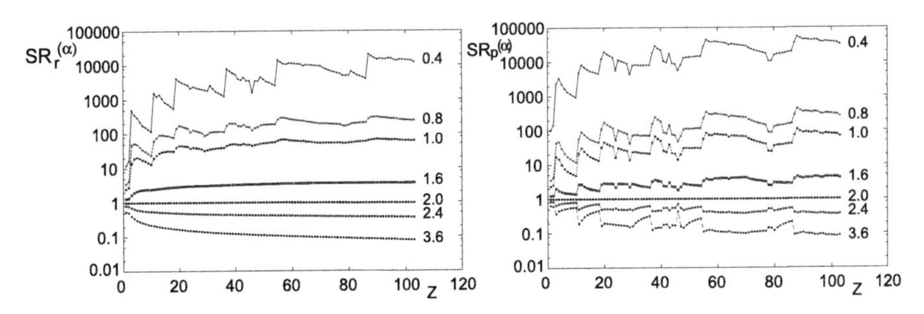

Fig. 6.5 Shape Rényi complexity $SR^{(\alpha)}$ with $\alpha = 0.4$, 0.8, 1.0, 1.6, 2.0, 2.4, 3.6 in (**a**) position space and (**b**) momentum space, for neutral atoms with $Z = 1$–103. Atomic units (a.u.) are used

different curves displayed. It is worthy to point out that for atomic systems the exponential long-range behavior of the position space density [106] allows any non-negative value $\alpha > 0$, while the momentum space one as p^{-8} [107] imposes the constraint $\alpha > 3/8 = 0.375$.

A first look at Fig. 6.5 allows to observe relevant differences between the structural characteristics of the Shape Rényi complexity $SR^{(\alpha)}$ after comparing the curves corresponding to both conjugated spaces. The position space measure $SR_r^{(\alpha)}$ (Fig. 6.5(a)) displays a much richer structure when dealing with very low values of α, reaching a higher smoothness and monotonicity as α increases. In those cases where the presence of local extrema is more apparent, a detailed analysis of their location reveals that they correspond either to closed shell systems or to atoms suffering the so called 'anomalous shell-filling'. These two characteristics depend on the occupation number of the outermost or the valence atomic subshell, where the aforementioned exponential behavior of $\rho(r)$ makes the density values to be very small as compared to those of the core region. Consequently, powering the density to a small α value enhances the contribution of the valence region, revealing the properties associated to the shell-filling process. Specially relevant is the strength for systems with 's' valence subshell as compared to other values of the angular momentum quantum number. It is additionally observed that changes of the $SR^{(\alpha)}$ in both spaces when increasing the nuclear charge (i.e. between consecutive systems) become smaller as far as considering heavier atoms, being much apparent for light ones.

The same study in momentum space (Fig. 6.5(b)) provides similar conclusions in what concerns the location of extrema and its interpretation in terms of the shell structure. The main difference when comparing to the position space curves is that such a structure is displayed independently of the α value consider, being much more apparent again for lower α's. Nevertheless, even for high α values that structure can be also observed under a much smaller scale. Again the reason for finding this behavior can be understood having in mind that the valence region is populated by low speed electrons, represented in terms of the momentum density $\gamma(p)$ by its value around the origin (i.e. close to $p = 0$). The momentum density in that region reaches high enough values in order to provide information on the valence electrons even without carrying out the enhancement operation by lowering the α parameter.

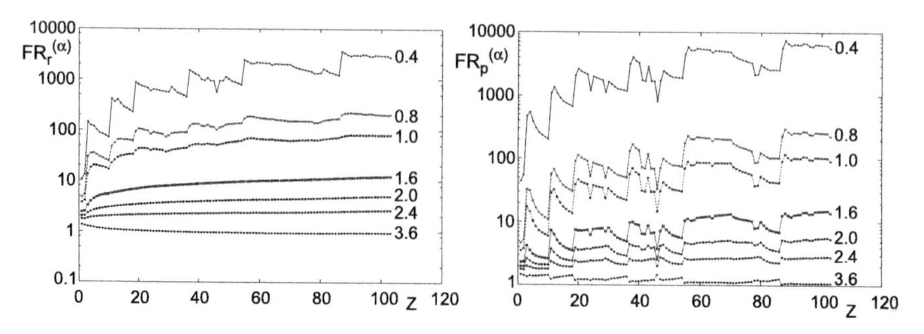

Fig. 6.6 Fisher-Rényi complexity $FR^{(\alpha)}$ with $\alpha = 0.4, 0.8, 1.0, 1.6, 2.0, 2.4, 3.6$ in (**a**) position space and (**b**) momentum space, for neutral atoms with $Z = 1$–103. Atomic units (a.u.) are used

Similar comments to those arising from the analysis of the figure corresponding to the Shape Rényi complexity $SR^{(\alpha)}$ in both conjugated spaces remain also valid for the Fisher-Rényi complexity $FR^{(\alpha)}$ as observed in the Fig. 6.6, at least in what concerns location of extrema and level of structure in each space. The Fig. 6.6 is composed similarly as the Fig. 6.5, i.e. position space (Fig. 6.6(a)) and momentum space (Fig. 6.6(b)). At this point it is worthy to remember the very different character of the factors involved as measures of order for each complexity, namely the disequilibrium and the Fisher information respectively. In spite of such a difference, the complexities themselves display a very similar structure for all the α values here considered. Nevertheless, a detailed analysis reveals the aforementioned 'local sensitivity' of the Fisher-Rényi complexity $FR^{(\alpha)}$ as compared to the Shape Rényi one $SR^{(\alpha)}$ in the magnitude of their variations for closed shells and anomalous shell-filling systems, specially in the momentum space, much less visible in the position one.

It should be pointed out the role played by the Rényi complexities SR and FR as compared to the individual factors composing them. It is well known the monotonic and structureless behavior of e.g. the disequilibrium D_r or the Fisher entropy I_r in position space [64], as also recently observed for the Rényi entropy $R_p^{(\alpha)}$ with $\alpha > 1$ [108].

The study of the Figs. 6.5 and 6.6 reveals not only the interest of considering different values of the Rényi parameter α in order to obtain a more complete information on the density structure in different atomic regions from the Rényi-like complexities, but also the usefulness of dealing simultaneously with both position and momentum spaces.

Far beyond the Shape Rényi and Fisher-Rényi atomic complexities as descriptors of the shell-filling pattern and information content, it appears also relevant the study of the contribution to the whole complexity of each of its composing factors, in order to analyse the location of all atomic systems here considered in the corresponding order-disorder plane. In this way, systems belonging to similar complexity values can be also classified attending to their disequilibrium/order on one hand, and to their uncertainty/disorder on the other.

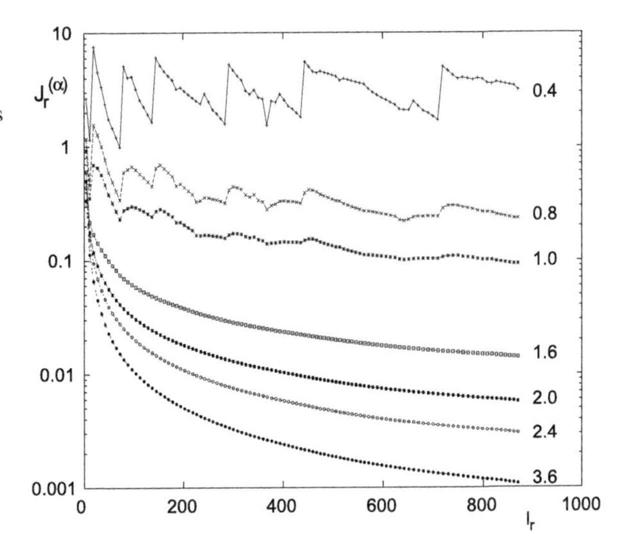

Fig. 6.7 Fisher-Rényi plane $I - J^{(\alpha)}$ in position space, with $\alpha = 0.4, 0.8, 1.0, 1.6, 2.0, 2.4, 3.6$, for neutral atoms with $Z = 1$–103. Atomic units (a.u.) are used

For illustration, the corresponding $I - J^{(\alpha)}$ and $D - L^{(\alpha)}$ planes are shown in Figs. 6.7 and 6.8, respectively, in the position space for the first case (i.e. $I_r - J_r^{(\alpha)}$ in Fig. 6.7) and in the momentum one for the other (i.e. $D_p - L_p^{(\alpha)}$ in Fig. 6.8). Similar conclusions are obtained for the other planes: for a given space, both planes look similar, the differences being mainly associated to the global and local character of the involved factors, as will be explained when discussing the Figs. 6.7 and 6.8 in detail. Nevertheless, it should be remarked that momentum space planes appear more involved than the position ones. As mentioned in the previous section, the information content of the atomic systems is mainly governed by the nuclear region in position space and by the valence subshells in the momentum one. Adding electrons to the atomic systems is a process which follows rules (shell-filling pattern) not as simple as merely increasing the nuclear charge. Such a difference is also displayed in the corresponding information planes.

Figure 6.7 displays the Fisher-Rényi plane in position space, for different values of the parameter α. The main two comments arising from the analysis of this figure are: (i) as previously observed for the position space complexities, the atomic shell structure is displayed, also in the information planes, for low α values, the curves being very smooth and almost monotonic for higher ones; the location of peaks corresponding to local extrema are associated to the characteristics of the atomic shell-filling, and (ii) all curves display a similar trend of large Fisher information and low power entropy for heavy atoms, which can be interpreted as a relevant increase of gradient at the origin as the electron cloud concentrates around the nuclear region when the nuclear charge increases, while in other regions the electron density spreads almost uniformly, increasing consequently the power entropy.

The aforementioned involvement in momentum space as a consequence of the shell-filling process is clearly observed in Fig. 6.8, where the location of the different atomic systems in the momentum $D - L^{(\alpha)}$ plane for a given value of the parameter α are displayed as a 'cloud', instead of a curve as in previous figures

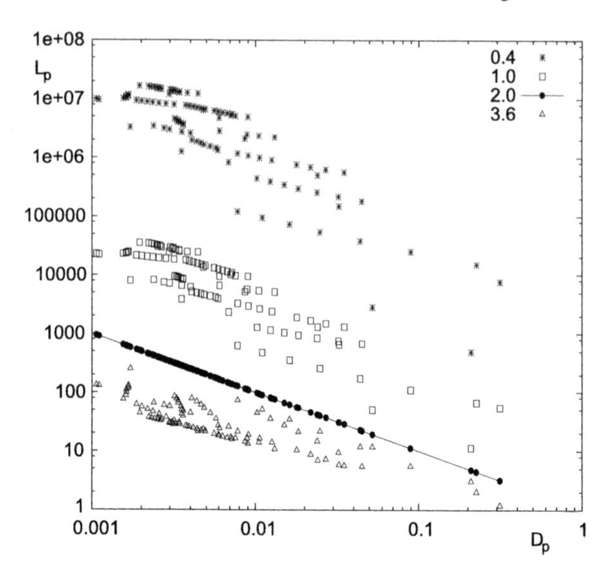

Fig. 6.8 Shape Rényi plane $D - L^{(\alpha)}$ in momentum space with $\alpha = 0.4, 1.0, 2.0, 3.6$, for neutral atoms with $Z = 1$–103. Atomic units (a.u.) are used

(apart from the trivial case $\alpha = 2$ with a constant $SR^{(\alpha)}$ product). Nevertheless, it is observed a general trend for each α value, in the sense that heavy systems concentrate around the upper-left region, corresponding to low disequilibrium and high exponential entropy (i.e. low order and high uncertainty). Additionally, the distance between consecutive systems becomes shorter as increasing their nuclear charge. In what concerns the dependence on α, it is observed that the clouds are ordered from above to below as increasing α, belonging to different bands, parallel to the unity product line.

A comparison between Figs. 6.7 and 6.8 perfectly shows the complementary character of the two conjugated spaces as well as that of the contributing individual factors to the whole complexity in both information planes. In this sense, it is worthy to remark that heavy systems are located, in the position space plane, in the lower right corner, corresponding to a high localization and a low entropy. Opposite trends, however, are observed in momentum space.

As in the complexity figures, it is also possible to distinguish the shell-filling patterns for low α in momentum space, more clearly for inner subshells (i.e. 1s, 2s, 2p). Nevertheless, the same can be also observed for additional subshells by employing an appropriate scale in the figure.

6.2.4 Bounds on Complexity

It should be expected that the higher $C(LMC)$ values are, the more far from uniformity the density is. To have an idea on the validity of this remark as well as on the comparison of complexity to unity, we show in Fig. 6.9 the values of LMC complexity of position and momentum densities, $\rho(\mathbf{r})$ and $\gamma(\mathbf{p})$ respectively, for neutral

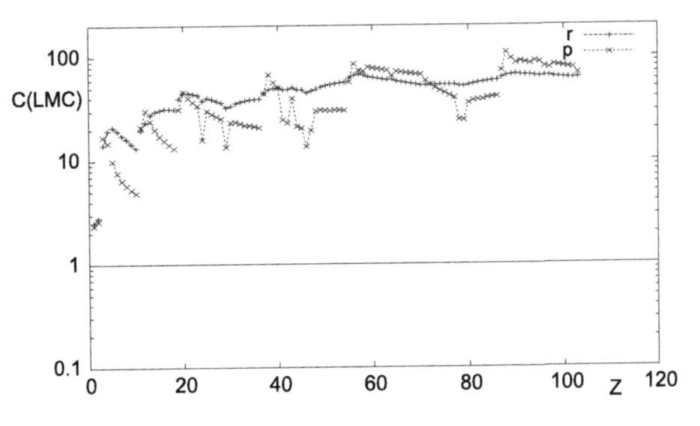

Fig. 6.9 LMC complexities $C(LMC)$ in position and momentum spaces for neutral atoms with nuclear charge from $Z=1$ to $Z=103$. Atomic units (a.u.) are used

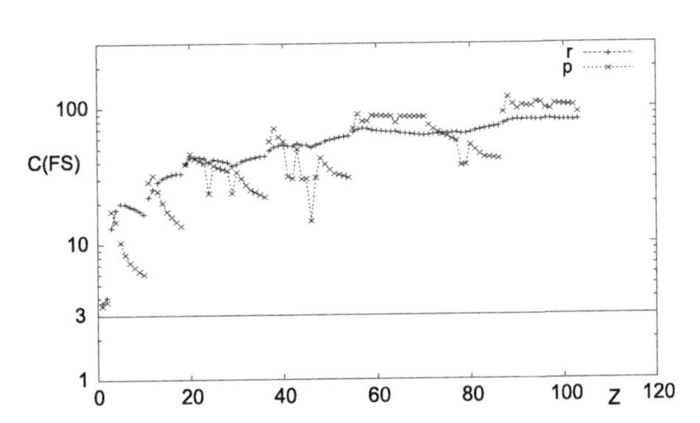

Fig. 6.10 Fisher-Shannon complexities $C(FS)$ in position and momentum spaces for neutral atoms with nuclear charge from $Z=1$ to $Z=103$. Atomic units (a.u.) are used

atoms with nuclear charge from $Z=1$ to $Z=103$. The discontinuities in the curves correspond to their decomposition according to the different periods conforming the whole Periodic Table, as also done for drawing the curves in other figures.

It is clearly observed that, apart from being all values in both spaces above unity, there appear different pieces in each curve corresponding to electronic filling of specific subshells, displaying monotonic behaviors which are opposite when comparing both conjugated spaces. This can be interpreted in terms of the uncertainty principle (which will be also analysed in next sections), in such a way that a higher delocalization in one space is associated to a higher localization in the conjugated one.

Additionally, higher values of complexity for heavy atoms are due to the *lost of uniformity* because of the increase in the level of shell structure, image with exactly corresponds to the intuitive notion of *complexity* of a system.

The numerical analysis of Fig. 6.10 for $C(FS)$ is now carried out similarly as done in Fig. 6.9 for $C(LMC)$, by considering exactly the same systems and spaces. Now, the lower bound is established by the three-dimensional ($n=3$) space as domain of the distributions. Similar comments to those of the previous figure, on the behavior in terms of the nuclear charge Z, can be also done.

In Fig. 6.11, a numerical computation of $C_p(CR)$ (momentum space) and the particular bound given by (6.31) including momentum expectation values is displayed for dimension $n=3$. It is clearly observed the similar trends followed by

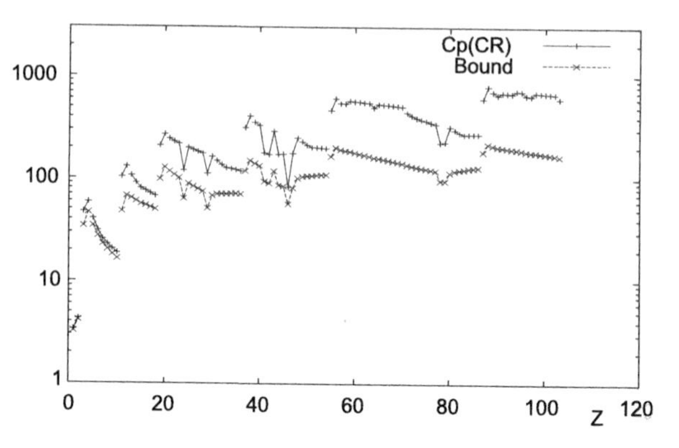

Fig. 6.11 Cramér-Rao complexity $C_p(CR)$ in momentum space, and lower bound in terms of radial expectation values, for neutral atoms with nuclear charge from $Z = 1$ to $Z = 103$. Atomic units (a.u.) are used

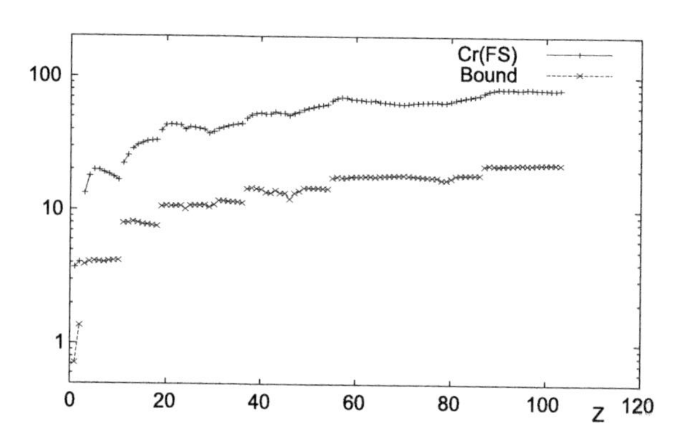

Fig. 6.12 Fisher-Shannon complexity $C_r(FS)$ in position space and lower bound in terms of Cramér-Rao complexity $C_r(CR)$, for neutral atoms with nuclear charge from $Z = 1$ to $Z = 103$. Atomic units (a.u.) are used

the exact complexity and its lower bound. Both curves display a structure strongly related to the shell-filling process. A similar figure is obtained for the corresponding quantities in position space, being consequently also valid the same comments as given above.

A numerical analysis of the inequality between FS and CR complexities is carried out (in position space) in Fig. 6.12, for the particular case $\alpha = 1$. As in previous figures, it is again observed the similar shape displayed by both the exact Fisher-Shannon complexity and its lower bound in terms of Cramér-Rao complexity and the chosen uncertainty product.

6.3 Ionization Processes and Atomic Complexity

This section is devoted to the analysis, by means of complexities, information planes and their basic ingredients, of specific kinds of ionization processes. Most usually, the study if performed by considering a neutral atom and by analyzing the evolution of the aforementioned quantities after the ionization of the initial system. Such an ionization can be performed in two different ways: (i) by adding or removing an electron of the neutral system while keeping fixed the nuclear charge, what gives rise to a singly-charged ion (Sect. 6.3.1), or (ii) by modifying the nuclear charge keeping fixed the number of electrons, the resulting system belonging, consequently, to the same isoelectronic series of the neutral atom (Sect. 6.3.2). Different behaviors will be discussed attending to the aforementioned ionization processes.

6.3.1 *Singly Charged Ions*

In this section the LMC, FS and CR complexities are analyzed for singly charged ions with a number of electrons up to $N = 54$, that is with a global charge $Q = Z - N = \pm 1$, Z being the nuclear charge. These quantities, together with the previously discussed values for neutral atoms within such N range, provide us with information on how complexity progresses in mono-ionization processes [67, 71]. In doing, we are considering a global of 150 systems (53 cations, 43 anions and 54 neutral atoms), the computations being performed by employing the accurate wavefunctions of [102].

6.3.1.1 LMC and FS Complexities of Singly-Charged Ions

A similar comparison between LMC and FS complexities as done previously for neutral atoms in both conjugated spaces has been also carried out for anions and cations in the two spaces. Conclusions raised by the analysis of these quantities for ions are almost identical to those provided when discussing the Fig. 6.1 for neutral atoms, in what concerns similarity between $C(LMC)$ and $C(FS)$ values as well as their connection with the shell-filling process by means of the location of their extrema, most minima of complexity corresponding to noble gases or the anomalous shell-filling set of atoms.

6.3.1.2 CR Complexity of Singly-Charged Ions and Neutral Atoms

Concerning the Cramér-Rao complexity C(CR), its evolution throughout the ionization is clearly displayed in Fig. 6.13, where its value is provided for the three considered species (anions, cations and neutrals) in order to determine to which extent the ionization processes (by adding or removing electrons keeping fixed the nuclear charge Z) modify the atomic complexity. For illustration, this comparison is carried out for the Cramér-Rao complexity $C_{rp}(CR)$ in the product space as shown in Fig. 6.13. Again, it is clearly observed the correlation of complexity with the atomic shell structure for all species. Additionally, it is appreciated that (i) complexity increases as the system loses an electron, and (ii) maxima are clearly associated to 's' valence subshells (those involved in ionization) while minima correspond to noble gases or some anomalous 'd' subshells filling.

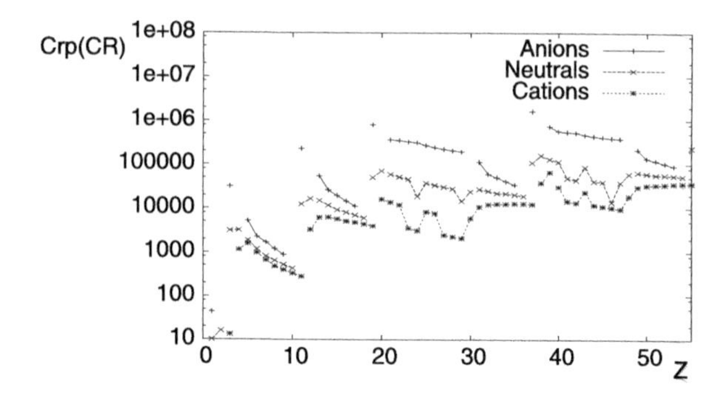

Fig. 6.13 CR complexity in product space for neutral atoms and singly charged ions with nuclear charge Z. Atomic units (a.u.) are used

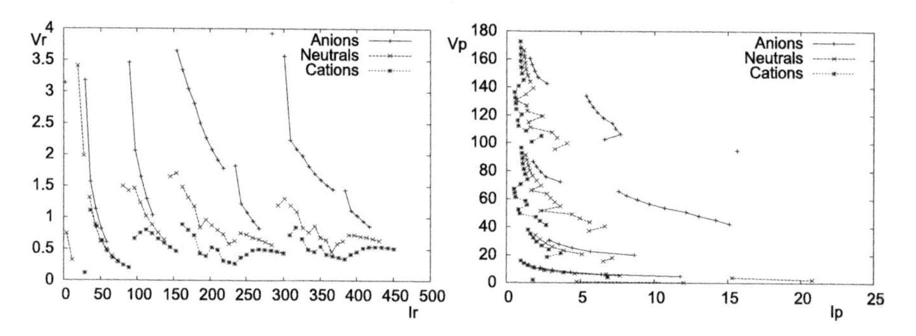

Fig. 6.14 Cramér-Rao plane (I, V) in position (*left*) and momentum (*right*) spaces, for neutral atoms and singly charged ions with nuclear charge Z. Atomic units (a.u.) are used

The Cramér-Rao informational plane subtended by the constituent factors (I, V) also provides interesting results interpreted according to the atomic shell structure. Figure 6.14 displays this plane in both conjugated spaces ((a) for position and (b) for momentum) for the systems here considered. Apart from the faithful reproduction of shell structure, it is worthy to remark that, as shown in Fig. 6.14(a), systems of large Z are highly localized and organized in position space while the light ones appear much more delocalized. Location at the position (I_r, V_r) plane after an ionization process slightly changes for heavy atoms as compared to the lighter ones. Additionally, for fixed nuclear charge Z complexity $C_r(CR)$ decreases following the sequence anion-neutral-cation, that is as losing electrons, being the changes associated to 's' electrons considerably higher to those of 'p' or 'd' subshells.

Exactly the opposite trends to those discussed in position space are observed in the momentum one, as shown in Fig. 6.14(b): large Z systems are now less localized and with a greater variance than the light ones, and losing electrons makes the variance to increase and Fisher information to decrease, just the reciprocal that happens in position space.

6.3.2 *Isoelectronic Series: Dependence of Complexity on the Nuclear Charge*

After carrying out the analysis of complexity dependence on the outermost subshells, as done in the previous section by considering ionization processes, let us now focus in the atomic core as source of the attractive forces and their effects on complexity values.

6.3.2.1 Composition and Number of Isoelectronic Series

We start by considering a neutral atom, that is a system with identical values of the nuclear charge Z and the number of electrons N, from which we give rise to a set of cations by progressively increasing one-by-one the nuclear charge Z keeping fixed the number of electrons (or, equivalently, starting from a global charge $Q \equiv Z - N = 0$ until reaching a maximum positive value, being $Q_{\max} = 20$ in the numerical application here considered). Such a set of cations together with the neutral atom is known as an 'isoelectronic series', characterized by the fixed number of electrons N as well as the maximum value $Q_{\max}$. Studying the previously considered complexity measures for a given isoelectronic series provides information on their dependence on the nuclear charge Z for fixed N electrons. In this section, such a study will be carried out for nine isoelectronic series, namely those corresponding to $N = 2$–10, within a Hartree-Fock framework [109]. Each series consists of 21 members (a neutral atom and 20 cations), giving rise consequently to analyze complexities of a global of 189 atomic systems.

6.3.2.2 LMC Complexity and Information Plane

In Fig. 6.15 the disequilibrium-Shannon plane (D, L) is shown in position, momentum and product spaces (Figs. 6.15(a), 6.15(b) and 6.15(c), respectively) for the isoelectronic series $N = 2$–10. For the individual spaces (position and momentum), each series roughly follows a linear trajectory in a double logarithmic scale. In fact, the Helium series ($N = 2$) displays an almost constant $C(LMC) = D \cdot L$ line in both spaces, what means that increasing the nuclear charge produces, as should be expected, a higher localization D and a lower uncertainty, both effects compensating each other proportionally and providing an almost constant product which defines LMC complexity. Concerning product space, the corresponding Disequilibrium-Shannon plane (D, L) is shown in Fig. 6.15(c). It is worthy to notice the strong changes in the slopes of all series as compared to those of the isolated spaces. While product entropy does not suffer drastic changes, localization appears very different within each series. Additionally, shell-filling patterns are clearly displayed, with systems having 2s as valence subshell having a higher complexity than those characterized by the 2p one. It is also remarkable that the $N = 2$ series display a very different behavior as compared to the other series. This can be interpreted by taking

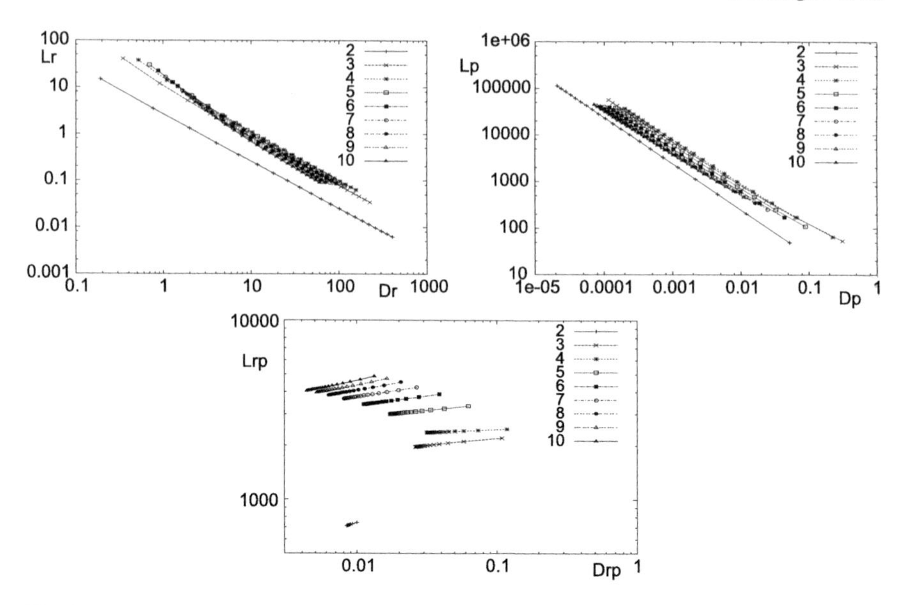

Fig. 6.15 Disequilibrium-Shannon plane (D, L) in position (*upper left*), momentum (*upper right*) and product (*lower*) spaces, for isoelectronic series with $N = 2$–10 electrons. Atomic units are used

into account that those systems are the unique ones here considered consisting only on a core shell. From all these comments it should be concluded that the product space plane is relevant in order to obtain an interpretation of the Disequilibrium-Shannon plane values in terms of shell structure.

In position space, systems with large nuclear charge Z for any isoelectronic series display a highly localized structure (large D) as shown in Fig. 6.15(a). In such a large D area, trajectories are almost linear which correspond to an almost constant product measure. Deviations from this linear shapes are better observed for low nuclear charge systems, possessing a greater complexity. Biggest position space complexities correspond to neutral systems, with a relatively lower localization and greater uncertainty as compared to its cations. All those comments are just the opposite ones in momentum spaces, as can be readily realized by observing Fig. 6.15(b). Heavy systems are characterized by a low localization and high entropy in momentum space, and neutrals deviate from isoproduct lines as possessing a higher level of structure. It is worthy to remark also that spacing between consecutive systems within each isoelectronic series decrease as increasing Z, because of a higher similarity between systems with large nuclear charge as compare to those with low Z, which progressively separate among themselves.

6.3.2.3 FS Complexity and Information Plane

A similar analysis has been also carried out for the Fisher-Shannon plane (I, J) in position, momentum and product spaces (Fig. 6.16). It is worthy to remember the

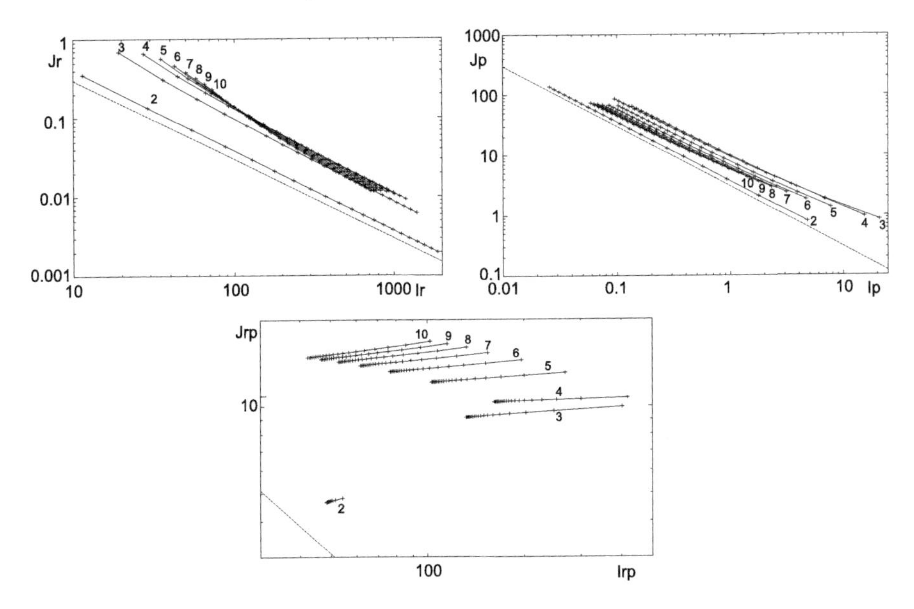

Fig. 6.16 Fisher-Shannon plane (I, J) in position (*upper left*), momentum (*upper right*) and product (*lower*) spaces, for isoelectronic series with $N = 2$–10 electrons. Atomic units are used

rigorous lower bound to the associated FS complexity $C(FS) = I \cdot J \geq$ constant (the constant being 3 for the conjugated spaces and $18\pi e$ for the product space) in order to verify such a bound for the systems here considered. The straight line $I \cdot J =$ constant drawn in the plane by using a double logarithmic scale divides it into an 'allowed' (upper) and a 'forbidden' (lower) parts. Parallel lines to that one represent isocomplexity points, and higher deviations from this frontier are associated to greater FS complexities. Such a parallel shape is roughly displayed by all isoelectronic series in both position and momentum spaces, as shown respectively in Figs. 6.16(a) and 6.16(b). Similar comments to those provided on discussing Fig. 6.15 in what concerns location areas of systems at the plane, distances within a series between consecutive systems, deviation from minimal complexity values and opposite behaviors in conjugated spaces are also valid for the position and momentum (I, J) planes as concluded by analyzing Figs. 6.16(a) and 6.16(b).

For the sake of brevity, results on the Cramér-Rao plane (I, V) are not displayed, but conclusions obtained from their values are the same as those just discussed for disequilibrium-Shannon and Fisher-Shannon planes.

6.4 Molecular Complexities and Chemical Properties

With the purpose of organizing and characterizing the complexity features of the molecular systems under study, several reactivity properties have been computed [73], such as the ionization potential (IP), the total dipole moment (μ), the hardness (η) and the electrophilicity index (ω). The ionization potential was obtained by use of the Koopmans theorem [110, 111] by calculating the HOMO and LUMO orbital energies at the B3LYP/6-311++G(3df, 2p) level of theory. The hardness η was obtained within the conceptual DFT framework [82, 112], through

$$\eta = \frac{1}{2S} = \frac{\varepsilon_{LUMO} - \varepsilon_{HOMO}}{2} \tag{6.40}$$

where ε denotes the frontier molecular orbital energies and 'S' the softness of the system. In general terms, hardness and softness are good descriptors of chemical reactivity, the former measures the global stability of the molecule (larger values of η means less reactive molecules), whereas the S index quantifies the polarizability of the molecule [113–116], thus soft molecules are more polarizable and possess predisposition to acquire additional electronic charge [117]. The chemical hardness η is a central quantity for use in the study of reactivity and stability, through the hard and soft acids and bases principle [118–120].

The electrophilicity index [121], ω, allows a quantitative classification of the global electrophilic nature of a molecule within a relative scale. Electrophilicity index of a system in terms of its chemical potential and hardness is given by the expression

$$\omega = \frac{\mu^2}{2\eta}. \tag{6.41}$$

The electrophilicity is also a good descriptor of chemical reactivity, which quantifies the global electrophilic power of the molecules [117].

The electronic structure calculations performed in the present study, for the whole set of molecules, were obtained with the Gaussian 03 suite of programs [122] at the CISD/6-311++G(3df, 2p) level of theory. The molecular information quantities (S, D, I) in both conjugated spaces were obtained by employing advanced software together with 3D numerical integration routines [123, 124] and the DGRID suite of programs [125]. All quantities calculated are given in atomic units throughout this work.

The molecular set chosen for the study includes different types of chemical organic and inorganic systems (aliphatic compounds, hydrocarbons, aromatic, alcohols, ethers, ketones). The set represents a variety of closed shell systems, radicals, isomers as well as molecules with heavy atoms such as sulphur, chlorine, magnesium and phosphorous. The geometries needed for the single points calculations above referred were obtained from standard databases [126]. The molecular set might be organized by isoelectronic groups as follows ('N' stands for the total number of electrons):

$N = 10$ NH_3 (ammonia)
$N = 12$ LiOH (lithium hydroxide)
$N = 14$ HBO (boron hydride oxide), Li_2O (dilithium oxide)
$N = 15$ HCO (formyl radical), NO (nitric oxide)
$N = 16$ H_2CO (formaldehyde), NHO (nitrosyl hydride), O_2 (oxygen)
$N = 17$ CH_3O (methoxy radical)
$N = 18$ CH_3NH_2 (methyl amine), CH_3OH (methyl alcohol), H_2O_2 (hydrogen peroxide), NH_2OH (hydroxylamine)

$N = 20$ NaOH (sodium hydroxide)
$N = 21$ BO_2 (boron dioxide), C_3H_3 (radical propargyl), MgOH (magnesium hydroxide), HCCO (ketenyl radical)
$N = 22$ C_3H_4 (cyclopropene), CH_2CCH_2 (allene), CH_3CCH (propyne), CH_2NN (diazomethane), CH_2CO (ketene), CH_3CN (acetonitrile), CH_3NC (methyl isocyanide), CO_2 (carbon dioxide), FCN (cyanogen fluoride), HBS (hydrogen boron sulfide), HCCOH (ethynol), HCNO (fulminic acid), HN_3 (hydrogen azide), HNCO (isocyanic acid), HOCN (cyanic acid), N_2O (nitrous oxide), NH_2CN (cyanamide)
$N = 23$ NO_2 (nitrogen dioxide), NS (mononitrogen monosulfide), PO (phosphorus monoxide)
$N = 24$ C_2H_4O (ethylene oxide), C_2H_5N (aziridine), C_3H_5 (allyl radical), C_3H_6 (cyclopropane), CF_2 (difluoromethylene), CH_2O_2 (dioxirane), CH_3CHO (acetaldehyde), CH_3CO (acetyl radical), $CHONH_2$ (formamide), FNO (nitrosyl fluoride), H_2CS (thioformaldehyde), HCOOH (formic acid), HNO_2 (nitrous acid) $NHCHNH_2$ (aminomethanimine), O_3 (ozone), SO (sulfur monoxide)
$N = 25$ $CH_2CH_2CH_3$ (npropyl radical), CH_3CHCH_3 (isopropyl radical), CH_3OO (methylperoxy radical), FO_2 (dioxygen monofluoride), NF_2 (difluoroamino radical), CH_3CHOH (ethoxy radical)
$N = 26$ C_3H_8 (propane), $CH_3CH_2NH_2$ (ethylamine), CH_3CH_2OH (ethanol), CH_3NHCH_3 (dimethylamine), CH_3OCH_3 (dimethyl ether), CH_3OOH (methyl peroxide), F_2O (difluorine monoxide)
$N = 27$ CH_3S (thiomethoxy)
$N = 30$ ClCN (chlorocyanogen), OCS (carbonyl sulfide), SiO_2 (silicon dioxide)
$N = 31$ PO_2 (phosphorus dioxide), PS (phosphorus sulfide)
$N = 32$ ClNO (nitrosyl chloride), S_2 (sulfur diatomic), SO_2 (sulfur dioxide)
$N = 33$ OClO (chlorine dioxide)
$N = 34$ CH_3CH_2SH (ethanethiol), CH_3SCH_3 (dimethyl sulfide), ClO_2 (chlorine dioxide), H_2S_2 (hydrogen sulfide), SF_2 (sulfur difluoride)
$N = 38$ CS_2 (carbon disulfide)
$N = 40$ CCl_2 (dichloromethylene), SO (sulfur monoxide)
$N = 46$ $MgCl_2$ (magnesium dichloride)
$N = 48$ S_3 (sulfur trimer), $SiCl_2$ (dichlorosilylene)
$N = 49$ ClS_2 (sulfur chloride)

For this set of molecules we have calculated different information and complexity measures, i.e. D, L, S (Shannon entropy), I, $C(LMC)$, $C(FS)$, in each of the conjugated spaces as well as in the product space.

In contrast with the atomic case, where the complexities possess a high level of natural organization provided by periodical properties [4, 64, 71, 72] the molecular case requires some sort of organization which could be affected by many factors (structural, energetic, entropic, etc.). So that several types of molecular systems, ordered according to the main chemical properties of interest, i.e., the total energy, the dipole moment, the ionization potential, the hardness and the electrophilicity, are here analyzed by establishing a chemical interpretation of the different complexity

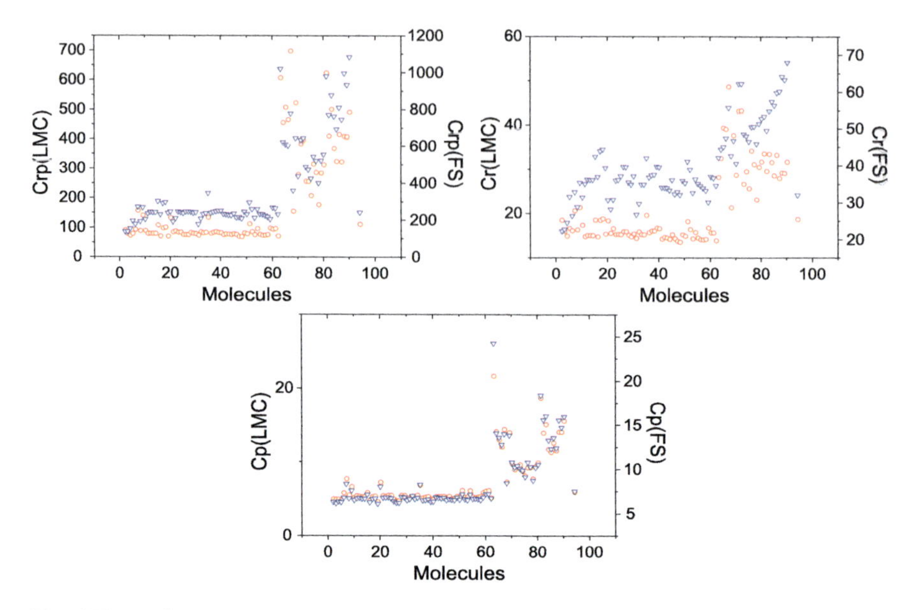

Fig. 6.17 LMC (*red circles*) and FS (*blue triangles*) complexities in product (*upper left*), position (*upper right*) and momentum (*lower*) spaces for the set of molecules 1 to 92 ordered as their energy increase. Atomic units (a.u.) are used

measures. From a numerical point of view, a common feature to be noticed in all cases studied below is that there exists a strong structural similarity between the $C_{rp}(LMC)$ and $C_{rp}(FS)$ measures.

6.4.1 Complexity and Molecular Energy

In Fig. 6.17(a) the $C_{rp}(FS)$ and $C_{rp}(LMC)$ values for all molecules are characterized by an increasing energetic behavior, i.e., molecules to the left side of the figure possess lower energies whereas molecules at the right side correspond with higher molecular energies. It may be observed that both complexity measures in the product space (rp) possess a similar behavior, i.e., both indicating a clear region of lower complexity (molecules with lower energies) and a region of higher complexity (molecules with higher energies). Then, we can establish for this particular set of molecules that the total energy governs the molecular complexity behavior in a simple manner.

In order to analyze the contribution of each conjugated space to the complexity measures LMC and FS previously discussed, the $C(LMC)$ and $C(FS)$ measures in position and momentum spaces are displayed in Figs. 6.17(b) and 6.17(c), respectively. The most remarkable feature that one may observe from these figures is the close resemblance between the two complexity measures in momentum space (Fig. 6.17(c)), whereas in position space the $C(FS)$ measure shows more complicated patterns of uncertainty (J) and organization (I). In contrast, the $C_r(LMC)$

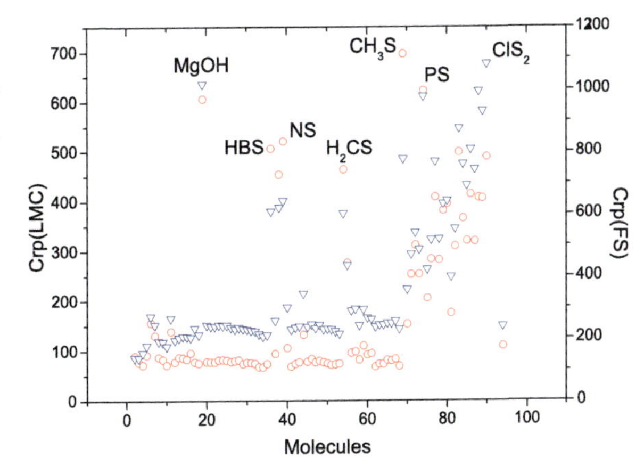

Fig. 6.18 LMC (*red circles*) and FS (*blue triangles*) complexities in product space for the set of molecules 1 to 92 ordered as their number of electrons increase. Atomic units (a.u.) are used

measure seems to be less sensitive to energetic changes, mainly in the region of lower energy molecules.

6.4.2 *Molecular Complexity and Number of Electrons*

In Fig. 6.18 we have plotted the complexity values for the $C_{rp}(LMC)$ and $C_{rp}(FS)$ measures constrained by the number of electrons (molecules are ordered from left to right according with their number of electrons in an increasing manner). We may observe from Fig. 6.18 that both complexity measures behave in a similar fashion, i.e. molecules with low number of electrons ($N < 26$) possess low complexities whereas molecules with larger number of electrons ($N > 26$) possess larger complexity values. A few exceptions may be noticed from Fig. 6.18, for molecules with low number of electrons and higher complexities which correspond to molecules containing heavier atoms (phosphorous, magnesium or sulphur); then, establishing that complexity also depends on the atomic number of the atom forming part of the molecular systems. This general observation will be analyzed below.

6.4.3 *Molecular Complexity and Hardness*

In Fig. 6.19 we have plotted the complexity values for the set of molecules constrained by increasing hardness values (molecules to the left posses lower hardness values and higher to the right). The main observation is again that both complexities behave in the same way, both indicating a clear relationship with the hardness and hence with the chemical reactivity of the molecules. We mentioned above that η is a central quantity for the study of reactivity and global molecular stability through the hard and soft acids and bases principle. Therefore we can observe from

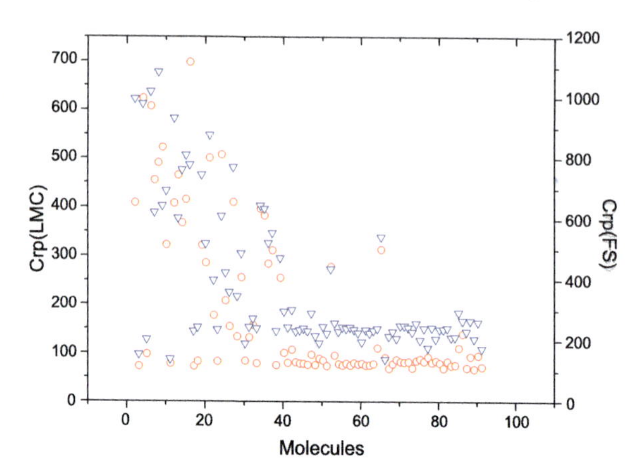

Fig. 6.19 LMC (*red circles*) and FS (*blue triangles*) complexities in product space for the set of molecules 1 to 92 ordered as their hardness increase. Atomic units (a.u.) are used

Fig. 6.19 that molecules that are more stable chemically (global stability) possess low complexity values, i.e. chemical reactivity seems to be directly related to the complexities in that higher values correspond with less stable molecules, with very few exceptions which correspond to molecules with heavier atoms as we mentioned before.

6.4.4 Molecular Complexity and Ionization Potential

Another way of assessing the chemical stability of the molecules in relation with their complexities may be analyzed through the ionization potential (IP) values and this might be observed from Fig. 6.20 where molecules with high IP values (more stable molecules) are located to the right of the figure. We note that stability is related with the molecular complexities in that higher complexity values correspond with more reactive molecules.

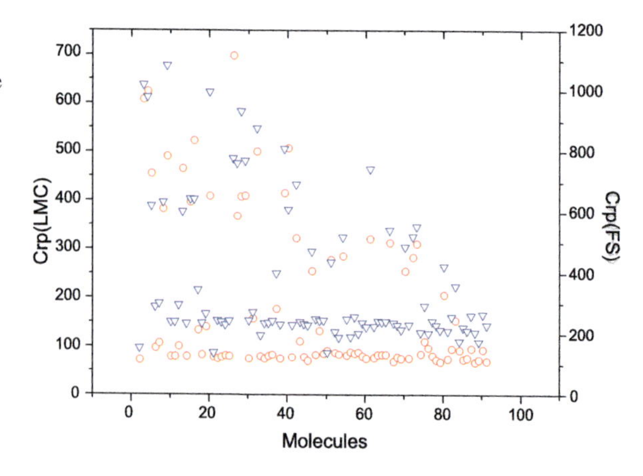

Fig. 6.20 LMC (*red circles*) and FS (*blue triangles*) complexities in product space for the set of molecules 1 to 92 ordered as their ionization potential increase. Atomic units (a.u.) are used

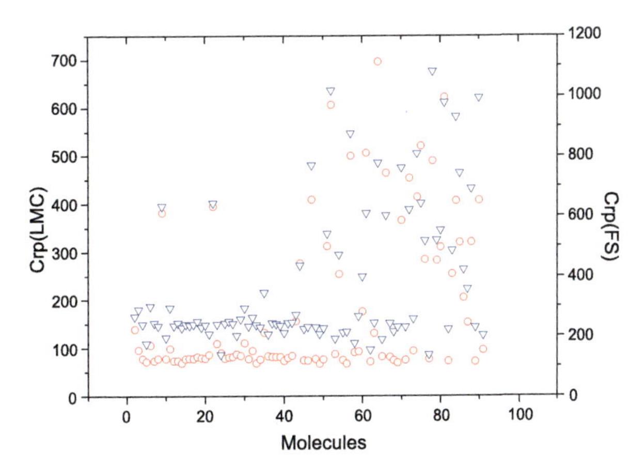

Fig. 6.21 LMC (*red circles*) and FS (*blue triangles*) complexities in product space for the set of molecules 1 to 92 ordered as their electrophilicity increase. Atomic units (a.u.) are used

6.4.5 *Molecular Complexity and Electrophilicity*

The electrophilicity index is other useful indicator of chemical reactivity which quantifies the global electrophilic power of the molecules (predisposition to acquire an additional electronic charge). In Fig. 6.21 we have plotted the complexity values in the product space for the set of molecules constrained by increasing electrophilicity values. We may observe that molecules with higher capacity to acquire molecular charge (less stable) possess higher complexity values.

6.4.6 *Information Planes in Molecules*

In the search of more pattern and organization we have found useful to impose one more constrain to the molecules through isoelectronic common features and so it is interesting to analyze the contribution of each one of the information measures D and L to the total LMC complexity. This is done in Figs. 6.22(a) and 6.22(b) in their respective conjugated spaces through the information plane (D-L) for some of the isoelectronic molecular series with $N = 22, 24, 25, 26$ electrons. Figure 6.22 depicts the complexity measures in r and p spaces in a double-logarithmic scale which show a division in the D-L plane into two regions. The left area is the forbidden region by inequality $D \cdot L \geq 1$ [84], and parallel lines to it represent isocomplexity lines showing that an increase (decrease) in uncertainty, L, along them is compensated by a proportional decrease (increase) of disequilibrium (order), and higher deviations from this frontier are associated to greater LMC complexities. Each isoelectronic series follows a trajectory in the D-L plane that can be easily analyzed.

For instance, the isoelectronic series corresponding to 26 electrons (green circles in both figures) shows an almost constant line in both spaces, indicating that the effect of increasing the energy (towards the left and upper region of the Fig. 6.22(a) in r-space and the right and lower region of the Fig. 6.22(b) in p-space) produces more uncertainty L and consequently less order (disequilibrium D). On the other

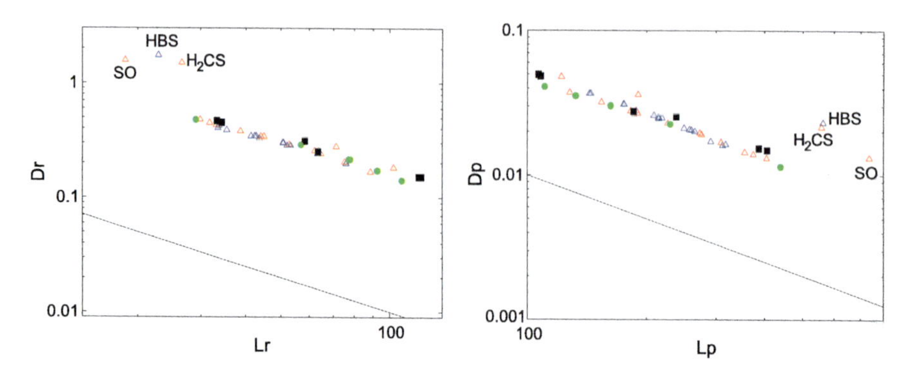

Fig. 6.22 Disequilibrium-Shannon plane (D, L) for isoelectronic series of 22 (*blue triangles*), 24 (*red triangles*), 25 (*black boxes*) and 26 (*green circles*) electrons, in position (*left*) and momentum (*right*) spaces. Double logarithmic scale. Lower bound $(DL = 1)$ is depicted by the *dashed line*. Atomic units (a.u.) are used

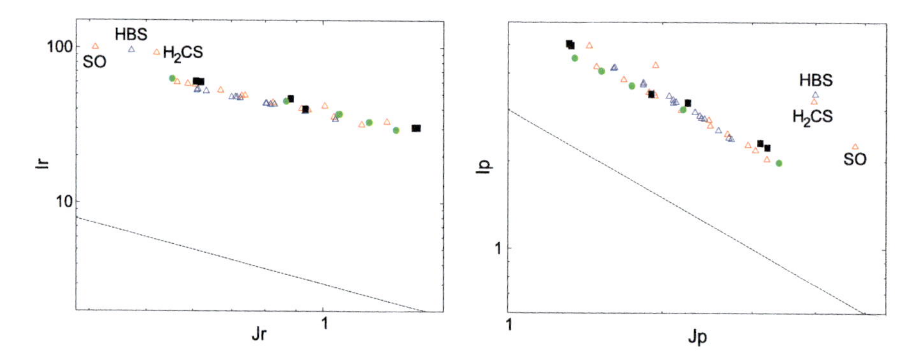

Fig. 6.23 Fisher-Shannon plane (I, J) for isoelectronic series of 22 (*blue triangles*), 24 (*red triangles*), 25 (*black boxes*) and 26 (*green circles*) electrons, in position (*left*) and momentum (*right*) spaces. Double logarithmic scale. Lower bound $(DL = 1)$ is depicted by the *dashed line*. Atomic units (a.u.) are used

hand, systems that are not in the isocomplexity lines belong to molecules with higher complexity values which possess heavier atoms, as we have discussed above. This behavior means that in position space, the higher complexity is due to higher disequilibrium values whereas in momentum space it is due to higher uncertainty values. It is also worthy to note that all isocomplexity lines representing the isoelectronic molecular series with $N = 22, 24, 25, 26$ electrons show large deviations (higher LMC complexities) from the rigorous lower bound as it may be observed from Fig. 6.22 in both conjugated spaces.

Continuing with the analysis of the pattern and organization of isoelectronic series we have analyzed the contribution of each one of the information measures I and J to the total FS complexity. This is done in Figs. 6.23(a) and 6.23(b) in their respective conjugated spaces through the information plane $(I\text{-}J)$ for some of the isoelectronic molecular series with $N = 22, 24, 25, 26$ electrons.

Figure 6.23 shows the results in r and p spaces which indicate a division of the I-J plane into two regions. At this point it is worth mentioning that the rigorous lower bound to the associated FS complexity $C(FS) = I \cdot J \geq$ constant (the constant being 3 for the conjugated spaces and $18\pi e$ for the product space) is verified for the systems here considered. The straight line $I \cdot J =$ constant drawn in the plane by using a double logarithmic scale divides it into an *allowed* (upper) and a *forbidden* (lower) parts. Parallel lines to that one represent isocomplexity points, and higher deviations from this frontier are associated to greater FS complexities. Over these lines an increase (decrease) in uncertainty (J) gets balanced by a proportional decrease (increase) of accuracy (I). Such a parallel shape is displayed by all isoelectronic series in both position and momentum spaces, as shown in Figs. 6.23(a) and 6.23(b) respectively.

Each isoelectronic series in Fig. 6.23 follows a trajectory in the I-J plane that can be easily analyzed. For instance, the isoelectronic series corresponding to 26 electrons (green circles in both figures) shows an almost constant line in both spaces, showing that the effect of increasing the energy (towards the left and upper region of the Fig. 6.23(a) in r-space and the right and lower region of the Fig. 6.23(b), in p-space, produces more uncertainty J and consequently less organization (accuracy I). On the other hand, systems that are not in the isocomplexity lines belong to higher complexity molecules as we have previously discussed, which possess heavier atoms as it may be observed from Fig. 6.23. It is also worth noting that all isocomplexity lines representing the isoelectronic molecular series with $N = 22$, 24, 25, 26 electrons show large deviations (higher FS complexities) from the rigorous lower bound ($I \cdot J = 3$) as it may be observed from Fig. 6.23 in both conjugated spaces.

In spite of the fact that not all information products are good candidates to form complexity measures that preserve the desirable properties of invariance under scaling, translation and replication, it appears interesting to analyze other planes which might be useful to analyze patterns of order-organization although the product fails to be invariant under scale transformation. Thus, in Fig. 6.24 we have plotted the (I, D) planes in their respective conjugated spaces for some of the isoelectronic molecular series with $N = 22$, 24, 25, 26 electrons. The results in r and p spaces show an interesting linear behavior with a positive slope for all isoelectronic molecular series, meaning that as the molecular order increases (higher D) their organization also increases (higher I). Interestingly, Fig. 6.24 shows that these planes are not only useful to detect molecular patterns of order-organization but also molecules of higher complexity (SO, HBS, H_2CS) which do not obey the above mentioned linear behavior. According to the results shown in these figures, molecules with heavier atoms possess more complexity patterns that do not fit with the simple description of order-organization. It is also interesting to note that whereas for these molecules the latter is true in position space (Fig. 6.24(a)), in momentum space the linear pattern of order-organization seems to be obeyed.

Beyond the practical use of employing information planes to study patterns of uncertainty-order (L-D), uncertainty-organization (J-I) or order-organization (D-I), it is chemically interesting to investigate the role of the chemical structure

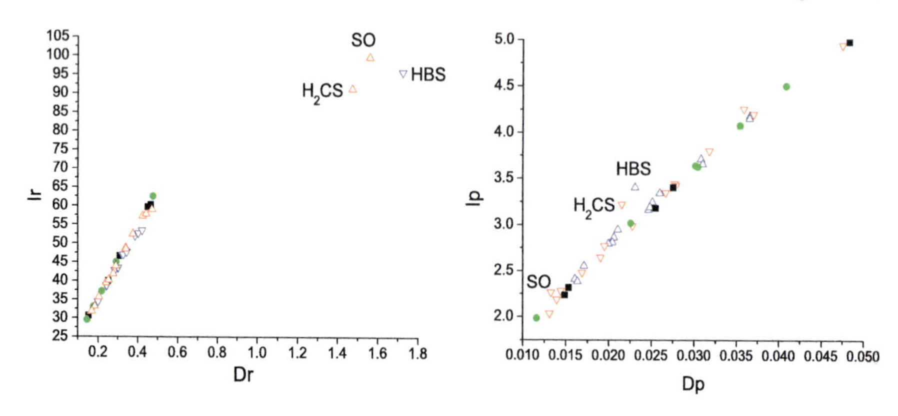

Fig. 6.24 Fisher-Disequilibrium plane (I, D) of the isoelectronic series of 22 (*blue triangles*), 24 (*red triangles*), 25 (*black boxes*) and 26 (*green circles*) electrons, in position (*left*) and momentum (*right*) spaces. Atomic units (a.u.) are used

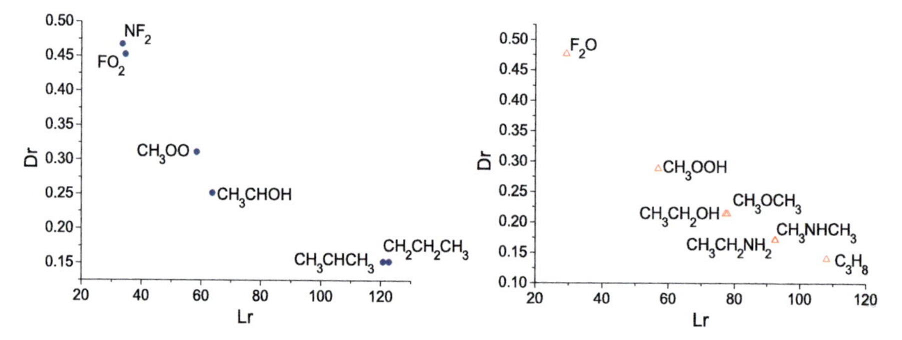

Fig. 6.25 Fisher-Disequilibrium plane (I, D) for the isoelectronic series of 25 (*left*) and 26 (*right*) electrons, in position space. Atomic units (a.u.) are used

of the analyzed molecules in terms of the different information planes and their corresponding interpretations. For instance, by representing the (D, L) plane in position space for the isoelectronic series of 25 and 26 electrons, we may analyze how the different chemical structures are linked to order and uncertainty in this case. Figure 6.25 shows such information for the 25 and 26 electrons series, and we may note that molecules that are much more ordered (less uncertainty) possess chemical structures that include heavier atoms. It so happens also that this order (increasing D_r) corresponds to an increase in energy. We may note also that isomeric molecules in both series possess the same information content in terms of order-uncertainty which provides a new chemical perspective on the relation between molecular energetic content and the order (disorder) of the corresponding chemical structures,

Finally, we have found useful to analyze the particular case of three isoelectronic isomers: HCNO (fulminic acid), HNCO (isocyanic acid) and HOCN (cyanic acid) in order to analyze their chemical properties with respect to their complexity values in product space. From an experimental side it is known that cyanic and isocyanic acids

Table 6.1 Chemical properties for the isomers HCNO, HNCO and HOCN in atomic units (a.u.)

Molecule	Name	Energy	Ionization Potential	Hardness	Electrophilicity
HCNO	Fulminic acid	−168.134	0.403	0.294	0.020
HNCO	Isocyanic acid	−168.261	0.447	0.308	0.031
HOCN	Cyanic	−168.223	0.453	0.305	0.036

Table 6.2 Complexity values in product space for the isomers HCNO, HNCO and HOCN in atomic units (a.u.)

Molecule	Name	I_{rp}	D_{rp}	$C_{rp}(LMC)$	$C_{rp}(FS)$
HCNO	Fulminic acid	135.999	0.0070	76.453	229.303
HNCO	Isocyanic acid	132.848	0.0068	74.159	223.889
HOCN	Cyanic	133.718	0.0069	75.922	225.656

are isomers of fulminic acid (H-C=N-O) which is an unstable compound [127]. From Tables 6.1 and 6.2 we may corroborate that this is indeed the case in that fulminic acid possess larger values for all complexity measures and according to our discussion above in terms of chemical properties this is indeed a more reactive molecule (unstable).

6.5 Concluding Remarks

Different information-theoretical quantities as well as complexities defined as the product of a couple of localization-delocalization factors have been shown to provide relevant information on the atomic and molecular systems. In the atomic case, not only on the shell structure and organization of a great variety of atomic systems, but also on ionization processes and their dependence on both the nuclear charge and the number of electrons. In the case of molecules, many different chemically relevant and/or experimental quantities have been shown their strong connection with the complexity values in both spaces. In doing so, it appears necessary to deal simultaneously with the conjugated position and momentum space electron densities, being also important to consider the product space in order to get a more detailed and complete description of such systems. The method here employed for carrying out the present study is also applyable to the analysis of additional multifermionic systems, as well as physical or chemical processes, such as reactions or polarization among others. Some of these subjects are now being studied and will be presented hopefully elsewhere.

It has been also shown the interest of studying the associated information planes subtended by two information functionals, which for the atomic case clearly display the characteristic shell-filling patterns throughout the whole periodic table, while in the molecular one their interpretation according to important quantities in a chemical description. It still remains open the question of the existence of additional functionals, planes and complexities providing further information on the atomic and molecular structure and/or chemical and physical processes, among others. Such is the case, for instance, of some generalizations of the complexity as those analyzed here based on the concept of Rényi entropy, which allows a deeper study of specific regions within the domain of the associated densities. Additional generalizations are actually being defined and analyzed, such as those arising from the Tsallis entropy in a similar fashion as done with the Rényi entropy.

A complementary concept to that of the complexity itself is the 'relative complexity', for which different definitions can also be considered. The main aim of introducing this concept is to define a double-density functional in order to compare, quantitatively, the levels of complexity associated to the systems under comparison. Such an application is carried out in a similar way as done with the divergences for quantifying the similarity/dissimilarity of different systems attending to the information content of their characteristics distributions. The present work has not been restricted to a purely numerical study of information measures and complexities, but also within a mathematical framework including numerous relevant properties and relationships among complexities, as well as universal bounds. Most of them are expressed as bounds, at times very tight, shown to be very useful for the study of many electron systems.

Acknowledgements We wish to thank Nelson Flores-Gallegos and Sheila López-Rosa for their kind help in the preparation of this chapter, and to Professor K.D. Sen for helpful discussions. This work was supported in part by the Spanish grants FIS-2008-02380 and FIS-2005-06237 (MICINN), FQM-1735 and P06-FQM-2445 (Junta de Andalucía), and the Mexican grants 08266 CONACyT, PIFI 3.3 PROMEP-SEP. We belong to the Andalusian research group FQM-0207.

References

1. Sears SB, Gadre SR (1981) An information theoretic synthesis and analysis of Compton profiles. J Chem Phys 75:4626
2. Nalewajski RF, Parr RG (2001) Information theory thermodynamics of molecules and their Hirshfeld fragments. J Phys Chem A 105:7391
3. Carbó-Dorca R, Arnau J, Leyda L (1980) How similar is a molecule to another? An electron density measure of similarity between two molecular structures. Int J Quant Chem 17:1185
4. Angulo JC, Antolín J (2007) Atomic quantum similarity indices in position and momentum spaces. J Chem Phys 126:044106
5. Antolín J, Angulo JC (2008) Quantum similarity indices for atomic ionization processes. Eur Phys J D 46:21
6. Cover TM, Thomas JA (1991) Elements of information theory. Wiley-Interscience, New York
7. Antolín J, Angulo JC, López-Rosa S (2009) Fisher and Jensen-Shannon divergences: quantitative comparisons among distributions. Application to position and momentum atomic densities. J Chem Phys 130:074110
8. López-Rosa S, Antolín J, Angulo JC, Esquivel RO (2009) Divergence analysis of atomic ionization processes and isoelectronic series. Phys Rev A 80:012505

9. Angulo JC, López-Rosa S, Antolín J (2010) Effect of the interelectronic repulsion on the information content of position and momentum atomic densities. Int J Quantum Chem 110:1738
10. Carbó-Dorca R, Girones X, Mezey PG (eds) (2001) Fundamentals of molecular similarity. Kluwer Academic/Plenum, Dordrecht/New York
11. Cioslowski J, Nanayakkara A (1993) Similarity of atoms in molecules. J Am Chem Soc 115:11213
12. Carbó-Dorca R, Amat L, Besalu E, Girones X, Robert D (2000) Quantum Mechanical Origin of QSAR: theory and applications. J Mol Struct, Theochem 504:181
13. Daudel R (1953) C R Acad Sci (Paris) 237:601
14. Aslangul C, Constanciel R, Daudel R, Kottis P (1972) Aspects of the localizability of electrons in atoms and molecules: loge theory and related methods. Adv Quantum Chem 6:94
15. Mezey PG, Daudel R, Csizmadia IG (1979) Dependence of approximate ab initio molecular loge sizes on the quality of basis functions. Int J Quant Chem 16:1009
16. Nalewajski RF (2003) Information principles in the loge theory. Chem Phys Lett 375:196
17. Wang L, Wang L, Arimoto S, Mezey PG (2006) Large-scale chirality measures and general symmetry deficiency measures for functional group polyhedra of proteins. J Math Chem 40:145
18. Avnir D, Meyer AY (1991) Quantifying the degree of molecular shape distortion. A chirality measure. J Mol Struct, Theochem 226:211
19. Cramér H (1946) Mathematical methods of statistics. Princeton University Press, Princeton
20. Shannon CE, Weaver W (1949) The mathematical theory of communication. University of Illinois Press, Urbana
21. Fisher RA (1925) Statistical methods for research workers. Proc Camb Philos Soc 22:700
22. Frieden BR (2004) Science from Fisher information. Cambridge University Press, Cambridge
23. Jaynes ET (1957) Information theory and statistical mechanics. Phys Rev A 106:620
24. Nagy A (2006) Fisher information in a two-electron entangled artificial atom. Chem Phys Lett 425:154
25. Nalewajski R (2003) Information principles in the theory of electronic structure. Chem Phys Lett 372:28
26. Reginatto M (1998) Derivation of the equations of nonrelativistic quantum mechanics using the principle of minimum Fisher information. Phys Rev A 58:1775
27. Romera E, Sánchez-Moreno P, Dehesa JS (2006) Uncertainty relation for Fisher information of D-dimensional single-particle systems with central potentials. J Math Phys 47:103504
28. Dehesa JS, González-Férez R, Sánchez-Moreno P (2007) The Fisher-information-based uncertainty relation, Cramér-Rao inequality and kinetic energy for the D-dimensional central problem. J Phys A 40:1845
29. Kolmogorov AN (1965) Three approaches to the quantitative definition of information. Probl Inf Transm 1:1
30. Chaitin GJ (1966) On the length of programs for computing finite binary sequences. J ACM 13:547
31. Crutchfield JP, Shalizi KL (1999) Thermodynamic depth of causal states: Objective complexity via minimal representations. Phys Rev E 59:275
32. Lempel A, Ziv J (1976) On the complexity of finite sequences. IEEE Trans Inf Theory 22:75
33. Grassberger P (1986) Toward a quantitative theory of self-generated complexity. Int J Theory Phys 25:907
34. Bennett CH (1988) Logical depth and physical complexity. In: The universal Turing machine: a half century survey. Oxford University Press, Oxford, pp 227–257
35. Lloyd SS, Pagels H (1988) Complexity as thermodynamic depth. Ann Phys NY 188:186
36. MacKay DJC (2003) Information theory, inference and learning algorithms. Cambridge University Press, Cambridge
37. Vitanyi PMB, Li M (2000) Minimum description length induction, Bayesianism, and Kolmogorov complexity. IEEE Trans Inf Theory 46:446

38. Shalizi CR, Shalizi KL, Haslinger R (2004) Quantifying self-organization with optimal predictors. Phys Rev Lett 93:118701
39. Rosso OA, Martin MT, Plastino A (2003) Brain electrical activity analysis using wavelet-based informational tools (II): Tsallis non-extensivity and complexity measures. Physica A 320:497
40. Chatzisavvas KCh, Moustakidis ChC, Panos CP (2005) Information entropy, information distances and complexity in atoms. J Chem Phys 123:174111
41. Feldman DP, Crutchfield JP (1998) Measures of statistical complexity: Why? Phys Lett A 238:244
42. Lamberti PW, Martin MP, Plastino A, Rosso OA (2004) Intensive entropic non-triviality measure. Physica A 334:119
43. López-Ruiz R, Mancini HL, Calbet X (1995) A statistical measure of complexity. Phys Lett A 209:321
44. Shiner JS, Davison M, Landsberg PT (1999) Simple measure for complexity. Phys Rev E 59:1459
45. Anteonodo C, Plastino A (1996) Some features of the López-Ruiz-Mancini-Calbet (LMC) statistical measure of complexity. Phys Lett A 223:348
46. Catalán RG, Garay J, López-Ruiz R (2002) Features of the extension of a statistical measure of complexity to continuous systems. Phys Rev E 66:011102
47. Martin MT, Plastino A, Rosso OA (2003) Statistical complexity and disequilibrium. Phys Lett A 311:126
48. López-Ruiz R (2005) Shannon information, LMC complexity and Rényi entropies: a straightforward approach. Biophys Chem 115:215
49. Yamano T (2004) A statistical complexity measure with nonextensive entropy and quasi-multiplicativity. J Math Phys 45:1974
50. Yamano T (2004) A statistical measure of complexity with nonextensive entropy. Physica A 340:131
51. Angulo JC (1994) Information entropy and uncertainty in D-dimensional many-body systems. Phys Rev A 50:311
52. Guevara NL, Sagar RP, Esquivel RO (2003) Shannon-information entropy sum as a correlation measure in atomic systems. Phys Rev A 67:012507
53. Romera E, Torres JJ, Angulo JC (2002) Reconstruction of atomic effective potentials from isotropic scattering factors. Phys Rev A 65:024502
54. Ho M, Smith VH Jr., Weaver DF, Gatti C, Sagar RP, Esquivel RO (1998) Molecular similarity based on information entropies and distances. J Chem Phys 108:5469
55. Zarzo A, Angulo JC, Antolín J, Yáñez RJ (1996) Maximum-entropy analysis of one-particle densities in atoms. Z Phys D 37:295
56. Antolín J, Zarzo A, Angulo JC, Cuchí JC (1997) Maximum-entropy analysis of momentum densities in diatomic molecules. Int J Quant Chem 61:77
57. Antolín J, Cuchí JC, Angulo JC (1999) Reciprocal form factors from momentum density magnitudes. J Phys B 32:577
58. Nagy A, Sen KD (2006) Atomic Fisher information versus atomic number. Phys Lett A 360:291
59. Sen KD, Panos CP, Chtazisavvas KCh, Moustakidis ChC (2007) Net Fisher information measure versus ionization potential and dipole polarizability in atoms. Phys Lett A 364:286
60. Hornyak I, Nagy A (2007) Phase-space Fisher information. Chem Phys Lett 437:132
61. Borgoo A, Godefroid M, Sen KD, de Proft F, Geerlings P (2004) Quantum similarity of atoms: a numerical Hartree-Fock and information theory approach. Chem Phys Lett 399:363
62. de Proft F, Ayers PW, Sen KD, Geerlings P (2004) On the importance of the density per particle (shape function) in the density functional theory. J Chem Phys 120:9969
63. Borgoo A, Godefroid M, Indelicato P, de Proft F, Geerlings P (2007) Quantum similarity study of atomic density functions: Insights from information theory and the role of relativistic effects. J Chem Phys 126:044102
64. Angulo JC, Antolín J, Sen KD (2008) Fisher-Shannon plane and statistical complexity of atoms. Phys Lett A 372:670

65. Onicescu O (1966) Énergie informationnelle. C R Acad Sci Paris A 263:841
66. Pipek J, Varga I (1992) Universal classification scheme for the spatial-localization properties of one-particle states in finite, d-dimensional systems. Phys Rev A 46:3148
67. Sen KD, Antolín J, Angulo JC (2007) Fisher-Shannon analysis of ionization processes and isoelectronic series. Phys Rev A 76:032502
68. Nagy A (2003) Fisher information in density functional theory. J Chem Phys 119:9401
69. Dembo A, Cover TA, Thomas JA (1991) Information theoretic inequalities. IEEE Trans Inf Theory 37:1501
70. Pearson JM (1997) A logarithmic Sobolev inequality on the real line. Proc Am Math Soc 125:3339
71. Antolín J, Angulo JC (2009) Complexity analysis of ionization processes and isoelectronic series. Int J Quant Chem 109:586
72. Angulo JC, Antolín J (2008) Atomic complexity measures in position and momentum spaces. J Chem Phys 128:164109
73. Esquivel RO, Angulo JC, Antolín J, Dehesa JS, López-Rosa S, Flores-Gallegos N (2009) Complexity analysis of selected molecules in position and momentum spaces. Preprint
74. Dehesa JS, Sánchez Moreno P, Yáñez RJ (2006) Cramér-Rao information plane of orthogonal hypergeometric polynomials. J Comput Appl Math 186:523
75. Calbet X, López-Ruiz R (2001) Tendency towards maximum complexity in a nonequilibrium isolated system. Phys Rev E 63:066116
76. Martin MT, Plastino A, Rosso OA (2006) Generalized statistical complexity measures: Geometrical and analytical properties. Physica A 369:439
77. Romera E, Nagy A (2008) Fisher-Rényi entropy product and information plane. Phys Lett A 372:6823
78. Antolín J, López-Rosa S, Angulo JC (2009) Rényi complexities and information planes: atomic structure in conjugated spaces. Chem Phys Lett 474:233
79. Rényi A (1961) On measures of entropy and information. In: Proc 4th Berkeley symposium on mathematics of statistics and probability, vol 1, pp 547–561
80. Tsallis C (1988) Possible generalization of Boltzmann-Gibbs statistics. J Stat Phys 52:479
81. Kendall MG, Stuart A (1969) The advanced theory of statistics, vol 1. Charles Griffin and Co Ltd, London
82. Parr RG, Yang W (1989) Density-functional theory of atoms and molecules. Oxford University Press, New York
83. Hall MJW (1999) Universal geometric approach to uncertainty, entropy, and information. Phys Rev A 59:2602
84. López-Rosa S, Angulo JC, Antolín J (2009) Rigorous properties and uncertainty-like relationships on product-complexity measures: Application to atomic systems. Physica A 388:2081
85. Dehesa JS, Gálvez FJ, Porras I (1989) Bounds to density-dependent quantities of D-dimensional many-particle systems in position and momentum spaces: Applications to atomic systems. Phys Rev A 40:35
86. Angulo JC, Dehesa JS (1992) Tight rigorous bounds to atomic information entropies. J Chem Phys 97:6485. Erratum 98:1 (1993)
87. Stam A (1959) Some inequalities satisfied by the quantities of information of Fisher and Shannon. Inf Control 2:101
88. Fraga S, Malli G (1968) Many electron systems: properties and interactions. Saunders, Philadelphia
89. Epstein IR (1973) Calculation of atomic and molecular momentum expectation values and total energies from Compton-scattering data. Phys Rev A 8:160
90. Bialynicky-Birula I, Mycielski J (1975) Uncertainty relations for information entropy in wave mechanics. Commun Math Phys 44:129
91. Heisenberg W (1927) Uber den anschaulichen inhalt der quanten-theoretischen kinematik und mechanik. Z Phys 443:172

92. Sánchez-Moreno P (2008) Medidas de Información de Funciones Especiales y sistemas mecano-cuánticos, y dinámica molecular en presencia de campos eléctricos homogéneos y dependientes del tiempo. PhD Thesis, University of Granada, Spain
93. Bialynicky-Birula I (2006) Formulation of the uncertainty relations in terms of the Rényi entropies 91. Phys Rev A 74:052101
94. Rajagopal AK (1995) The Sobolev inequality and the Tsallis entropic uncertainty relation. Phys Lett A 205:32
95. Angulo JC (1993) Uncertainty relationships in many-body systems. J Phys A 26:6493
96. Gadre SR (1984) Information entropy and Thomas-Fermi theory. Phys Rev A 30:620
97. Gadre SR, Bendale RD (1985) Information entropies in quantum chemistry. Curr Sci (India) 54:970
98. Panos CP, Chatzisavvas KCh, Moustakidis ChC, Kyhou EG (2007) Comparison of SDL and LMC measures of complexity: Atoms as a testbed. Phys Lett A 363:78
99. Gadre SR, Bendale RD, Gejji SP (1985) Refinement of electron momentum densities of ionic solids using an experimental energy constraint. Chem Phys Lett 117:138
100. Sagar RP, Guevara NL (2006) Mutual information and electron correlation in momentum space. J Chem Phys 124:134101
101. Romera E, Dehesa JS (2004) The Fisher-Shannon information plane, an electron correlation tool. J Chem Phys 120:8906
102. Koga T, Kanayama K, Watanabe S, Thakkar AJ (1999) Analytical Hartree-Fock wave functions subject to cusp and asymptotic constraints: He to Xe, Li^+ to Cs^+, H^- to I^-. Int J Quant Chem 71:491
103. Koga T, Kanayama K, Watanabe S, Imai S, Thakkar AJ (2000) Analytical Hartree-Fock wave functions for the atoms Cs to Lr. Theor Chem Acc 104:411
104. Borgoo A, de Proft F, Geerlings P, Sen KD (2007) Complexity of Dirac-Fock atom increases with atomic number. Chem Phys Lett 444:186
105. Szabo JB, Sen KD, Nagy A (2008) The Fisher-Shannon information plane for atoms. Phys Lett A 372:2428
106. Hoffmann-Ostenhof M, Hoffmann-Ostenhof T (1977) "Schrödinger inequalities" and asymptotic behavior of the electron density of atoms and molecules. Phys Rev A 16:1782
107. Benesch R, Smith VH Jr (1973) Wave mechanics: the first fifty years. Butterworth, London
108. Romera E, Nagy A (2008) Rényi information of atoms. Phys Lett A 372:4918
109. Koga T, Omura M, Teruya H, Thakkar AJ (1995) Improved Roothaan-Hartree-Fock wavefunctions for isoelectronic series of the atoms He to Ne. J Phys B 28:3113
110. Koopmans TA (1933) Über die Zuordnung von Wellenfunktionen und Eigenwerten zu den Einzelnen Elektronen Eines Atoms. Physica 1:104
111. Janak JF (1978) Proof that $\partial E/\partial n_i = \varepsilon$ in density-functional theory. Phys Rev B 18:7165
112. Parr RG, Pearson RG (1983) Absolute hardness: companion parameter to absolute electronegativity. J Am Chem Soc 105:7512
113. Ghanty TK, Ghosh SK (1993) Correlation between hardness, polarizability, and size of atoms, molecules, and clusters. J Phys Chem 97:4951
114. Roy R, Chandra AK, Pal S (1994) Correlation of polarizability, hardness, and electronegativity: polyatomic molecules. J Phys Chem 98:10447
115. Hati S, Datta D (1994) Hardness and electric dipole polarizability. Atoms and clusters. J Phys Chem 98:10451
116. Simon-Manso Y, Fuentealba E (1998) On the density functional relationship between static dipole polarizability and global softness. J Phys Chem A 102:2029
117. Chattaraj PK, Sarkar U, Roy DR (2006) Electrophilicity index. Chem Rev 106:2065
118. Pearson RG (1963) Hard and soft acids and bases. J Am Chem Soc 85:3533
119. Pearson RG (1973) Hard and soft acids and bases. Dowen, Hutchinson and Ross, Stroudsberg
120. Pearson RG (1997) Chemical hardness. Wiley-VCH, New York
121. Parr RG, Szentpály LV, Liu S (1999) Electrophilicity index. J Am Chem Soc 121:1922

122. Frisch MJ, Trucks GW, Schlegel HB, Scuseria GE, Robb MA, Cheeseman JR, Montgomery JA Jr, Vreven T, Kudin KN, Burant JC, Millam JM, Iyengar SS, Tomasi J, Barone V, Mennucci B, Cossi M, Scalmani G, Rega N, Petersson GA, Nakatsuji H, Hada M, Ehara M, Toyota K, Fukuda R, Hasegawa J, Ishida M, Nakajima T, Honda Y, Kitao O, Nakai H, Klene M, Li X, Knox JE, Hratchian HP, Cross JB, Bakken V, Adamo C, Jaramillo J, Gomperts R, Stratmann RE, Yazyev O, Austin AJ, Cammi R, Pomelli C, Ochterski JW, Ayala PY, Morokuma K, Voth GA, Salvador P, Dannenberg JJ, Zakrzewski VG, Dapprich S, Daniels AD, Strain MC, Farkas O, Malick DK, Rabuck AD, Raghavachari K, Foresman JB, Ortiz JV, Cui Q, Baboul AG, Clifford S, Cioslowski J, Stefanov BB, Liu G, Liashenko A, Piskorz P, Komaromi I, Martin RL, Fox DJ, Keith T, Al-Laham MA, Peng CY, Nanayakkara A, Challacombe M, Gill PMW, Johnson B, Chen W, Wong MW, González C, Pople JA (2004) Gaussian 03, Revision D.01, Gaussian Inc, Wallingford
123. Pérez-Jordá JM, San-Fabián E (1993) A simple, efficient and more reliable scheme for automatic numerical integration. Comput Phys Commun 77:46
124. Pérez-Jordá JM, Becke AD, San-Fabián E (1994) Automatic numerical integration techniques for polyatomic molecules. J Chem Phys 100:6520
125. Kohout M (2007) Program DGRID, version 4.2
126. Computational Chemistry Comparison and Benchmark DataBase, http://cccbdb.nist.gov/
127. Kurzer F (2000) Fulminic acid in the history of organic chemistry. J Chem Educ 77:851

Chapter 7
Rényi Entropy and Complexity

Á. Nagy and E. Romera

Abstract Several important properties of the Rényi entropy and Rényi entropy power are presented. Uncertainty relations for the Rényi entropy including uncertainty relations for single particle densities of many particle systems in position and momentum spaces are discussed. Connection between Fisher information and Rényi entropy is studied. The Fisher-Rényi information plane and entropic product are presented.

Position and momentum space Rényi entropies of order α are presented for ground-state neutral atoms with atomic numbers $Z = 1$–103. It is emphasized that the values of $\alpha \leq 1$ ($\alpha \geq 1$) stress the shell structure for position-space (momentum-space) Rényi entropies. Position and momentum space relative Rényi entropies of order α are presented for ground-state neutral atoms with atomic numbers $Z = 1$–103. Simple hydrogen-like model densities are used as the reference. A relationship with the atomic radius and quantum capacitance is also discussed.

The relationship between the statistical complexity and the Rényi entropy is studied. A recently introduced, one-parameter extension of the LMC complexity is presented.

The maximum Rényi entropy principle is used to generalize the Thomas-Fermi model. A simple relation between the dimension and the Rényi parameter is emphasized.

7.1 Introduction

Recently, there has been a growing interest in using information concepts in several fields of science. Shannon information [1] has been applied in describing atomic and molecular properties [2–14] for several decades. Recently, Fisher information [15] has proved to be a very useful tool in analyzing atoms and molecules [16–31].

Á. Nagy (✉)
Department of Theoretical Physics, University of Debrecen, 4010 Debrecen, Hungary
e-mail: anagy@madget.atomki.hu

K.D. Sen (ed.), *Statistical Complexity*,
DOI 10.1007/978-90-481-3890-6_7, © Springer Science+Business Media B.V. 2011

Although Rényi entropy has been introduced by Rényi [32] as early as 1961, this kind of entropy has only recently obtained a wide range of application e.g. in the analysis of quantum entanglement [33–35], quantum communication protocols [36, 37], quantum correlations [38], quantum revivals [39] or localization properties [40].

This chapter summarizes recent results on Rényi entropy. Section 7.2 contains the definition of the Rényi entropy and the relative Rényi entropy and presents important uncertainty relations. Application of Rényi information in studying atoms and molecules has started only recently. The present authors published the first study of Rényi entropy and the relative Rényi entropy for atoms [41, 42]. These results are summarized in Sect. 7.3. Section 7.4 presents the Fisher-Rényi product [43]. Rényi information is a component of different complexity measures. Complexity measures are considered a hot topic, an efficient tool for analyzing systems. The relationship between complexity [44] and Rényi entropy is discussed in Sect. 7.5. In Sect. 7.6 the Thomas-Fermi model is generalized using the maximum Rényi entropy principle [45]. It also gives a physical meaning of the Rényi parameter. The last section is devoted to a summary and proposition of future directions.

7.2 Rényi Entropy and Rényi Entropy Power

Rényi entropy is a one-parameter extension of Shannon entropy. The Shannon entropy has the form

$$S_f = -\int f(x) \ln f(x) dx \tag{7.1}$$

for a continuous distribution $f(x)$. The probability distribution $f(x)$ can be associated with a wave function $\psi(x)$ as $f(x) = |\psi(x)|^2$. The Shannon entropy can be written as

$$S_g = -\int g(p) \ln g(p) dp, \tag{7.2}$$

where the probability distribution $g(p)$ is given by the momentum space wave function $\phi(p)$, the Fourier transform of the wave function $\psi(x)$ as $g(p) = |\phi(p)|^2$.

There exists an entropic uncertainty relation in the form

$$S_f + S_g \geq \ln(e\pi). \tag{7.3}$$

This relation was conjectured by Hirschman [46] and proved by Bialynicki-Birula and Mycielski [47] and by Beckner [48]. The above equations can be readily generalized to arbitrary dimensions.

Rényi entropy of order α for a D dimensional probability density function $f(r_1, \ldots, r_D)$ normalized to one is defined by

$$R_f^\alpha \equiv \frac{1}{1-\alpha} \ln \int f^\alpha(\mathbf{r}) d\mathbf{r} \quad \text{for } 0 < \alpha < \infty,\ \alpha \neq 1, \tag{7.4}$$

where $\mathbf{r}$ stands for $r_1, \ldots, r_D$. The Rényi entropy tends to the Shannon entropy when $\alpha \to 1$:

$$S_f = -\int f(\mathbf{r}) \ln f(\mathbf{r}) d\mathbf{r}. \tag{7.5}$$

Rényi entropy is a nonincreasing function of α [49].

Consider a D-variable function $\Psi(r_1, \ldots, r_D)$ and its conjugate Fourier transform $\Phi(p_1, \ldots, p_D)$ and the corresponding distribution functions $|\Psi|^2$ and $|\Phi|^2$. Bialynicki-Birula derived an uncertainty relation [50] for the Rényi entropy sum:

$$R^{\alpha}_{|\Psi|^2} + R^{\beta}_{|\Phi|^2} \geq f(\alpha, \beta), \quad \frac{1}{\alpha} + \frac{1}{\beta} = 2, \tag{7.6}$$

$$f(\alpha, \beta) = \frac{D}{2}\left[\frac{1}{\alpha - 1} \ln\left(\frac{\alpha}{\pi}\right) + \frac{1}{\beta - 1} \ln\left(\frac{\beta}{\pi}\right)\right]. \tag{7.7}$$

This uncertainty relation reaches the Shannon entropic uncertainty relation (7.3) in the limit $\alpha \to 1$.

The Rényi entropy power of index α is defined by

$$\begin{aligned} N^{\alpha}_f &\equiv \left(\frac{\alpha}{2\alpha - 1}\right)^{\frac{2\alpha-1}{\alpha-1}} \frac{1}{2\pi} \exp\left(\frac{2}{D} R^{\alpha}_f\right) \\ &= \beta^{1/(1-\beta)} \frac{1}{2\pi} \exp\left(\frac{2}{D} R^{\alpha}_f\right) \end{aligned} \tag{7.8}$$

where β satisfies the equation $\alpha^{-1} + \beta^{-1} = 2$.

The Rényi uncertainty relation (7.6) can be expressed in terms of Rényi entropy power in a compact form:

$$N^{\alpha}_{|\Psi|^2} N^{\beta}_{|\Phi|^2} \geq 1/4. \tag{7.9}$$

The Rényi entropy power is an extension of Shannon entropy power, that is, $N^{\alpha}_f \to N_f = \frac{1}{2\pi e} e^{\frac{2}{D} S_f}$ if $\alpha \to 1$. In this limit the uncertainty relation (7.6) reaches the Shannon entropic uncertainty relation [47] $N_{|\Psi|^2} N_{|\Phi|^2} \geq 1/4$.

Scaling of the function as $\Psi_\lambda(r_1, \ldots, r_D) = \lambda^{D/2} \Psi(\lambda r_1, \ldots, \lambda r_D)$, the Rényi entropy power scales

$$N^{\alpha}_{|\Psi_\lambda|^2} = \lambda^{-2} N^{\alpha}_{|\Psi|^2}. \tag{7.10}$$

The Rényi entropy power has also the property [49]

$$N^{\alpha}_f > N^{\alpha'}_f, \quad \alpha < \alpha'. \tag{7.11}$$

As special cases the Rényi and Shannon information powers satisfy the inequalities [43]:

$$N^{\alpha}_f \geq N_f, \quad \frac{1}{2} < \alpha \leq 1 \tag{7.12}$$

and

$$N^{\alpha}_f \leq N_f, \quad \alpha \geq 1. \tag{7.13}$$

The inequality (7.11) can be readily proved by taking into account that the Rényi entropy is a nonincreasing function of α [49] and $c(\alpha) = (\frac{\alpha}{2\alpha-1})^{\frac{2\alpha-1}{\alpha-1}}$ is a nonincreasing function for $\alpha > 1/2$. Then using the definition of the Rényi entropy power (7.8) we immediately obtain that the Rényi entropy power is a nonincreasing function of α. We also notice that $c(\alpha) > e^{-1}$ and $R_f^\alpha > S_f$ for $\frac{1}{2} < \alpha < 1$, and that $c(\alpha) < e^{-1}$ and $R_f^\alpha < S_f$ for $\alpha > 1$.

7.3 Atomic Rényi and Relative Rényi Entropies

The atomic Rényi entropies were numerically calculated [41] with the density obtained from ground-state Roothaan-Hartree-Fock (RHF) wave functions [51, 52]. The calculations were performed for atomic numbers $Z = 1$–103 and for several values of α. The upper panels of Figs. 7.1 and 7.2 present position-space Rényi entropy for different values of α including the case $\alpha = 1$ (Shannon) and $\alpha = 0.5$ vs atomic number of neutral atoms. We can clearly observe the shell structure of atoms for $\alpha \leq 1$. For $\alpha \geq 2$ there is no sign of a shell structure [41].

Calculation were performed also in momentum space. The upper panels of Figs. 7.3 and 7.4 present the momentum-space Rényi entropy for different values of α including the case $\alpha = 1$ (Shannon) vs atomic number of neutral atoms. Complementary to the position-space results, we can clearly see the shell structure of atoms for $\alpha \geq 1$. For $\alpha < 1$ no shell structure can be found [41].

To explain the results presented on Figs. 7.1–7.4 simple analytic model calculations were done [41]. The limit $\alpha \to 1$ gives the Shannon case for which Sagar et al. [53] used cusp- and asymptotic-constrained model densities. Consider the model density

$$\rho_{mod}(r) = \frac{b^3}{8\pi} \exp(-br). \tag{7.14}$$

Then the Rényi entropy has the form

$$R^\alpha_{\rho,mod} = \frac{3\ln\alpha}{\alpha - 1} - \ln\frac{b^3}{8\pi}. \tag{7.15}$$

The choice $b = 2\sqrt{2I}$ gives a model density with correct asymptotic behaviour, while taking $b = 2Z$ gives a model density with a correct cusp. I is the ionization energy. The first term in both $R^\alpha_{\rho,asymp}$ and $R^\alpha_{\rho,cusp}$ are the same. The difference is in the second term: the second term contains the information on the asymptotic behaviour and the cusp, respectively. It is remarkable that the order α appears only in the first term, the term which is the same in both $R^\alpha_{\rho,asymp}$ and $R^\alpha_{\rho,cusp}$.

The limit $\alpha \to 1$ gives the Shannon information:

$$S_{\rho,asymp} = 3 - \ln\frac{2\sqrt{2}I^{3/2}}{\pi} \tag{7.16}$$

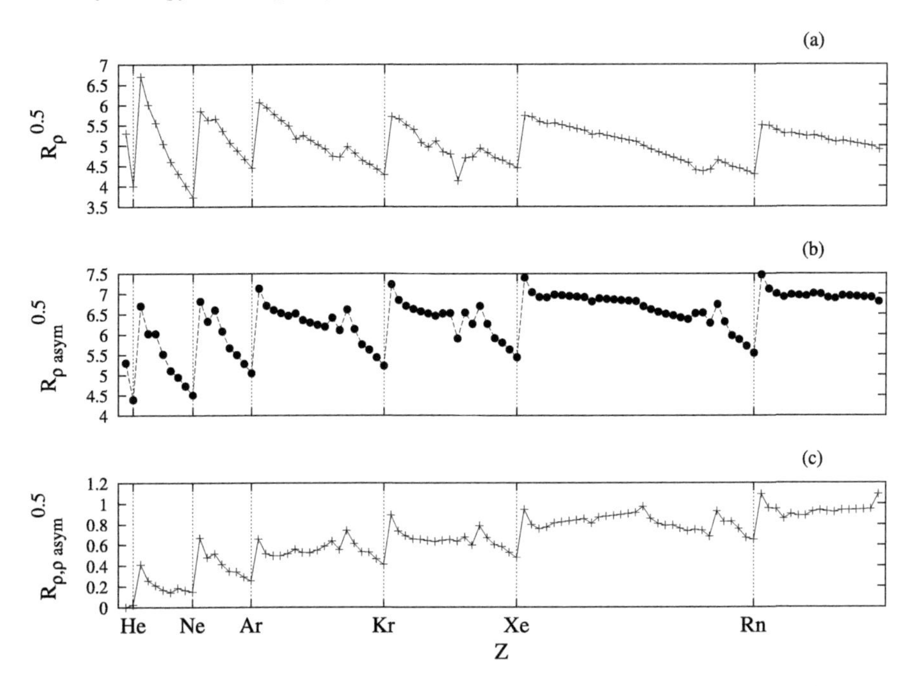

Fig. 7.1 Position space (**a**) Rényi entropy R_ρ^α, (**b**) model asymptotic Rényi entropy $R_{\rho_{asym}}^\alpha$ and (**c**) relative Rényi entropy with model density ρ_{asym}; with $\alpha = 0.5$ for the ground state of neutral atoms with $Z = 1$–103. Noble gasses are indicated by *vertical dotted lines*

and

$$S_{\rho,cusp} = 3 - \ln \frac{Z^3}{\pi}. \tag{7.17}$$

The results in (7.16) and (7.17) were first derived in [53]. If $\alpha \to \infty$ the first term in (7.15) goes to zero in both $R_{\rho,asymp}^\alpha$ and $R_{\rho,cusp}^\alpha$ and neither of them depends on α any more. Expression (7.15) suggests that the α-dependence is stronger for small values of α.

In the momentum space Sagar et al. [53] applied the density

$$\gamma(p) = \frac{8}{\pi^2} \frac{a^5}{(a^2 + p^2)^4}, \tag{7.18}$$

which is obtained from the Fourier transform of the model wave function

$$\psi(r) = \frac{a^{3/2}}{\pi^{1/2}} \exp(-ar). \tag{7.19}$$

Then the Rényi entropy in the momentum space

$$R_\gamma^\alpha = \frac{1}{1-\alpha} \ln \int \gamma^\alpha(\mathbf{p}) d\mathbf{p} \tag{7.20}$$

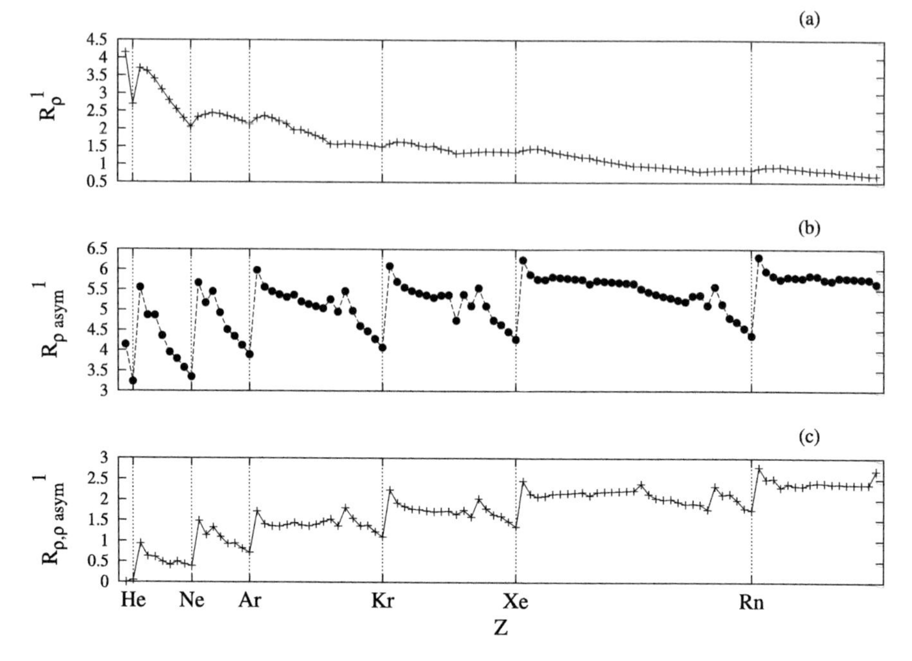

Fig. 7.2 Position space (**a**) Rényi entropy R_ρ^α, (**b**) model asymptotic Rényi entropy $R_{\rho_{asym}}^\alpha$ and (**c**) relative Rényi entropy with model density ρ_{asym}; with $\alpha = 1$ for the ground state of neutral atoms with $Z = 1$–103. Noble gasses are indicated by *vertical dotted lines*

has the form

$$R_\gamma^\alpha = \frac{1}{1-\alpha} \ln\left[\frac{8^\alpha}{\pi^{2\alpha}} 4\pi \int_0^\infty \frac{a^{5\alpha}}{(a^2+p^2)^{4\alpha}} p^2 dp\right]. \tag{7.21}$$

The integration leads to the result

$$R_\gamma^\alpha = \frac{1}{1-\alpha} \ln\left[\frac{8^{\alpha+1}\pi^{\frac{3}{2}-2\alpha}\Gamma(4\alpha+1/2)\Gamma(8\alpha-3)}{\Gamma(8\alpha)\Gamma(4\alpha-1)}\right] + 3\ln a. \tag{7.22}$$

Substituting

$$a = \sqrt{2I} \tag{7.23}$$

into (7.22) we arrive at the expression

$$R_{\gamma,asymp}^\alpha = \frac{1}{1-\alpha} \ln\left[\frac{8^{\alpha+1}\pi^{\frac{5}{2}-3\alpha}\Gamma(4\alpha+1/2)\Gamma(8\alpha-3)}{\Gamma(8\alpha)\Gamma(4\alpha-1)}\right] + \ln\frac{2\sqrt{2}I^{3/2}}{\pi}. \tag{7.24}$$

Applying $a = Z$ we are led to the result

$$R_{\gamma,cusp}^\alpha = \frac{1}{1-\alpha} \ln\left[\frac{8^{\alpha+1}\pi^{\frac{5}{2}-3\alpha}\Gamma(4\alpha+1/2)\Gamma(8\alpha-3)}{\Gamma(8\alpha)\Gamma(4\alpha-1)}\right] + \ln\frac{Z^3}{\pi}. \tag{7.25}$$

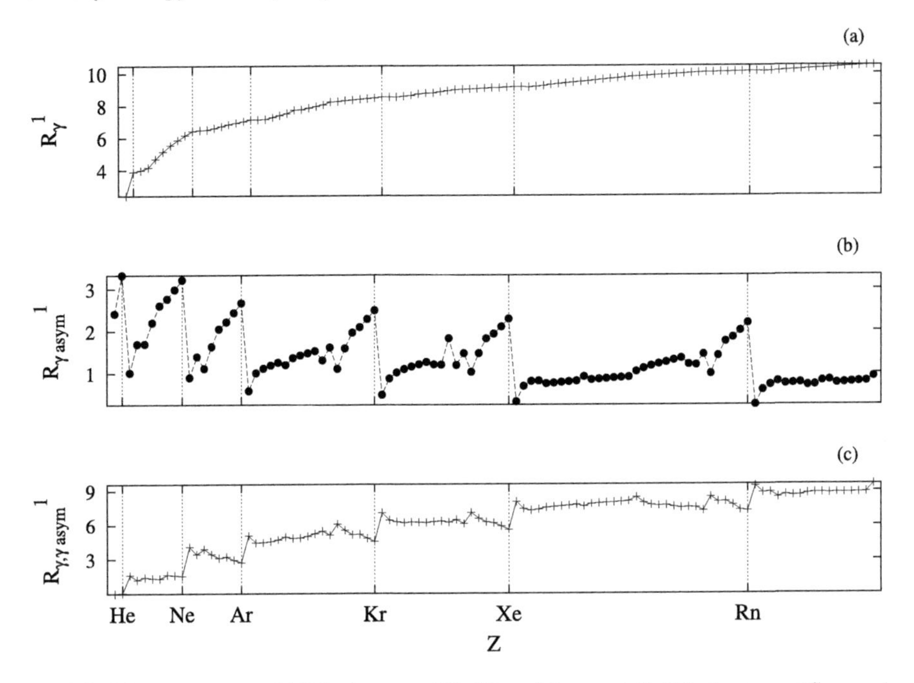

Fig. 7.3 Momentum space (**a**) Rényi entropy R_γ^α, (**b**) model asymptotic Rényi entropy $R_{\gamma_{asym}}^\alpha$ and (**c**) relative Rényi entropy with model density γ_{asym}; with $\alpha = 1$ for the ground state of neutral atoms with $Z = 1$–103. Noble gasses are indicated by *vertical dotted lines*

Here we again see that the first term in both $R_{\gamma,asymp}^\alpha$ and $R_{\gamma,cusp}^\alpha$ are the same. The difference is in the second term: the second term contains the information on the asymptotic behaviour and the cusp, respectively. The order α appears only in the first term, the term which is common in both $R_{\gamma,asymp}^\alpha$ and $R_{\gamma,cusp}^\alpha$.

The limit $\alpha \to 1$ gives the Shannon entropy:

$$S_{\gamma,asymp} = \ln \frac{\pi^3}{8} + 4\left(2\ln 2 - \frac{5}{6}\right) + \ln \frac{2\sqrt{2}I^{3/2}}{\pi} \tag{7.26}$$

and

$$S_{\gamma,cusp} = \ln \frac{\pi^3}{8} + 4\left(2\ln 2 - \frac{5}{6}\right) + \ln \frac{Z^3}{\pi}. \tag{7.27}$$

The results in (7.26) and (7.27) were first derived in [53]. If $\alpha \to \infty$ we obtain

$$R_{\gamma,asymp}^\infty = \ln \frac{\pi^3}{8} + \ln \frac{2\sqrt{2}I^{3/2}}{\pi} \tag{7.28}$$

and

$$R_{\gamma,cusp}^\infty = \ln \frac{\pi^3}{8} + \ln \frac{Z^3}{\pi}. \tag{7.29}$$

The middle panels of Figs. 7.1–7.2 and 7.3–7.4 present the Rényi entropies $R_{\rho,asymp}^\alpha$ and $R_{\gamma,asymp}^\alpha$. The values obtained from the model densities show a very

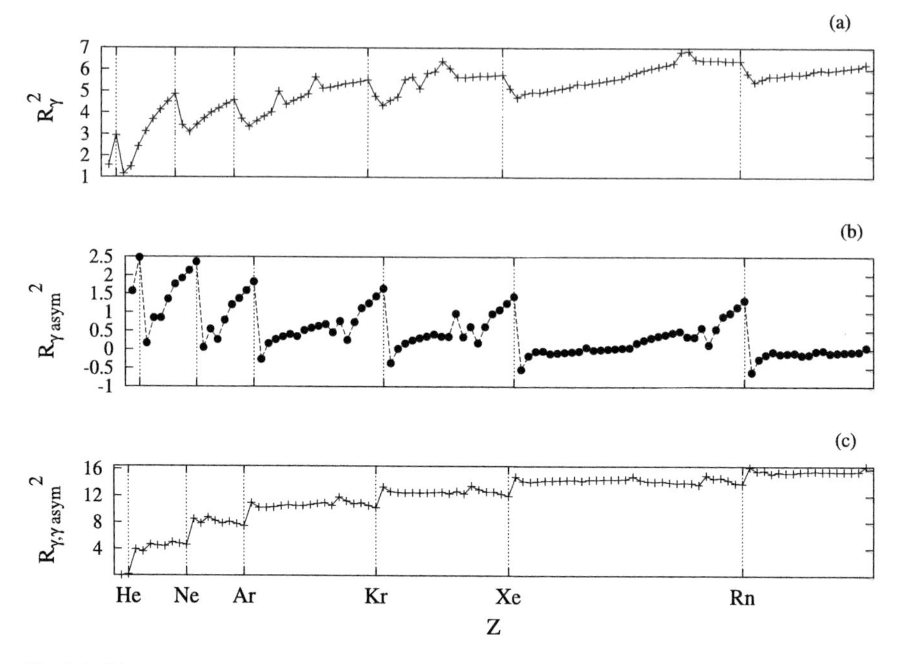

Fig. 7.4 Momentum space (**a**) Rényi entropy R_γ^α, (**b**) model asymptotic Rényi entropy $R_{\gamma_{asym}}^\alpha$ and (**c**) relative Rényi entropy with model density γ_{asym}; with $\alpha = 2$ for the ground state of neutral atoms with $Z = 1$–103. Noble gasses are indicated by *vertical dotted lines*

clear shell structure for $\alpha \leq 1$ (position space) and $\alpha \geq 1$ (momentum space), respectively. The agreement is better for smaller α in the position space and for larger α in the momentum space.

The fact that the results obtained from these simple analytical models give such a good description of the accurate values suggests that this model wave functions and densities include the main information, that is, the most important aspects of the true accurate wave functions and densities, namely, the cusp and asymptotic behaviour.

It is very interesting that the limit value of α where the shell structure appears (or disappears) is $\alpha = 1$, that is, the Shannon case.

The shell structure is more dominant for very small α in case of the position-space Rényi entropies and very large α in case of the momentum-space Rényi entropies. Also the agreement between the accurate Rényi entropies and results obtained from the model density with correct asymptotic behaviour is better for smaller α in the position space and for larger α in the momentum space. It means that small α in the position space (or large α in the momentum space) stress the asymptotic part of the density. On the other hand, in case of large α in the position space (or small α in the momentum space) the cusp part of the density dominates. That is different values of α amplify the shell structure in different manner. We can conjecture that other properties are amplified by other value of α.

We calculated the position-space and momentum-space Rényi entropies from accurate and model densities for ground-state anions and cations [41]. We found excellent agreement between the values arising from the accurate densities and the model densities with correct asymptotic behaviour and arrived at the same conclusion for ground-state anions and cations: the values of $\alpha \leq 1$ ($\alpha \geq 1$) stress the shell structure for position-space Rényi entropies (momentum-space Rényi entropies).

To gain more insight into the relationship of the model and accurate densities and the Rényi entropies corresponding to them, we studied relative Rényi entropies [42].

The relative Rényi entropy of order α is associated with two probability density functions $g(\mathbf{r})$ and $f(\mathbf{r})$:

$$R^{\alpha}_{f,g} = \frac{1}{\alpha - 1} \ln \int \frac{f^{\alpha}(\mathbf{r})}{g^{\alpha-1}(\mathbf{r})} d\mathbf{r} \tag{7.30}$$

provided that the integral on the right exists and $\alpha > 0$. The limit $\alpha \to 1$ gives the relative or Kullback-Leibler entropy:

$$I_{KL}(f,g) = \int f(\mathbf{r}) \ln \frac{f(\mathbf{r})}{g(\mathbf{r})} d\mathbf{r}. \tag{7.31}$$

The relative entropy is a measure of the deviation of $f(\mathbf{r})$ from $g(\mathbf{r})$, which is called the reference density. In a recent paper [54] relative entropy were studied in atomic systems using hydrogen-like model density as reference. We extended their investigation to the relative Rényi entropy [42].

The relative Rényi entropy is zero if $f = g$, and it measures the deviation of the density $\rho(r)$ from $\rho_{asymp}(r)$ or $\rho_{cusp}(r)$ and $\gamma(p)$ from $\gamma_{asymp}(p)$ or $\gamma_{cusp}(p)$.

We calculated the position space relative Rényi entropy with the model densities ρ_{cusp} for $\alpha \leq 1$ [42]. We can observe a shell structure only for $\alpha = 1$, no shell structure can be found if $\alpha < 1$. The relative Rényi entropy is close to 0 for small atomic numbers, that is, the deviation of the density from the cusp model density is not large. For larger atomic numbers the relative Rényi entropy is also larger showing a larger deviation from the cusp model density. The calculations cannot be performed for $\alpha > 1$, because the integral (7.30) does not exist for cusp model densities [42].

The bottom panels of Figs. 7.1–7.2 present the position space relative Rényi entropy with the model densities ρ_{asymp} for the values of $\alpha = 0.5$ and 1. One can see the shell effect for all values of α. The position space relative Rényi entropy is much smaller for the model densities ρ_{asymp} than for the model densities ρ_{cusp}. That is, in the position space the deviation of the density from the asymptotic model density is much smaller than from the cusp model density.

The bottom panels of Figs. 7.1–7.3 present the momentum space relative Rényi entropy with the model densities γ_{asymp} for the values of $\alpha = 1$ and 2. We can see the shell effect for all values of α. In the momentum space there is not much difference in the relative Rényi entropies obtained by the two model densities.

We mention in passing that there is a relationship between the relative Rényi entropies and the atomic radius and the quantum capacitance. The quantity

$$\langle r \rangle_a = \int r \rho_{HOMO}(\mathbf{r}) d\mathbf{r} \tag{7.32}$$

has turned to be a good measure of the atomic radius [54], where ρ_{HOMO} is the unity normalized density of the highest occupied orbital. The quantum capacitance is defined as [55–57]

$$C_I = \frac{1}{I - A} = \frac{1}{\eta}, \tag{7.33}$$

that is, the inverse of the hardness η. I and A are the ionization potential and the electron affinity, respectively.

Recently, Sagar et al. [54] demonstrated a linear relation between Shannon entropy and relative Shannon entropy with atomic radii and capacitance. We have numerically checked [42] that a linear regression of position space Rényi entropy and relative Rényi entropy for the values $\alpha = 0.5, 0.82, 0.86$ and 0.98 with the capacitance produce excellent fits for Alkali metals (Li, Na, K, Rb), Alkali earths (Be, Mg, Ca, Sr) and second row P-state (Al, Si, S, Cl). So this facts supports the consideration [54] of the capacitance in terms of entropies and consequently in terms of localization properties of the atomic valence density.

7.4 Fisher-Rényi Product

The Fisher information of the probability density function f is given by

$$I_f \equiv \int \frac{|\nabla f(\mathbf{r})|^2}{f(\mathbf{r})} d\mathbf{r}. \tag{7.34}$$

There hold the Stam uncertainty relations

$$I_{|\Psi|^2} \leq 4\sigma^2_{|\Phi|^2}, \tag{7.35}$$

where $\sigma^2_{|\Phi|^2}$ is the variance of the distribution in the momentum space and

$$I_{|\Phi|^2} \leq 4\sigma^2_{|\Psi|^2}, \tag{7.36}$$

where $\sigma^2_{|\Psi|^2}$ is the variance of the distribution in the position space.

When the D dimensional probability density $f(\mathbf{r})$ is scaling by a scalar factor λ as $f_\lambda(\mathbf{r}) = \lambda^D f(\lambda \mathbf{r})$, the Rényi entropy power (see (7.10)) and Fisher information [58] transform as follows:

$$N^\alpha_{f_\lambda} = \lambda^{-2} N^\alpha_f \quad \text{and} \quad I_{f_\lambda} = \lambda^2 I_\lambda. \tag{7.37}$$

Now, we take the product of the Fisher information and the Rényi entropy power

$$P^\alpha_f = \frac{1}{D} N^\alpha_f I_f \quad \text{with } \alpha \in (1/2, 1] \tag{7.38}$$

as an extension of the Fisher-Shannon product [25] with the important properties

(i) $P^\alpha_{f_\gamma}$ is invariant under scaling transformation $f_\lambda = \lambda^D f(\lambda \mathbf{r})$, i.e. $P^\alpha_{f_\lambda} = P^\alpha_f$,
(ii) there holds an uncertainty property

$$P^\alpha_f \geq 1, \tag{7.39}$$

(iii) P_f^α is a nonincreasing function of α

$$P_f^\alpha > P_f^{\alpha'} \geq 1 \quad \text{with } \alpha \in (1/2, 1] \tag{7.40}$$

for any probability density.

Now, we can combine the Stam uncertainty relations (7.35) and (7.36) to obtain a Stam type uncertainty relation for Rényi information

$$N_{|\Psi|^2}^\alpha \geq \frac{D}{4\sigma_{|\Phi|^2}^2}, \qquad \alpha \in (1/2, 1] \tag{7.41}$$

in the position space, and

$$N_{|\Phi|^2}^\alpha \geq \frac{D}{4\sigma_{|\Psi|^2}^2}, \qquad \alpha \in (1/2, 1] \tag{7.42}$$

in the momentum space.

Inequalities (7.41) and (7.42) can straightforwardly be extended to different parameters:

$$N_{|\Psi|^2}^\alpha > N_{|\Psi|^2}^{\alpha'} \geq \frac{D}{4\sigma_{|\Phi|^2}^2}, \qquad 1/2 < \alpha < \alpha' \leq 1 \tag{7.43}$$

and

$$N_{|\Phi|^2}^\alpha > N_{|\Phi|^2}^{\alpha'} \geq \frac{D}{4\sigma_{|\Psi|^2}^2}, \qquad 1/2 < \alpha < \alpha' \leq 1. \tag{7.44}$$

Let us point out that taking into account the Shannon uncertainty relation $N_{|\Psi|^2} N_{|\Phi|^2} \geq 1/4$ and inequality (7.12) we obtain a Rényi uncertainty relation:

$$N_{|\Psi|^2}^\alpha N_{|\Phi|^2}^\alpha \geq 1/4, \quad \alpha \in (1/2, 1]. \tag{7.45}$$

This inequality has been recently found by different means by Zozor et al. [59].

Except in the case $\alpha = 1$ (for which above inequalities saturates for Gaussians) the relations given by (7.39), (7.41) and (7.45) are not sharp.

Consider now an N particle system and denote the single particle densities in position and momentum spaces by

$$\rho(\mathbf{r}) = \int |\Psi(\mathbf{r}, \mathbf{r}_2, \ldots, \mathbf{r}_N)|^2 d\mathbf{r}_2 \ldots d\mathbf{r}_N \tag{7.46}$$

and

$$\gamma(\mathbf{p}) = \int |\Phi(\mathbf{p}, \mathbf{p}_2, \ldots, \mathbf{p}_N)|^2 d\mathbf{p}_2 \ldots d\mathbf{p}_N, \tag{7.47}$$

respectively. Note that the density is normalized to one. We also introduce the position space single particle Fisher-Rényi product

$$P_\rho^\alpha = \frac{1}{D} N_\rho^\alpha I_\rho \quad \text{with } \alpha \in (1/2, 1] \tag{7.48}$$

and the momentum space single particle Fisher-Rényi product

$$P_\gamma^\alpha = \frac{1}{D} N_\gamma^\alpha I_\gamma \quad \text{with } \alpha \in (1/2, 1]. \tag{7.49}$$

The relations (7.39)–(7.45) can readily be transcribed in terms of the corresponding one-particle probability densities $\rho(\mathbf{r})$ and $\gamma(\mathbf{p})$ by [43]

$$P_\rho^\alpha \geq 1, \quad P_\gamma^\alpha \geq 1 \quad \text{with } \alpha \in (1/2, 1], \tag{7.50}$$

and for different parameters:

$$P_\rho^\alpha > P_\rho^{\alpha'} \geq 1 \quad \text{with } 1/2 < \alpha < \alpha' \leq 1 \tag{7.51}$$

and

$$P_\gamma^\alpha > P_\gamma^{\alpha'} \geq 1 \quad \text{with } 1/2 < \alpha < \alpha' \leq 1. \tag{7.52}$$

On the other hand, the superadditivity and subadditivity of the Fisher and Shannon informations, respectively, have been used to prove the validity of Stam's uncertainty [60, 61] relation and Shannon uncertainty [4, 61] relation for single-particle densities in position and momentum spaces. Using the one-particle densities in position and momentum spaces, the inequalities take the form:

$$I_\rho \leq 4\sigma_\gamma^2, \qquad I_\gamma \leq 4\sigma_\rho^2 \tag{7.53}$$

and

$$N_\rho N_\gamma \geq 1/4. \tag{7.54}$$

Note, that the relationship between the single particle and N-particle variance is $\sigma_{|\Psi|^2}^2 = N\sigma_\rho^2$ in coordinate space and there is a similar expression for momentum space. Now, using the uncertainty relations and the inequalities (7.41),(7.42) and (7.12) it is fulfilled that

$$N_\rho^\alpha \geq \frac{D}{4\sigma_\gamma^2}, \quad N_\gamma^\alpha \geq \frac{D}{4\sigma_\rho^2}, \quad \alpha \in (1/2, 1] \tag{7.55}$$

and

$$N_\rho^\alpha N_\gamma^\alpha \geq 1/4, \quad \alpha \in (1/2, 1]. \tag{7.56}$$

We remark that to our best knowledge, relation (7.56) is the first Rényi uncertainty relation valid for monoparticle densities in conjugate spaces for $N \geq 2$ [43]. For $\alpha = 1$ we recover the Shannon uncertainty relation which is sharp, and which saturates for Gaussian functions.

As we have just seen there exists a link between Fisher and Rényi information. It is the generalization of the Fisher-Shannon connection proposed by Vignat et al. [58]. We also defined the Fisher-Rényi information plane and suggested that we can characterize density probabilities considering the location in this information plane [43]. Moreover, the information product scaling property shows that densities under a scaling transformation belong to the same curve in that plane.

7.5 Complexity and Rényi Entropy

Now we turn to study the relationship between the complexity and Rényi entropy. There are several measures of complexity (see other chapters of this monograph). Here we consider the LMC complexity [44].

Consider the Rényi entropy with $\alpha = 2$. From (7.4) follows that

$$R_\rho^2 = -\ln\left(\int \rho^2 d\mathbf{r}\right) = -\ln\langle\rho\rangle. \tag{7.57}$$

This quantity is closely related to the position-space disequilibrium

$$D_\rho = \int \rho^2 d\mathbf{r} = \langle\rho\rangle, \tag{7.58}$$

namely,

$$R_\rho^2 = -\ln D_\rho \tag{7.59}$$

and there is a similar expression for the momentum space:

$$R_\gamma^2 = -\ln D_\gamma. \tag{7.60}$$

The quantity D, also known as quantum self-similarity [62–64], information energy [65] or linear entropy [66, 67] is a measurable quantity [68]. D expresses the distance from equilibrium.

Consider the complexity measure defined by [44]

$$C_{LMC} = HD, \tag{7.61}$$

where

$$H = e^S \tag{7.62}$$

and S is the Shannon entropy. Making use of (7.59) expression (7.61) can also be written as [69]

$$\ln C_{LMC} = S + \ln D = S - R^2. \tag{7.63}$$

From the expressions of Rényi entropy obtained from the model densities (7.15), (7.24) and (7.25) we immediately see that the logarithm of the complexity is a constant (does not depend on the atomic number or the ionization potential). The accurate values, however, are not constant and show a clear shell structure. This reflects that the model densities do not provide a good approximation when the values of α is around 1 or 2. We mention that in recent papers [31, 70] the Z-dependence of the complexity is studied and a fit has been presented as a function of atomic number.

Romera et al. [71] have recently introduced a generalized, α-dependent measure of complexity. The new measure $C_f^{(\alpha)}$ is defined by

$$C_f^{(\alpha)} = H_f^{(\alpha)} Q_f, \quad \text{with } H_f^{(\alpha)} = e^{R_f^{(\alpha)}}. \tag{7.64}$$

In the limit $\alpha \to 1$ $C_f^{(\alpha)}$ tends to the original LMC complexity.

$C_f^{(\alpha)}$ has the properties:

(i) $C_f^{(\alpha=2)} = 1$,
(ii) $C_f^{(\alpha)}$ is invariant under scaling transformation, $f_\lambda = \lambda^D f(\lambda \mathbf{r})$, i.e. $C_{f_\lambda}^{(\alpha)} = C_f^{(\alpha)}$,
(iii) $C_f^{(\alpha)} \geq 1$ for $\alpha < 2$,
(iv) $C_f^{(\alpha)} \leq 1$ for $\alpha > 2$, and
(v) $C_f^{(\alpha)}$ is a nonincreasing function of α.

One can easily see that when α goes to 1, a lower bound for the original LMC complexity is obtained.

We emphasize that the Fisher-Rényi entropy can also be considered a measure of complexity. The hydrogen atom has recently been used as a test system [71] to study the different measures of statistical complexity: the Fisher-Rényi entropy product and the generalized statistical complexity. It was found that for each level of energy, both indicators take their minimum values on the orbitals that correspond to the highest orbital angular momentum. Hence, in the same way as happens with the Fisher-Shannon and the statistical complexity, these generalized Rényi-like statistical magnitudes break the energy degeneration in the H-atom.

Finally, we want to point out that a two-parameter family of complexity measures $\tilde{C}^{(\alpha,\beta)}$

$$\tilde{C}_f^{(\alpha,\beta)} = e^{R_f^{(\alpha)} - R_f^{(\beta)}}, \quad 0 < \alpha, \beta < \infty, \tag{7.65}$$

based on the Rényi entropies was introduced and characterized by a detailed study of its mathematical properties in [89]. The whole family is identified by two parameters, α and β. For the special case of β going to infinity, it is remarkable that the new complexity measure is the product of a global quantity by a local information of the density distribution.

7.6 Maximum Rényi Entropy Principle and the Generalized Thomas-Fermi Model

In this section we explore how the maximum Rényi entropy principle can be applied to generalize the Thomas-Fermi model [45]. Earlier the principle of extreme physical information [16] was used to derive the Euler and the Kohn-Sham equations of the density functional theory [19, 20]. First, the derivation of the Euler equation of the density functional theory is summarized. Consider a system of N electrons moving in a local external potential v. Imagine a system of non-interacting electrons having the same density $\rho(\mathbf{r})$ as that of the original interacting electrons. The kinetic energy $T_s[\rho]$ of the non-interacting system is defined by the Levy-Lieb [72, 73] constraint search as

$$T_s[\rho] = \min_{\Phi \to \rho} \langle \Phi | \hat{T} | \Phi \rangle = \langle \Phi[\rho] | \hat{T} | \Phi[\rho] \rangle, \tag{7.66}$$

where $\hat{T}$ is the kinetic energy operator and the minimum is searched over the non-interacting wave functions Φ having the density ρ. Then the Euler equation can be derived minimizing the non-interacting kinetic energy $T_s[\rho]$ with the constraints that the density is fixed (the same as the interacting one). In this section the density is normalized to the number of electrons N:

$$T_s[\rho] + \int v_{KS}(\mathbf{r})\rho(\mathbf{r})d\mathbf{r} - \mu \int \rho(\mathbf{r})d\mathbf{r}. \tag{7.67}$$

The Lagrange multipliers $v_{KS}(\mathbf{r})$ and μ are the Kohn-Sham potential and the chemical potential, respectively. The variation of this expression with respect to the density ρ leads to the Euler equation of the density functional theory:

$$\frac{\delta T_s[\rho]}{\delta \rho(\mathbf{r})} + v_{KS}(\mathbf{r}) = \mu. \tag{7.68}$$

Though (7.66) gives the formal definition of the non-interacting kinetic energy, its exact form as a functional of the density is unknown. There are, however, several approximations for it. The simplest one is the Thomas-Fermi functional [74, 75]:

$$T_{TF}[\rho] = C_{TF} \int \rho(\mathbf{r})^{5/3} d\mathbf{r}, \tag{7.69}$$

with

$$C_{TF} = \frac{3}{10}(3\pi^2)^{2/3}. \tag{7.70}$$

The Thomas-Fermi functional was generalized for arbitrary dimension [76–80]. The kinetic energy of a D dimensional system has the form

$$T_D[\rho] = C_D \int \rho(\mathbf{r})^{1+2/D} d\mathbf{r}, \tag{7.71}$$

with

$$C_D = \frac{D}{2(D+2)} \left(\frac{D}{2K_D} \right)^{2/D}. \tag{7.72}$$

The factor K_D can be given by the recurrence relation

$$K_{D+2} = \frac{K_D}{2\pi D}, \tag{7.73}$$

with

$$K_1 = \frac{1}{\pi} \tag{7.74}$$

and

$$K_2 = \frac{1}{2\pi}. \tag{7.75}$$

The maximum Rényi entropy principle is a generalization of the maximum entropy principle developed by Jaynes [81, 82]. In Jaynes' principle the Shannon entropy is maximized under proper conditions including the normalization of the probability density. The maximum Rényi entropy principle has been used recently to generalize the Thomas-Fermi model [45]. The extremum of the Rényi entropy of order α is searched under the conditions:

1. The density is normalized to 1. A Lagrange multiplicator ν is introduced.
2. The density is kept fixed. This requirement is taken into account by a local potential $w(\mathbf{r})$. This constraint is used to ensure that the density of the non-interacting system be equal to that of the interacting one.

Then, the extremum of the functional

$$R_f^\alpha - \nu \int f(\mathbf{r})d\mathbf{r} + \int w(\mathbf{r}) f(\mathbf{r})d\mathbf{r} \tag{7.76}$$

is considered. The variation leads to the equation:

$$\frac{\delta R_f^\alpha}{\delta f} - \nu + w(\mathbf{r}) = 0. \tag{7.77}$$

The functional derivative of the Rényi entropy (7.4) is

$$\frac{\delta R_f^\alpha}{\delta f} = \frac{\alpha}{1-\alpha}\frac{f^{\alpha-1}}{\int f^\alpha d\mathbf{r}}. \tag{7.78}$$

Then (7.77) takes the form:

$$\frac{\alpha}{1-\alpha}\frac{f^{\alpha-1}}{\int f^\alpha d\mathbf{r}} + w(\mathbf{r}) = \nu \tag{7.79}$$

or it can also be written as

$$cf^{\alpha-1} + \tilde{w}(\mathbf{r}) = \tilde{\nu} \tag{7.80}$$

with

$$\tilde{w} = \frac{c(1-\alpha)}{\alpha} w \int f^\alpha d\mathbf{r}, \tag{7.81}$$

$$\tilde{\nu} = \frac{c(1-\alpha)}{\alpha} \nu \int f^\alpha d\mathbf{r} \tag{7.82}$$

and c is any constant. Comparing (7.80) with (7.69) we immediately see that the Thomas-Fermi case recovers with the choice $\alpha = 5/3$ and $c = c_{TF}$. Equation (7.80) is a generalization of the Thomas-Fermi model derived from the maximum Rényi entropy principle. The Thomas-Fermi model was worked out for the free-electron gas and can be applied in systems with slowly changing densities. It is frequently used even today, for example to study the structure of the new artificial atoms or quantum dots [83], Bose-Einstein condensates in traps [84], systems under pressure [85], simulations in chemical and biological systems [86].

Here we use the Thomas-Fermi kinetic energy functional not the original Thomas-Fermi model itself, that is, the Thomas-Fermi model in a more general sense. Comparing (7.80) with (7.71) we can recover the D-dimensional Thomas-Fermi model. We are led to the simple relationship between the dimension D and the parameter α

$$\alpha = 1 + \frac{2}{D}. \tag{7.83}$$

In this case there is a physical meaning of the Rényi parameter α as it is related to the dimension.

The relationship between the kinetic energy and Shannon information entropy (that is, $\alpha = 1$) was studied earlier. Massen and Panos [87] fitted numerical values for several nuclei and atomic clusters and found that

$$S = aT + bT\ln(cT), \tag{7.84}$$

where a, b and c are constants.

Recently, we have extended this kind of investigation to cover several other values of α. The Thomas-Fermi model is a crude approximation for atoms (though it is correct in the limit of large atomic numbers). Applying the generalization presented here, we studied the relation between the kinetic energy functional and the Rényi entropy of order α for neutral atoms with $Z = 1$–36. The Rényi entropy can be approximated with the kinetic energy as

$$R_\rho^\alpha = a(\alpha)\ln T + b(\alpha), \tag{7.85}$$

where the parameters can be found in [45].

Finally, we utilize the fact that $\lim_{\alpha\to\infty} R_\rho^\alpha = -\ln\|\rho\|_\infty = -\ln\rho(0)$. It follows from $\|\rho\|_p \equiv (\int \rho^p d\mathbf{r})^{1/p}$ and $\|\rho\|_\infty \equiv \sup|\rho|$ [88], where we made use of the fact that the ground state density of atoms is a non-increasing function of r. Consequently, one finds that when α goes to infinity expression (7.85) gives us $-\ln\rho(0) = a(\infty)\ln T + b(\infty)$. Thus, there is a relation between the value of the density at the nucleus $\rho(0)$ and the kinetic energy $\rho(0) = ZT^{-a(\infty)}e^{-b(\infty)}$.

7.7 Summary and Future Directions

This chapter is devoted to summarize several properties of the Rényi entropy. It is emphasized that there exist uncertainty relations for the Rényi entropy sum. There are relations for the Rényi entropy sum arising both from the many particle and for single particle densities. The Fisher information and the Rényi entropy can be combined to define the Fisher-Rényi product and information plane. Several properties, including inequalities are derived.

Position and momentum space Rényi entropies of order α are presented for ground-state neutral atoms. It is found that the values of $\alpha \leq 1$ ($\alpha \geq 1$) stress the shell structure for position-space (momentum-space) Rényi entropies. Position and momentum space relative Rényi entropies of order α are also presented. A relationship with the atomic radius and quantum capacitance is also discussed.

The maximum Rényi entropy principle is used to generalize the Thomas-Fermi model. In this case there is a physical meaning of the Rényi parameter α as it is related to the dimension. An approximate expression between the Rényi entropy and the logarithm of the kinetic energy of atoms and as a special case an approximate relation between the density at the nucleus and the kinetic energy are presented.

The link between complexity and Rényi entropy is emphasized. A generalized statistical complexity based on the Rényi entropy is defined. This is a one-parameter extension of the original LMC complexity. A further, two-parameter extension of the complexity measure has just taken place [89].

We can conclude that the Rényi entropy is an important quantity to characterize complex systems. It is a fundamental ingredient of two essential measures of complexity: the LMC complexity and the Fisher-Rényi product. Moreover, properly defined combinations of two Rényi entropies give a one-parameter (or even a two-parameter) family of complexity measures. Further studies in these directions are in progress.

Acknowledgements E.R. acknowledges the Spanish project FQM-165/0207 (Junta de Andalucía) and No. FIS2008-01143. Á.N. acknowledges grant OTKA No. T67923. The work was also supported by the TAMOP 4.2.1/B-09/1/KONV-2010-0007 project. The project is co-financed by the European Union and the European Social Fund.

References

1. Shannon CE (1948) A mathematical theory of communication. Bell Syst Tech J 27:379–423
2. Sears SB, Parr RG, Dinur U (1980) On the quantum-mechanical kinetic-energy as a measure of the information in a distribution. Isr J Chem 19:165–173
3. Gadre SR (1984) Information entropy and Thomas-Fermi theory. Phys Rev A 30:620–621
4. Gadre SR, Sears SB, Chakravorty SJ, Bendale RD (1985) Some novel characteristics of atomic information entropies. Phys Rev A 32:2602–2606
5. Tripathi AN, Sagar RP, Esquivel RO, Smith VH Jr (1992) Electron correlation in momentum space—the beryllium-atom isoelectronic sequence. Phys Rev A 45:4385–4392
6. Yánez RJ, Van Assche W, Dehesa JS (1994) Position and momentum information entropies of the D-dimensional harmonic-oscillator and hydrogen-atom. Phys Rev A 32:3065–3079
7. Hó M, Sagar RP, Smith VH Jr, Esquivel RO (1994) Atomic information entropies beyond the Hartree-Fock limit. J Phys B 27:5149–5157
8. Hó M, Sagar RP, Pérez-Jordá JM, Smith VH Jr, Esquivel RO (1994) A molecular study of molecular information entropies. Chem Phys Lett 219:15–20
9. Nagy Á, Parr RG (1996) Information entropy as a measure of the quality of an approximate electronic wave function. Int J Quant Chem 58:323–327
10. Guevara NL, Sagar RP, Esquivel RO (2003) Shannon-information entropy sum as a correlation measure in atomic systems. Phys Rev A 67:012507
11. Guevara NL, Sagar RP, Esquivel RO (2003) Information uncertainty-type inequalities in atomic systems. J Chem Phys 119:7030–7036
12. Guevara NL, Sagar RP, Esquivel RO (2005) Local correlation measures in atomic systems. J Chem Phys 122:084101
13. Moustakidis ChC, Massen SE (2005) Dependence of information entropy of uniform Fermi systems on correlations and thermal effects. Phys Rev B 71:045102
14. Sen KD (2005) Characteristic features of Shannon information entropy of confined atoms. J Chem Phys 123:074110
15. Fisher RA (1925) Theory of statistical estimation. Proc Camb Philos Soc 22:700–725
16. Frieden BR (1989) Fisher information as the basis for the Schrodinger wave-equation. Am J Phys 57:1004–1008
17. Reginatto M (1998) Derivation of the equations of nonrelativistic quantum mechanics using the principle of minimum Fisher information. Phys Rev A 58:1775–1778
18. Frieden BR (1998) Physics from Fisher information. A unification. Cambridge University Press, Cambridge
19. Nalewajski R (2003) Information principles in the theory of electronic structure. Chem Phys Lett 372:28–34

20. Nagy Á (2003) Fisher information in density functional theory. J Chem Phys 119:9401–9405
21. Romera E, Sánchez-Morena P, Dehesa JS (2005) The Fisher information of single-particle systems with a central potential. Chem Phys Lett 414:468–472
22. Nagy Á (2006) Fisher information in a two-electron entangled artificial atom. Chem Phys Lett 425:154–156
23. Nagy Á, Sen KD (2006) Atomic Fisher information versus atomic number. Phys Lett A 360:291–293
24. Hornyák I, Nagy Á (2007) Phase-space Fisher information. Chem Phys Lett 437:132–137
25. Romera E, Dehesa JS (2004) The Fisher-Shannon information plane, an electron correlation tool. J Chem Phys 120:8906–8912
26. Romera E (2002) Stam's principle D-dimensional uncertainty-like relationships and some atomic properties. Mol Phys 100:3325–3329
27. Nagy Á, Sen KD (2006) Atomic Fisher information versus atomic number. Phys Lett A 360:291–293
28. Liu SB (2007) On the relationship between densities of Shannon entropy and Fisher information for atoms and molecules. J Chem Phys 126:191107
29. Nagy Á (2007) Fisher information and Steric effect. Chem Phys Lett 449:212–215
30. Nagy Á, Liu SB (2008) Local wave-vector, Shannon and Fisher information. Phys Lett A 372:1654–1656
31. Szabó JB, Sen KD, Nagy Á (2008) The Fisher-Shannon information plane for atoms. Phys Lett A 372:2428–2430
32. Rényi A (1961) In: Proceedings of fourth Berkeley symp on mathematics, statistics and probability, vol 1. Univ California Press, Berkeley, p 547
33. Gühne O, Lewenstein M (2004) Entropic uncertainty relations and entanglement. Phys Rev A 70:022316
34. Adesso G, Serafini A, Illuminati F (2004) Extremal entanglement and mixedness in continuous variable systems. Phys Rev A 70:022318
35. Bovino A, Castagnolli G, Ekert A, Horodecki P, Alves CM, Serfienko AV (2005) Direct measurement of nonlinear properties of bipartite quantum states. Phys Rev Lett 95:240407
36. Renner R, Gisin N, Kraus B (2005) Information-theoretic security proof for quantum-key-distribution protocols. Phys Rev A 72:012332
37. Giovannetti V, Lloyd S (2004) Additivity properties of a Gaussian channel. Phys Rev A 69:062307
38. Lévay P, Nagy S, Pipek J (2005) Elementary formula for entanglement entropies of fermionic systems. Phys Rev A 72:022302
39. Romera E, de los Santos F (2008) Fractional revivals through Renyi uncertainty relations. Phys Rev A 78:013837
40. Arbo DG, Reinhold CO, Burgdörfer J, Pattanayak AK, Stokely CL, Zhao W, Lancaster JC, Dunning FB (2003) Pulse-induced focusing of Rydberg wave packets. Phys Rev A 67:063401
41. Romera E, Nagy Á (2008) Rényi information of atoms. Phys Lett A 372:4918–4922
42. Nagy Á, Romera E (2009) Relative Rényi entropy for atoms. Int J Quant Chem 109:2490–2494
43. Romera E, Nagy Á (2008) Fisher-Rényi entropy product and information plane. Phys Lett A 372:6823–6825
44. López-Ruiz R, Mancini HL, Calbet X (1995) A statistical measure of complexity. Phys Lett A 209:321–326
45. Nagy Á, Romera E (2009) Maximum Renyi entropy principle and the generalized Thomas-Fermi model. Phys Lett A 373:844–846
46. Hirschman IJ (1957) A note on entropy. Am J Math 79:152–156
47. Bialynicki-Birula I, Mycielski I (1975) Uncertainty relations for information entropy in wave mechanics. Commun Math Phys 44:129–132
48. Beckner W (1975) Inequalities in Fourier-analysis. Ann Math 102:159–182
49. Dembo A, Cover TM, Thomas JA (1991) Information theoretic inequalities. IEEE Trans Inf Theory 37:1501–1518

50. Bialynicki-Birula I (2006) Formulation of the uncertainty relations in terms of the Renyi entropies. Phys Rev A 74:052101
51. Koga T, Kanayama K, Watanabe S, Thakkar AJ (1999) Analytical Hartree-Fock wave functions subject to cusp and asymptotic constraints: He to Xe, Li^+ to Cs^+, H^- to I^-. Int J Quant Chem 71:491–497
52. Koga T, Kanayama K, Watanabe S, Imai T, Thakkar AJ (2000) Analytical Hartree-Fock wave functions for the atoms Cs to Lr. Theor Chem Acc 104:411–413
53. Sagar RP, Ramirez JC, Esquivel RO, Ho M, Smith VH Jr (2001) Shannon entropies and logarithmic mean excitation energies from cusp- and asymptotic-constrained model densities. Phys Rev A 63:022509
54. Sagar RB, Guevara NL (2008) Relative entropy and atomic structure. J Mol Struct, Theochem 857:72–77
55. Ellenbogen JC (2006) Neutral atoms behave much like classical spherical capacitors. Phys Rev A 74:034501
56. Iafrate GJ, Hess K, Krieger JB, Macucci M (1995) Capacitive nature of atomic-sized structures. Phys Rev B 52:10737–10739
57. Perdew JP (1988) Correction. Phys Rev B 37:4267–4267
58. Vignat C, Bercher JF (2003) Analysis of signals in the Fisher-Shannon information plane. Phys Lett A 312:27–33
59. Zozor S, Portesi M, Vignat C (2008) Some extensions of the uncertainty principle. Physica A 387:19–20
60. Hoffmann Ostenhof M, Hoffmann Ostenhof T (1977) "Schrödinger inequalities" and asymptotic behavior of the electron density of atoms and molecules. Phys Rev A 16:1782–1785
61. Carlen EA (1991) Superadditivity of Fisher information and logarithmic Sobolev inequalities. J Funct Anal 101:194–211
62. Carbó R, Arnau J, Leyda L (1980) How similar is a molecule to another—an electron-density measure of similarity between 2 molecular-structures. Int J Quant Chem 17:1185–1189
63. Borgou A, Godefroid M, Indelicato P, De Proft F, Geerlings P (2007) Quantum similarity study of atomic density functions: insights from information theory and the role of relativistic effects. J Chem Phys 126:044102
64. Angulo JC (2007) Atomic quantum similarity indices in position and momentum spaces. J Chem Phys 126:044106
65. Oniescu O (1966) C R Acad Sci Paris A 263:25
66. Hall MJW (1999) Universal geometric approach to uncertainty, entropy, and information. Phys Rev A 59:2602–2615
67. Pennini F, Plastino A (2007) Localization estimation and global vs local information measures. Phys Lett A 365:263–267
68. Hyman AS, Yaniger SI, Liebman JL (1978) Interrelations among X-ray-scattering, electron-densities, and ionization-potentials. Int J Quant Chem 19:757–766
69. Pipek J, Varga I (1997) Statistical electron densities. Int J Quant Chem 64:85–93
70. Borgou A, De Proft F, Geerlings P, Sen KD (2007) Complexity of Dirac-Fock atom increases with atomic number. Chem Phys Lett 44:186–191
71. Romera E, López-Ruiz R, Sanudo J, Nagy Á (2009) Generalized statistical complexity and Fisher-Rényi entropy product in the H-atom. Int Rev Phys (IREPHY) 3:207–211
72. Levy M (1979) Universal variational functionals of electron-densities, 1st-order density-matrices, and natural spin-orbitals and solution of the v-representability problem. Proc Natl Acad Sci USA 76:6062–6065
73. Lieb EH (1983) Density functional for coulomb systems. Int J Quant Chem 24:243–277
74. Thomas LH (1927) The calculation of atomic fields. Proc Camb Philos Soc 23:542–548
75. Fermi E (1928) Eine statistische Methode zur Bestimmung einiger Eigenschaften des Atoms und ihre Anwendung auf die Theorie des periodischen Systems der Elemente. Z Phys 48:73–79
76. Kventsel GF, Katriel J (1981) Thomas-Fermi atom in N-dimensions. Phys Rev A 24:2299–2301

77. March NH (1985) Scaling properties of total energy of heavy positive-ions in d-dimensions. J Math Phys 26:554–555
78. Holas A, March NH (1994) Perturbation and density-gradient expansions in d-dimensions. Philos Mag 69:787–798
79. March NH, Kais S (1997) Kinetic energy functional derivative for the Thomas-Fermi atom in D dimensions. Int J Quant Chem 65:411–413
80. Shivamoggi BK (1998) Thomas-Fermi theory in an n-dimensional space. Physica A 248:195–206
81. Janes ET (1957) Information theory and statistical mechanics. Phys Rev 106:620–630
82. Janes ET (1957) Information theory and statistical mechanics. II. Phys Rev 108:171–190
83. Sanudo J, Pacheco AF (2006) Electrons in a box: Thomas-Fermi solution. Can J Phys 84:833–844
84. Schuck P, Vinas X (2000) Thomas-Fermi approximation for Bose-Einstein condensates in traps. Phys Rev A 61:043603
85. Cappelluti E, Delle Site L (2002) Generalized Thomas-Fermi approach for systems under pressure. Physica A 303:481–492
86. Hodak M, Lu W, http://meetings.aps.org/link/BAPS.2006.MAR.V27.8
87. Massen, Panos (2001) A link of information entropy and kinetic energy for quantum many-body systems. Phys Lett A 280:65–69
88. Debnath L, Mikusinski P (2005) Introduction to Hilbert spaces. Academic Press, San Diego
89. López-Ruiz R, Nagy Á, Romera E, Sanudo J (2009) A generalized statistical complexity measure: applications to quantum systems. J Math Phys 50:123528

Chapter 8
Scaling Properties of Net Information Measures for Bound States of Spherical Model Potentials

K.D. Sen and S.H. Patil

Abstract Using dimensional analysis of (a) the position and momentum variances which define the quantum mechanical Heisenberg uncertainty product, and (b) several *composite* information theoretical measures, the scaling properties of the various information theoretical *uncertainty-like* relationships are derived for the bound states corresponding to a set of non-relativistic spherical model potentials, $V(r)$. The potentials considered are described by (1) adding a/r^2 term to (i) the isotropic harmonic oscillator, $V_1(r) = \frac{1}{2}kr^2 + \frac{a}{r^2}$ and to (ii) the Coulombic hydrogen-like potentials, $V_2(r) = -\frac{Z}{r} + \frac{a}{r^2}$ (2) the exponential cosine screened Coulomb potentials generated by multiplying the *superposition* of (i) Yukawa-like, $-Z\frac{e^{-\mu r}}{r}$, and (ii) Hulthén-like, $-Z\mu\frac{1}{e^{\mu r}-1}$, potentials by $\cos(b\mu r)$ followed by addition of the term a/r^2 where a and $b \geq 0$, μ is the screening parameter and Z, in case of atoms, represents the nuclear charge, along with their generalized forms, $V_3(r) = -[Z\frac{e^{-\mu r}}{r} + \sum_i c_i Z\frac{e^{-s_i \mu r}}{r}]\cos(b\mu r) + \frac{a}{r^2}$, and $V_4(r) = -[Z\frac{\mu}{e^{\mu r}-1} + \sum_i c_i Z\frac{s_i \mu}{e^{s_i \mu r}-1}]\cos(b\mu r) + \frac{a}{r^2}$, (3) the Pöschl-Teller, $V_5(r) = -Z\,\mathrm{sech}^2(\mu r)$, and the Morse potentials, $V_6(r) = Z[e^{-2\mu(r-r_0)} - 2e^{-\mu(r-r_0)}]$, (4) two soft-Coulomb potentials which describe an electron moving in the central field due to a smeared nuclear charge described by Z and $\beta > 0$ according to $V_7(r) = -\frac{Z}{(r^n+\beta^n)^{1/n}}$, (5) the spherically confined Hydrogenic-potential, $V_8(r) = -\frac{Z}{r}$ for $r < R$; and $= \infty$ for $r \geq R$, (6) the superposition of the power potential of the form $V_9(r) = Zr^n + \sum_i Z_i r^{n_i}$, where Z, Z_i, n, n_i are parameters, in the free state as well as in the additional presence of a spherical *penetrable* and impenetrable boundary wall located at radius R, and (7) the Hydrogen-like atoms in the presence of strong parallel and perpendicular magnetic (B) and electric (F) fields, $V_{10}(\mathbf{r}) = -\frac{Z}{r} + \frac{1}{2}B^2(x^2+y^2) + Fz$. As an illustrative example we consider the superposition of the power potential of the form $V(r) = Zr^n + \sum_i Z_i r^{n_i}$ where Z, Z_i, n, n_i are parameters yielding *bound states* for a particle of mass M. The uncertainty product and *all* other *net* information

K.D. Sen (✉)
School of Chemistry, University of Hyderabad, Hyderabad 500 046, India
e-mail: kalidas.sen@gmail.com

K.D. Sen (ed.), *Statistical Complexity*,
DOI 10.1007/978-90-481-3890-6_8,

measures are shown here to depend only on the parameters $[s_i]$ defined by the ratios $Z_i/Z^{(n_i+2)/(n+2)}$. Under the imposition of a spherical impenetrable boundary of radius R over the polynomial potential, parametric dependence becomes $[s_i, t_1]$ where t_1 is given by $RZ^{1/(n+2)}$. Introduction of a finite potential, V_c at the radial distance $r \geq R$ results in a complete set of scaling parameters given by $[s_i, t_1, t_2]$, where $t_2 = V_c/(Z)^{2/(n+2)}$. Analogous results on the scaling property of the uncertainty-like information theoretical measures are tabulated for the chosen set of spherical model potentials. Significance of the scaling properties will be discussed. Illustrative numerical tests of the scaling behavior will be presented.

8.1 Introduction

The uncertainty relations are the basic properties of quantum mechanics [1, 2], in particular, we have the Heisenberg uncertainty principle for the product of the uncertainties in position and momentum,

$$\begin{aligned} &\sigma_x \sigma_p \geq \frac{1}{2}\hbar, \\ &\sigma_x^2 = \langle (x - \langle x \rangle)^2 \rangle, \qquad \sigma_p^2 = \langle (p_x - \langle p_x \rangle)^2 \rangle, \end{aligned} \tag{8.1}$$

in terms of Planck's constant. The uncertainty product has many interesting properties for different potentials, for example, the product for bound states in homogeneous, power potentials is independent of the strength of the potentials. While the uncertainty principle is by definition expressed in terms of the quantum mechanical expectation values, there exist other interesting uncertainty-like relationships [3, 4] expressed in terms of the information theoretical measures [5–16] defined by the probability. Several numerical studies have been carried out using the continuous electron probability density distributions in the atomic and molecular structure theory [17–48]. Here we will consider some general properties for the bound states in superpositions of power potentials with a finite barrier. It is observed that the dimensionality and scaling properties lead to interesting properties of the uncertainty product, and densities with implications for entropies and information. The scaling properties corresponding to a test-set of spherical potentials listed in the abstract have been studied earlier [54–68] by us following a similar dimensional analysis. A comprehensive summary of the results so obtained are also compiled in a tabular form. The aim of such a study is to ascertain how the parameters in the potential describe the various information-theoretical uncertainty-like measures expressed through the electron probability density as the key parameter. Finally, illustrative numerical tests of the scaling behavior will be presented in the case of spherically confined hydrogen atom inside an impenetrable cavity with interesting results.

8.2 Heisenberg Uncertainty Relations

Here we analyze some dimensionality properties and their implications for the uncertainty relations for the bound states in superpositions of power potentials with a finite barrier.

8.2.1 Superpositions of Power Potentials with a Finite Barrier

Consider a potential of the form

$$V(r) = \left[Zr^n + \sum_i Z_i r^{n_i} \right] \theta(R-r) + V_c \theta(r-R) \tag{8.2}$$

where Z, Z_i, n, n_i, R, V_c are parameters (n, n_i may not be integers) with θ being the Heaviside theta function, in which there are bound states for a particle of mass M. Specifically, we have

$$V_1(r) = [-kr^2 + \lambda r^4]\theta(R-r) + V_c\theta(r-R) \tag{8.3}$$

for a symmetric double well potential,

$$V_2(r) = \left[\frac{1}{2}kr^2 + \frac{a}{r^2} \right] \theta(R-r) + V_c\theta(r-R) \tag{8.4}$$

for a modified s.h.o., and

$$V_3(r) = \left[-\frac{Z}{r} + \lambda r \right] \theta(R-r) + V_c\theta(r-R) \tag{8.5}$$

for a confined hydrogenic system. The Schrödinger equation for the potential in (8.2) is

$$-\frac{\hbar^2}{2M}\nabla^2\psi + \left[\left(Zr^n + \sum Z_i r^{n_i} \right) \theta(R-r) + V_c\theta(r-R) \right] \psi = E\psi. \tag{8.6}$$

8.2.2 Dimensionality and Uncertainty Relations

The basic dimensional parameters in our Schrödinger equation are $\hbar^2/M$, Z, Z_i, R, V_c. Of these,

$$s_i = \frac{M}{\hbar^2} Z_i \left(\frac{\hbar^2}{MZ} \right)^{\left(\frac{n_i+2}{n+2}\right)}, \qquad t_1 = R(MZ/\hbar^2)^{1/(n+2)},$$
$$t_2 = V_c(M/\hbar^2)(\hbar^2/MZ)^{2/(n+2)} \tag{8.7}$$

are the dimensionless parameters. Now we consider the deviations

$$\sigma_{\mathbf{r}}^2 = \langle (\mathbf{r} - \langle \mathbf{r} \rangle)^2 \rangle, \qquad \sigma_{\mathbf{p}}^2 = \langle (\mathbf{p} - \langle \mathbf{p} \rangle)^2 \rangle. \tag{8.8}$$

For our potential in (8.6), the dimensionality properties imply that the deviations are of the form

$$\sigma_{\mathbf{r}} = (\hbar^2/MZ)^{1/(n+2)}\, g_1(s_i, t_i), \qquad \sigma_{\mathbf{p}} = \hbar(MZ/\hbar^2)^{1/(n+2)}\, g_2(s_i, t_i), \tag{8.9}$$

so that the uncertainty product is

$$\sigma_{\mathbf{r}}\sigma_{\mathbf{p}} = \hbar g_1(s_i, t_i) g_2(s_i, t_i), \qquad s_i = \frac{M}{\hbar^2} Z_i \left(\frac{\hbar^2}{MZ}\right)^{\left(\frac{n_i+2}{n+2}\right)},$$
$$t_1 = R(MZ/\hbar^2)^{1/(n+2)}, \qquad t_2 = V_c(M/\hbar^2)(\hbar^2/MZ)^{2/(n+2)}. \tag{8.10}$$

This implies that the uncertainty product depends only on the dimensionless parameters s_i, t_i. It may also be noted that the bound state energies are of the form

$$E = (\hbar^2/M)(MZ/\hbar^2)^{2/(n+2)} g_3(s_i, t_i). \tag{8.11}$$

These results follow from just the dimensionality properties of the parameters.

It is interesting to note that one can analyze the minimum uncertainty in terms of the scaled parameters. For example, in the case of a hydrogen atom with a finite barrier potential, one can obtain the minimum uncertainty by keeping V_c fixed but varying RZ. Alternatively, one can keep R fixed but vary V_c. These considerations can be extended to the superpositions of power potentials with a finite barrier potential in terms of the scaled variables s_i and t_i.

8.3 Scaling Properties and Entropies

We will now consider some scaling properties for bound states in a superposition of power potentials, and their implications for Shannon entropy [5, 6] and other properties.

8.3.1 Scaling Properties

For the Schrödinger equation in (8.6), the energy E and eigenfunction ψ are functions of the form

$$E : E(\hbar^2/M, Z, Z_i, R, V_c), \qquad \psi : \psi(\hbar^2/M, Z, Z_i, R, V_c, r). \tag{8.12}$$

Multiplying (8.6) by $M/\hbar^2$, and introducing a scale transformation

$$\mathbf{r} = \lambda \mathbf{r}' \tag{8.13}$$

one gets

$$-\frac{1}{2}\nabla'^2\psi + (M/\hbar^2)\left[Z\lambda^{n+2}r'^n + \sum Z_i\lambda^{n_i+2}r'^{n_i}\right]\theta(R/\lambda - r')\psi$$
$$+ (M/\hbar^2)\lambda^2 V_c\theta(r' - R/\lambda)\psi = (M/\hbar^2)\lambda^2 E\psi. \tag{8.14}$$

Taking

$$\lambda = \left(\frac{\hbar^2}{MZ}\right)^{1/(n+2)}, \tag{8.15}$$

it leads to

$$-\frac{1}{2}\nabla'^2\psi + \left[r'^n + \frac{M}{\hbar^2}\sum Z_i\left(\frac{\hbar^2}{MZ}\right)^{\left(\frac{n_i+2}{n+2}\right)} r'^{n_i}\right]\theta\left(R(MZ/\hbar^2)^{1/(n+2)} - r'\right)\psi$$
$$+ \frac{M}{\hbar^2}(\hbar^2/MZ)^{2/(n+2)}V_c\theta\left(r' - R(MZ/\hbar^2)^{1/(n+2)}\right)\psi$$
$$= \frac{M}{\hbar^2}\left(\frac{\hbar^2}{MZ}\right)^{2/(n+2)}E\psi. \tag{8.16}$$

Comparing this with (8.6), we obtain

$$E\left(\frac{\hbar^2}{M}, Z, Z_i, R, V_c\right) = (\hbar^2/M)\lambda^{-2}E(1, 1, Z_i\lambda^{n_i+2}M/\hbar^2, R/\lambda, V_c\lambda^2 M/\hbar^2),$$
$$\psi\left(\frac{\hbar^2}{M}, Z, Z_i, R, V_c, r\right) = A\psi(1, 1, Z_i\lambda^{n_i+2}M/\hbar^2, R/\lambda, V_c\lambda^2 M/\hbar^2, r'),$$
$$r = \lambda r',\ \lambda = (\hbar^2/MZ)^{1/(n+2)}. \tag{8.17}$$

Taking $\psi(1, 1, Z_i\lambda^{n_i+2}M/\hbar^2, R/\lambda, V_c\lambda^2/\hbar^2, r')$ to be normalised, the normalization of the wave function $\psi(\hbar^2/M, Z, Z_i, R, V_c, r)$ leads to

$$1 = A^2\int |\psi(1, 1, Z_i\lambda^{n_i+2}M/\hbar^2, R/\lambda, V_c\lambda^2 M/\hbar^2, r')|^2 d^3r$$
$$= A^2\lambda^3 \quad\Rightarrow\quad A = \lambda^{-3/2} = (MZ/\hbar^2)^{\frac{3}{2(n+2)}}, \tag{8.18}$$

so that

$$\psi\left(\frac{\hbar^2}{M}, Z, Z_i, R, V_c, r\right) = \lambda^{-3/2}\psi(1, 1, Z_i\lambda^{n_i+2}M/\hbar^2, R/\lambda, V_c\lambda^2 M/\hbar^2, r'),$$
$$\lambda = (\hbar^2/MZ)^{1/(n+2)},\ r' = r/\lambda. \tag{8.19}$$

For obtaining the wave function in the momentum space, we take the Fourier transform of the wave function in (8.19), leading to

$$f(\hbar^2/M, Z, Z_i, R, V_c, p) = \frac{1}{(2\pi\hbar)^{3/2}}\int d^3r e^{-i\mathbf{p}\cdot\mathbf{r}/\hbar}\psi(\hbar^2/M, Z, Z_i, R, V_c, r). \tag{8.20}$$

Using the relation in (8.19) and changing the integration variable to r', we get

$$f(\hbar^2/M, Z, Z_i, R, V_c, p) = \lambda^{3/2}f(1, 1, Z_i\lambda^{n_i+2}M/\hbar^2, R/\lambda, V_c\lambda^2 M/\hbar^2, p'),$$
$$\lambda = (\hbar^2/MZ)^{1/(n+2)},\ p' = \lambda p. \tag{8.21}$$

From the relations in (8.19) and (8.21), one has for the corresponding position and momentum densities,

$$\rho(\hbar^2/M, Z, Z_i, R, V_c, r) = \lambda^{-3}\rho(1, 1, s_i, t_1, t_2, r'),$$
$$\gamma(\hbar^2/M, Z, Z_i, R, V_c, p) = \lambda^3\gamma(1, 1, s_i, t_1, t_2, p'),$$
$$\lambda = (\hbar^2/MZ)^{1/(n+2)},\ s_i = \lambda^{n_i+2}MZ_i/\hbar^2,\ t_1 = R/\lambda,$$
$$t_2 = V_c\lambda^2M/\hbar^2,\ r' = r/\lambda,\ p' = \lambda p, \tag{8.22}$$

with s_i, t_i being the scaled parameters.

8.3.2 Shannon Entropy Sum

The Shannon entropies [5, 6] in the position space and momentum space, are

$$S_r = -\int \rho(r)[\ln\rho(r)]d^3r, \qquad S_p = -\int \gamma(p)[\ln\gamma(p)]d^3p. \tag{8.23}$$

Using the relations in (8.22), we get for these entropies

$$\begin{aligned} S_r(\hbar^2/M, Z, Z_i, R, V_c) &= 3\ln\lambda + S_r(1, 1, s_i, t_1, t_2), \\ S_p(\hbar^2/M, Z, Z_i, R, V_c) &= -3\ln\lambda + S_p(1, 1, s_i, t_1, t_2), \end{aligned} \tag{8.24}$$

which imply that the Shannon entropy sum $S_T = S_r + S_p$ satisfies the relation

$$S_T(\hbar^2/M, Z, Z_i, R, V_c) = S_T(1, 1, s_i, t_1, t_2), \quad s_i = \frac{M}{\hbar^2}Z_i\left(\frac{\hbar^2}{MZ}\right)^{(\frac{n_i+2}{n+2})},$$
$$t_1 = R(MZ/\hbar^2)^{1/(n+2)},\ t_2 = V_c(M/\hbar^2)(\hbar^2/MZ)^{2/(n+2)}. \tag{8.25}$$

Therefore, for given values of the parameters Z, Z_i, R, V_c, the Shannon entropy sum depends only on $Z_i/Z^{(n_i+2)/(n+2)}, t_1, t_2$. This result is significant in relation to the uncertainty-like relationship [3]

$$S_T = S_r + S_p \geq N(1 + \ln\pi), \tag{8.26}$$

which is the strongest statement of Uncertainty Principle known for an N-dimensional system in quantum mechanics.

8.3.3 Fisher Information

The Fisher information measures [7–9] for position and momentum are

$$I_r = \int \frac{[\nabla\rho(r)]^2}{\rho(r)}d^3r, \qquad I_p = \int \frac{[\nabla\gamma(p)]^2}{\gamma(p)}d^3p. \tag{8.27}$$

Using the relations in (8.22), one obtains

$$\begin{aligned} I_r(\hbar^2/M, Z, Z_i, R, V_c) &= \frac{1}{\lambda^2}I_r(1, 1, s_i, t_1, t_2), \\ I_p(\hbar^2/M, Z, Z_i, R, V_c) &= \lambda^2 I_p(1, 1, s_i, t_1, t_2), \end{aligned} \tag{8.28}$$

which together imply that the Fisher information product $I_r I_p$ satisfies the relation

$$\begin{aligned} &I_{rp}(\hbar^2/M, Z, Z_i, R, V_c) = I_{rp}(1, 1, s_i, t_1, t_2), \quad I_{rp} = I_r I_p, \\ &s_i = \frac{M}{\hbar^2} Z_i \left(\frac{\hbar^2}{MZ}\right)^{(\frac{n_i+2}{n+2})}, \ t_1 = R(MZ/\hbar^2)^{1/(n+2)}, \\ &t_2 = V_c(M/\hbar^2)(\hbar^2/MZ)^{2/(n+2)}. \end{aligned} \tag{8.29}$$

Here, for given values of the parameters Z, Z_i, R, V_c, the Fisher information product depends only on $Z_i/Z^{(n_i+2)/(n+2)}, t_1, t_2$.

8.3.4 Rényi Entropy

The Rényi entropies [12–14] in position and momentum spaces are

$$H_\alpha^{(r)} = \frac{1}{1-\alpha} \ln \int [\rho(r)]^\alpha d^3 r, \qquad H_\alpha^{(p)} = \frac{1}{1-\alpha} \ln \int [\gamma(p)]^\alpha d^3 p. \tag{8.30}$$

With the relations in (8.22), we get for these entropies,

$$\begin{aligned} H_\alpha^{(r)}(\hbar^2/M, Z, Z_i, R, V_c) &= 3 \ln \lambda + H_\alpha^{(r)}(1, 1, s_i, t_1, t_2), \\ H_\alpha^{(p)}(\hbar^2/M, Z, Z_i, R, V_c) &= -3 \ln \lambda + H_\alpha^{(p)}(1, 1, s_i, t_1, t_2), \end{aligned} \tag{8.31}$$

which imply that the Rényi entropy sum $H_\alpha^{(T)} = H_\alpha^{(r)} + H_\alpha^{(p)}$ satisfies the relation

$$\begin{aligned} &H_\alpha^{(T)}\left(\frac{\hbar^2}{M}, Z, Z_i, R, V_c\right) = H_\alpha^{(T)}(1, 1, s_i, t_1, t_2), \quad s_i = \frac{M}{\hbar^2} Z_i \left(\frac{\hbar^2}{MZ}\right)^{(\frac{n_i+2}{n+2})}, \\ &t_1 = R(MZ/\hbar^2)^{1/(n+2)}, \ t_2 = V_c(M/\hbar^2)(\hbar^2/MZ)^{2/(n+2)}. \end{aligned} \tag{8.32}$$

Therefore, as in other cases, for given values of the parameters Z, Z_i, R, V_c, the Rényi entropy sum depends on $Z_i/Z^{(n_i+2)/(n+2)}, t_1, t_2$.

8.3.5 Onicescu Energies

The Onicescu energies [15] in position and momentum spaces are

$$E_r = \int [\rho(r)]^2 d^3 r, \qquad E_p = \int [\gamma(p)]^2 d^3 p. \tag{8.33}$$

Using the relations in (8.22), we get

$$\begin{aligned} E_r(\hbar^2/M, Z, Z_i, R, V_c) &= \frac{1}{\lambda^3} E_r(1, 1, s_i, t_1, t_2), \\ E_p(\hbar^2/M, Z, Z_i, R, V_c) &= \lambda^3 E_p(1, 1, s_i, t_1, t_2), \end{aligned} \tag{8.34}$$

which imply that the Onicescu energy product $E_{rp} = E_r E_p$ satisfies the relation

$$E_{rp}(\hbar^2/M, Z, Z_i, R, V_c) = E_{rp}(1, 1, s_i, t_1, t_2), \quad s_i = \frac{M}{\hbar^2} Z_i \left(\frac{\hbar^2}{MZ}\right)^{(\frac{n_i+2}{n+2})},$$
$$t_1 = R(MZ/\hbar^2)^{1/(n+2)}, \; t_2 = V_c(M/\hbar^2)(\hbar^2/MZ)^{2/(n+2)}. \tag{8.35}$$

In this case also, for given values of the parameters Z, Z_i, R, V_c, the Onicescu energy product depends on $Z_i/Z^{(n_i+2)/(n+2)}, t_1, t_2$.

8.3.6 Tsallis Entropy

The Tsallis entropies [16] in position and momentum spaces are

$$T_r = \frac{1}{q-1}\left[1 - \int [\rho(r)]^q d^3 r\right], \quad T_p = \frac{1}{m-1}\left[1 - \int [\gamma(p)]^m d^3 p\right],$$
$$\frac{1}{q} + \frac{1}{m} = 2. \tag{8.36}$$

We consider the integral terms

$$J_r(\hbar^2/M, Z, Z_i, R, V_c) = \int [\rho(r)]^q d^3 r,$$
$$J_p(\hbar^2/M, Z, Z_i, R, V_c) = \int [\gamma(p)]^m d^3 p. \tag{8.37}$$

Using the relations in (8.22), we get

$$J_r(\hbar^2/M, Z, Z_i, R, V_c) = \lambda^{3-3q} J_r(1, 1, s_i, t_1, t_2),$$
$$J_p(\hbar^2/M, Z, Z_i, R, V_c) = \lambda^{3m-3} J_p(1, 1, s_i, t_1, t_2). \tag{8.38}$$

Then one obtains for the ratio,

$$J_{p/r}(\hbar^2/M, Z, Z_i, R, V_c) = J_{p/r}(1, 1, s_i, t_1, t_2), \quad J_{p/r} = \frac{J_p^{1/2m}}{J_r^{1/2q}}, \; \frac{1}{m} + \frac{1}{q} = 2,$$
$$s_i = \frac{M}{\hbar^2} Z_i \left(\frac{\hbar^2}{MZ}\right)^{(\frac{n_i+2}{n+2})}, \; t_1 = R(MZ/\hbar^2)^{1/(n+2)},$$
$$t_2 = V_c(M/\hbar^2)(\hbar^2/MZ)^{2/(n+2)}. \tag{8.39}$$

Therefore in this case also, for given values of the parameters Z, Z_i, R, V_c, the ratio of Tsallis entropies depends on $Z_i/Z^{(n_i+2)/(n+2)}, t_1, t_2$.

8.3.7 Statistical Complexity Measure

A couple of statistical measures of complexity [49–52] has been used to study the complexity of electronic systems. A given measure becomes significant when a rigorous bound on it is known to exist. In this letter, we focus on the LMC (López-Ruiz-Mancini-Calbet) complexity [49, 50], C_{LMC}, with the aim of deriving a general lower bound for it.

Consider a D-dimensional distribution function $f(\mathbf{r})$, with $f(\mathbf{r})$ nonnegative and $\int f(\mathbf{r})d\mathbf{r} = 1$; $\mathbf{r}$ stands for $r_1, \ldots, r_D$. The Shannon entropy [5, 6] and the Shannon entropy power are defined as

$$S_f = -\int f(\mathbf{r}) \ln f(\mathbf{r}) d\mathbf{r}, \tag{8.40}$$

$$H_f = e^{S_f}, \tag{8.41}$$

respectively. The so called disequilibrium D has the form

$$D_f = \int f^2(\mathbf{r}) d\mathbf{r}. \tag{8.42}$$

The definition of the LMC complexity measure is [49, 50]

$$C_{LMC} = H.D. \tag{8.43}$$

The form of C_{LMC} is designed such that it vanishes for the two extreme probability distributions corresponding to perfect order ($H = 0$) and maximum disorder ($D = 0$). It is known [49, 50] that the complexity corresponding to probability distributions given by rectangular, triangular, Gaussian and exponential functions in one-dimensional position space is given by 1, $(2/3)(e^{1/2})$, $(e^{1/2})/2$, and $e/2$, respectively. The rectangular probability distribution, by definition, corresponds to the minimum statistical complexity. We shall now derive the lower bound for C_{LMC} corresponding to a given one-electron density.

8.3.8 Lower Bound on LMC Complexity

To derive a lower bound for the LMC complexity we cite Theorem 2 in the reference [53]. The position-space entropy $\tilde{S}_\varrho$ of an N-electron system in a physical state characterized by the (normalized to N) one-electron density $\varrho(\mathbf{r})$ fulfills the inequality

$$\tilde{S}_\varrho + \langle \ln g(\mathbf{r}) \rangle \leq N \ln\left(\frac{\int g(\mathbf{r}) d\mathbf{r}}{N} \right), \tag{8.44}$$

where $g(\mathbf{r})$ is an arbitrary positive function. From the relationship between the Shannon entropies coming from densities normalized to 1 and N [53]:

$$S_f = \frac{\tilde{S}_\varrho}{N} + \ln N \tag{8.45}$$

Table 8.1 A compilation of the definitions of quantum mechanical uncertainty principle, information theoretical uncertainty-like relationships and LMC statistical complexity measure

Heisenberg Uncertainty Measure	$\sigma_{\mathbf{r}}^2\sigma_{\mathbf{p}}^2 = \langle(\mathbf{r}-\langle\mathbf{r}\rangle)^2\rangle\langle(\mathbf{p}-\langle\mathbf{p}\rangle)^2\rangle,$
Shannon Information Entropy Sum	$S_r = -\int \rho(\mathbf{r})\ln\rho(\mathbf{r})\,d\mathbf{r}$
	$S_p = -\int \rho(\mathbf{p})\ln\rho(\mathbf{p})\,d\mathbf{p}$
	$S_T = S_r + S_p \geq D(1+\ln\pi)$
Fisher Information Measure Product	$I_r = \int \frac{\vert\nabla\rho(\mathbf{r})\vert^2}{\rho(\mathbf{r})}\,d\mathbf{r},\ I_p = \int \frac{\vert\nabla\rho(\mathbf{p})\vert^2}{\rho(\mathbf{p})}\,d\mathbf{p}$
Onicescu information product	$E_r = \int[\rho(\mathbf{r})]^2 d\mathbf{r},\ E_p = \int[\gamma(\mathbf{p})]^2 d\mathbf{p}$
Rényi entropy sum	$H_\alpha^{(r)} = \frac{1}{1-\alpha}\ln\int[\rho(\mathbf{r})]^\alpha d\mathbf{r}$
	$H_\alpha^{(p)} = \frac{1}{1-\alpha}\ln\int[\gamma(\mathbf{p})]^\alpha d\mathbf{p}$
Tsallis entropy ratio	$T_n^{(r)} = \frac{1}{n-1}[1-J_n^{(r)}],\ T_q^{(p)} = \frac{1}{q-1}[1-J_q^{(p)}]$
	$J_n^{(r)} = \int[\rho(\mathbf{r})]^n d\mathbf{r},\ J_q^{(p)} = \int[\gamma(\mathbf{p})]^q d\mathbf{p}$
	$J_{p/r} = \frac{(J_q^{(p)})^{1/2q}}{(J_n^{(r)})^{1/2n}}$
Fisher Shannon Plane	$N_r = \frac{1}{\pi e}e^{2S_r/3},\ N_p = \frac{1}{\pi e}e^{2S_p/3},\ N_r I_r$ and $N_p I_p$
Statistical Complexity Measure	$C_r = e^{S_r}E_r,\ C_p = e^{S_p}E_p$

and taking $g = f^2$, we obtain the inequality

$$S_f + \ln\left(\int f^2(\mathbf{r})d\mathbf{r}\right) \geq 0. \tag{8.46}$$

From the definition of the LMC complexity (8.43) we obtain the upper bound

$$\ln C_{LMC} \geq 0 \tag{8.47}$$

or

$$C_{LMC} \geq 1. \tag{8.48}$$

It is to be noted that the scaling properties of C_{LMC} in either the position or momentum space are identical to that of the Heisenberg Uncertainty product and the other composite information theoretical measures such as the sum of Shannon entropy, S_T and the Fisher information product $I_r I_p$.

The various information theoretical measures considered by us [54–68] have been compiled in Table 8.1.

In Table 8.2 we present the scaling properties of the Heisenberg uncertainty product corresponding to a selected group of potentials studied till date by us [54–68] using the procedure described under Sect. 8.3 wherein the new results have been presented for the spherically symmetric polynomial potential with finite barrier heights.

Table 8.2 A compilation of the scaling properties of the various measures of uncertainty and complexity as listed under Table 8.1

Potential	Scaling property
$V_{\text{Hydrogenic}}(r) = -\frac{Z}{r} + \frac{a}{r^2}$	$\sigma_{\mathbf{r}}^2\sigma_{\mathbf{p}}^2 = \hbar^2\, g_1(Ma/\hbar^2)g_2(Ma/\hbar^2)$
$V_{\text{Mod. Yukawa}}(r) = -[Z\frac{e^{-\mu r}}{r} + \sum_i c_i Z\frac{e^{-s_i\mu r}}{r}]\cos(b\mu r) + \frac{a}{r^2}$	$f_1(\hbar^2\mu/MZ, Ma/\hbar^2, c_i, s_i, b) \times f_2(\hbar^2\mu/MZ, Ma/\hbar^2, c_i, s_i, b)$
$V_{\text{Mod. Hulthen}}(r) = -[Z\frac{\mu}{e^{\mu r}-1} + \sum_i c_i Z\frac{s_i\mu}{e^{s_i\mu r}-1}]\cos(b\mu r) + \frac{a}{r^2}$	$\sigma_{\mathbf{r}}^2\sigma_{\mathbf{p}}^2 = \hbar^2 f_1(\hbar^2\mu/MZ, Ma/\hbar^2, c_i, s_i, b) \times f_2(\hbar^2\mu/MZ, Ma/\hbar^2, c_i, s_i, b)$
$V_{\text{Truncated Coulomb}}(r) = -[\frac{Z}{(r^q+\beta^q)^{\frac{1}{q}}}]$	$\sigma_{\mathbf{r}}\sigma_{\mathbf{p}} = \hbar g_1(s)g_2(s),\ s = \frac{MZ\beta}{\hbar^2}$
$V_{\text{3D HO}}(r) = \frac{kr^2}{2} + \frac{a}{r^2}$	$\sigma_{\mathbf{r}}\sigma_{\mathbf{p}} = \hbar(2n + l' + 3/2) \times [1 - \frac{2Ma/\hbar^2}{(l'+1/2)(2n+l'+3/2)}]^{1/2}$
$V_{\text{Pöschl-Teller}}(r) = -Z\,\text{sech}^2(\mu r)$	$\sigma_{\mathbf{r}}\sigma_{\mathbf{p}} = \hbar g_1(s)g_2(s),\ s = \frac{\hbar\mu}{\sqrt{MZ}}$
$V_{\text{Morse}}(r) = Z[e^{-2\mu(r-r_0)} - 2e^{-\mu(r-r_0)}]$	$g_4(s_1, s_2)g_5(s_1, s_2),\ s_1 = \hbar\mu/\sqrt{MZ},\ s_2 = r_0\sqrt{MZ}/\hbar$
$V_{\text{Sum of Power}}(r) = Zr^n + \sum_i Z_i r^{n_i}$	$g_1(s_i)g_2(s_i),\ s_i = \frac{M}{\hbar^2}Z_i(\frac{\hbar^2}{MZ})^{(\frac{n_i+2}{n+2})}$
$V_{\text{BH}}(r) = -\frac{Ze^2}{4\pi\epsilon_0 r} + \frac{eB}{m}L_z + \frac{e^2B^2}{2m}r^2\sin^2\theta + eFr\cos\theta$	$f_1(s_1)f_2(s_2),\ s_1 = \frac{B\hbar^3(4\pi\epsilon_0)^2}{Z^2m^2e^3},\ s_2 = \frac{F\hbar^4(4\pi\epsilon_0)^3}{Z^3e^5m^2}$

8.4 Conclusions

Using the dimensional analysis, we have studied the scaling properties of the Heisenberg Uncertainty principle and a set of information theoretical uncertainty-like relationships for the bound states corresponding to several useful model potentials in chemical physics. Such results enrich our understanding of the potential → electron probability density → uncertainty-like bounds defined in the position and momentum spaces describing the physical systems. Further, except in a very limited number of potentials where the exact solutions are known, the presently derived results can be used to test the accuracy of the approximate wave functions.

Acknowledgements One of the authors, S.H.P. acknowledges support from A.I.C.T.E. as emeritus fellow. K.D.S. is grateful to the Department of Science and Technology for the award of J.C. Bose National Fellowship during 2009–2014. He is thankful to Professor H.E. Montgomery for constant encouragement.

References

1. Heisenberg W (1927) Z Phys 43:172
2. Kennard EH (1927) Z Phys 44:326
3. Bialynicki-Birula I, Mycielski J (1975) Commun Math Phys 44:129

4. Bialynicki-Birula I (2006) Phys Rev A 74:052101
5. Shannon CE (1948) Bell Syst Tech J 27:379
6. Shannon CE (1948) Bell Syst Tech J 27:623
7. Fisher RA (1925) Proc Camb Philos Soc 22:700
8. Frieden BR (2004) Science from Fisher information. Cambridge University Press, Cambridge
9. Rau ARP, arXiv:0910.4163v1 [physics.bio-phys] 21 Oct 2009. For a lucid historical account of the development of Design Theory
10. Rao CR (1965) Linear statistical interference and its applications. Wiley, New York
11. Stam A (1959) Inf Control 2:101
12. Rényi A (1960) Some fundamental questions of information theory. MTA III Oszt Kzl 10:251
13. Rényi A (1960) On measures of information and entropy. In: Proceedings of the fourth Berkeley symposium on mathematics, statistics and probability. Berkeley University Press, Berkeley, p 547
14. Rényi A (1970) Probability theory. North-Holland, Amsterdam
15. Onicescu O (1966) C R Acad Sci Paris A 263:25
16. Tsallis C (1988) J Stat Phys 52:479
17. Tsapline B (1970) Chem Phys Lett 6:596
18. Majernik V, Richterek L (1997) J Phys A, Math Gen 30:L49
19. Grypeoos ME, Koutroulos CG, Oyewumi KJ, Petridou Th (2004) J Phys A, Math Gen 37:7895
20. Kuo CD (2005) Ann Phys 316:431
21. Gadre SR, Sears SB, Chakravorty SG, Bendale RD (1985) Phys Rev A 32:2602
22. Gadre SR (1984) Phys Rev A 30:620
23. Gadre SR, Bendale RD (1985) Int J Quant Chem 28:311
24. Gadre SR, Kulkarni SA, Shrivastava IH (1990) Chem Phys Lett 16:445
25. Gadre SR, Bendale RD, Gejji SP (1985) Chem Phys Lett 117:138
26. Gadre SR (2002) Reviews of modern quantum chemistry. World Scientific, Singapore, p 108. Edited: KD Sen
27. Yáñez RJ, Van Assche W, Dehesa JS (1994) Phys Rev A 50:3065
28. Dehesa JS, Martinez-Finkelstein A, Sanchez-Ruiz J (2001) J Comput Appl Math 133:23
29. Romera E, Sanchez-Moreno P, Dehesa JS (2005) Chem Phys Lett 414:468
30. Romera E, Dehesa JS (2004) J Chem Phys 120:8906
31. Dehesa JS, Martinez-Finkelstein A, Sorokin VN (2006) Mol Phys 104:613
32. Dehesa JS, Lopez-Rosa S, Olmos B, Yanez RJ (2006) J Math Phys 47:052104
33. Dehesa JS, Olmos B, Yáñez RJ (2006) arXiv:math.CA/0606133v1
34. Dehesa JS, López-Rosa S, Olmos B (2006) J Math Phys 47:052104
35. Romera E, Sánchez-Moreno P, Dehesa JS (2006) J Math Phys 47:103504
36. Sánchez-Moreno P, González-Férez R, Dehesa JS (2006) New J Phys A 8:1
37. Dehesa JS, González-Férez R, Sánchez-Moreno P (2007) J Phys A 40:1845
38. Dehesa JS, López-Rosa S, Yáñez RJ (2007) J Math Phys 48:043503
39. Tripathi AN, Smith VH Jr, Sagar RP, Esquivel RO (1998) Phys Rev A 54:1877
40. Ho M, Weaver DF, Smith VH Jr, Sagar RP, Esquivel RO (1998) Phys Rev A 57:4512
41. Ho M, Smith VH Jr, Weaver DF, Gatti C, Sagar RP, Esquivel RO (1998) J Chem Phys 108:5469
42. Ramirez JC, Perez JMH, Sagar RP, Esquivel RO, Ho M, Smith VH Jr (1998) Phys Rev A 58:3507
43. Ho M, Weaver DF, Smith VH Jr, Sagar RP, Esquivel RO, Yamamoto S (1998) J Chem Phys 109:10620
44. Sagar RP, Ramirez JC, Esquivel RO, Ho M, Smith VH Jr (2001) Phys Rev A 63:022509
45. Guevara NL, Sagar RP, Esquivel RO (2003) J Chem Phys 119:7030
46. Guevara NL, Sagar RP, Esquivel RO (2005) J Chem Phys 122:084101
47. Shi Q, Kais S (2004) J Chem Phys 121:5611
48. Shi Q, Kais S (2005) J Chem Phys 309:127
49. López-Ruiz R, Mancini HL, Calbet X (1995) Phys Lett A 209:321

50. Catalan RG, Garay J, López-Ruiz R (2002) Phys Rev E 66:011102
51. Chatzisavvas KCh, Moustakidis ChC, Panos CP (2005) J Chem Phys 123:174111
52. Panos CP, Chatzisavvas KCh, Moustakidis ChC, Kyrkou EG (2007) Phys Lett A 363:78
53. Yáñez RJ, Angulo JC, Dehesa JS (1998) Int J Quant Chem 56:489
54. Patil SH, Sen KD (2007) Uncertainty relations for modified isotropic harmonic oscillator and Coulomb potentials. Phys Lett A 363:109
55. Patil SH, Sen KD (2007) Net information measures for modified Yukawa and Hulthn potentials. Int J Quant Chem 107:1864
56. Patil SH, Sen KD, Watson NA, Montgomery HE Jr (2007) Characteristic features of net information measures for modified constrained Coulomb potentials. J Phys B 40:2147
57. Patil SH, Sen KD (2007) Scaling properties of net information measures for superpositions of power potentials: free and spherically confined cases. Phys Lett A 370:354
58. Borgoo A, De Proft F, Geerlings P, Sen KD (2007) Complexity of Dirac-Fock atom increases with atomic number. Chem Phys Lett 44:186
59. Sen KD, Angulo JC, Antolin JC (2007) Fisher-Shannon analysis of ionization processes and isoelectronic series. Phys Rev A 76:032502
60. Sen KD, Angulo JC, Antolin JC (2008) Fisher-Shannon plane and statistical complexity of atoms. Phys Lett A 372:5
61. Montgomery HE Jr, Sen KD (2008) Statistical complexity and Fisher-Shannon information measure of H_2^+. Phys Lett A 372:2271
62. Szabo JB, Sen KD, Nagy A (2008) The Fisher-Shannon information plane for atoms. Phys Lett A 372:2428
63. Borgoo A, Geerlings P, Sen KD (2008) Electron density and Fisher information of Dirac-Fock atoms. Phys Lett A 372:5106
64. Howard IA, Sen KD, Borgoo A, Geerlings P (2008) Characterization of the Chandrasekhar correlated two-electron wave functions using Fisher, Shannon and statistical complexity information measures. Phys Lett A 372:6321
65. Nagy A, Sen KD, Montgomery HE Jr (2009) LMC complexity for the ground states of different quantum systems. Phys Lett A 373:2552
66. Howard IA, Borgoo A, Sen KD, Geerlings P (2009) Comparative characterization of two-electron wavefunctions using information-theory measures. Phys Lett A 373:3277
67. Gonzales-Ferez R, Dehesa JS, Patil SH, Sen KD (2009) Scaling properties of composite information measures and shape complexity for hydrogenic atoms in parallel magnetic and electric fields. Physica A 388:4919
68. Katriel J, Sen KD (2010) Relativistic effects on information measures for hydrogen-like atoms. J Comput Appl Math 233:1399

Chapter 9
Chemical Information from Information Discrimination and Relative Complexity

Alex Borgoo, Paul Geerlings, and K.D. Sen

Abstract Following the spirit of the Hohenberg-Kohn theorems of DFT, the atomic and molecular density functions are used as the key parameters to extract chemical information. The functional form of the Kullback-Leibler relative information measure permitted the construction of density functionals, which along with the choice of a good reference density, for example, noble gas elements, is shown to reveal the periodicity in Mendeleev's table. Similarly, the specific choice of the transition state as reference, leads to valuable information on the chemically interesting reaction profiles. The investigation of atomic complexity measures is reported in terms of the non-relativistic and relativistic shape electron densities. Such an analysis leads to the interesting result that the atomic complexity derived from the relativistic electron density increases with atomic number whereas a saturation value is attained when the non-relativistic density is employed. It is shown that a relative complexity measure can be defined which reflects the diversity of electron density functions with respect to a reference atom as the sub shells are filled across the periodic table.

9.1 Introduction

The density functional theory (DFT) forms the basis of the most widely used computational tools in quantum chemistry today. The DFT is based on the Hohenberg-Kohn theorems [1]. According to its most common understanding, the N-electronic system is completely specified by its ground state electron density function. In other words an atom's or a molecule's energy—and in fact any other physical property—can be determined by evaluating a density functional. However, the construction of a functional,which corresponds to a given physical property, has proven to be very difficult. Moreover, to the present day no general and systematic way for constructing such functionals has been established. Although energy functionals, which are

P. Geerlings (✉)
Vrije Universiteit Brussel, Brussels, Belgium
e-mail: pgeerlin@vub.ac.be

K.D. Sen (ed.), *Statistical Complexity*,
DOI 10.1007/978-90-481-3890-6_9, © Springer Science+Business Media B.V. 2011

accurate enough for numerous practical purposes, have been around for some time now, the complicated rationale and the everlasting search for even more accurate energy functionals are proof of the difficulties encountered when constructing such functionals. In the domain of conceptual DFT, where chemical reactivity is investigated, a scheme for the construction of functionals, based on derivatives of the energy with respect to the number of electrons and/or the external potential, has proven very successful [2, 3]. In such studies, an intimate connection between the electronic probability density and the statistical thermodynamics principles is established. To the extent that the quantum mechanical uncertainty principle dictates the electronic probability distribution of atoms and molecules, it is hardly surprising that the ideas for the construction of chemically interesting functionals have also originated from information theory and statistical mathematics. This connection is generally to be attributed to the existence of several information theoretical uncertainty-like measures defined in terms of the electron density function, rather than the quantum mechanical expectation values. Information theoretical concepts found their way into chemistry during the seventies of last century. They were introduced to investigate experimental and computed energy distributions from molecular collision experiments. The purpose of the information theoretical approach was to measure the significance of theoretical models and conversely to decide which parameters should be investigated to gain the best insight into the actual distribution. For an overview of this approach to molecular reaction dynamics, we refer to Levine's work [4]. Although the investigated energy distributions have little relation with electronic wave functions and density functions, the same ideas and concepts found their way to quantum chemistry and the chemical reactivity studies which are an important study field of it. As mentioned above, this was stimulated by the fact that atoms and molecules can be described by their density function, which is ultimately a probability distribution. Since the first applications of information theoretical concepts in quantum chemistry described in the pioneering work of Sears, Parr and Dinur [5] many applications of information theoretical concepts to investigate wave functions and density functions, have been reported. In [6] Gadre gives a detailed review of the original ideas behind and the literature on "Information Theoretical Approaches to Quantum Chemistry". To motivate our work in this field I paraphrase the author's concluding sentence:

> "Thus it is felt that the information theoretical principles will continue to serve as powerful guidelines for predictive and interpretive purposes in quantum chemistry."

One of the most important and basic cornerstones of chemistry is the periodic table of elements revealing the periodicity of atomic properties starting with e.g. the shell structure of atoms. Its recovery on the basis of the electron density alone can be considered a significant result. In an information theoretical context, the periodicity revealing functional can be interpreted as a quantification of the amount of information in a given atom's density function, missing from the density function of the noble gas atom which precedes it in the periodic table. In the same spirit, the concept of complexity has been taken under consideration for the investigation of electron density functions. Complexity has appeared in many fields of scientific inquiry e.g. physics, statistics, biology, computer science end economics [7]. At present there does not exist a general definition which quantifies complexity, however several attempts have been made. For us, two of these stands out due to its functional form and their link with information theory. In this chapter we present a review of the work done in the Brussels group which illustrates how a variety of chemically interesting information theoretical functionals of electron density could be constructed. In all cases, extensive quantitative tests are presented in order to test the suitability of such constructions.

Throughout this chapter it becomes clear that different information and complexity measures can be used to distinguish electron density functions. Their evaluation and interpretation for atomic and molecular density functions gradually gives us a better understanding of how the density function carries physical and chemical information. *This exploration of the density function using information measures teaches us to read this information.*

Before going into more details about our research [8–11] several concepts should be formally introduced. For our research, which deals with applications of functional measures to atomic and molecular density functions, a brief discussion of these measures should suffice. The theoretical sections are followed by an in depth discussion of our results. In the concluding section we formulate general remarks and some perspectives.

9.2 Shannon's Measure: An Axiomatic Definition

In 1948 Shannon constructed his information measure—also referred to as "entropy"—for probability distributions according to a set of characterizing axioms [12]. A subsequent work showed that, to obtain the desired characterization, Shannon's original four axioms should be completed with a fifth one [13]. Different equivalent sets of axioms exist which yield Shannon's information measure. The original axioms, with the necessary fifth, can be found in [14]. Here we state the set of axioms described by Kinchin [6, 15].

For a stochastic event with a set of n possible outcomes (called the event space) $\{A_1, A_2, \ldots, A_n\}$ where the associated probability distribution $P = \{P_1, P_2, \ldots, P_n\}$ with $P_i \geq 0$ for all i and $\sum_{i=1}^{n} P_i = 1$, the measure S should satisfy:

1. the entropy functional S is a continuous functional of P;
2. the entropy is maximal when P is the uniform distribution i.e. $P_i = 1/n$;
3. the entropy of independent schemes are additive i.e. $S(P_A + P_B) = S(P_A) + S(P_B)$ (a weaker condition for dependent schemes exists);
4. adding any number of impossible events to the event space does not change the entropy i.e. $S(P_1, P_2, \ldots, P_n, 0, 0, \ldots, 0) = S(P_1, P_2, \ldots, P_n)$.

It can be proven [15] that these axioms suffice to uniquely characterize Shannon's entropy functional

$$S = -k \sum_i P_i \log P_i, \tag{9.1}$$

with k a positive constant. The sum runs over the event space i.e. the entire probability distribution. In physics, expression (9.1) also defines the entropy of a given macro-state, where the sum runs over all micro-states and where P_i is the probability corresponding to the i-th micro-state. The uniform distribution possesses the largest entropy indicating that the measure can be considered as a measure of randomness or uncertainty, or alternatively, it indicates the presence of information.

When Shannon made the straightforward generalization for continuous probability distributions $P(x)$

$$S[P(x)] = -k \int P(x) \log P(x)\, dx, \tag{9.2}$$

he noticed that the obtained functional depends on the choice of the coordinates. This is easily demonstrated for an arbitrary coordinate transformation $y = g(x)$, by employing the transformation rule for the probability distribution $p(x)$

$$q(y) = p(x) J^{-1} \tag{9.3}$$

and the integrandum

$$dy = J\, dx, \tag{9.4}$$

where J is the Jacobian of the coordinate transformation and J^{-1} its inverse. The entropy hence becomes

$$\int q(y) \log(q(y))\, dy = \int p(x) \log(p(x) J^{-1})\, dx, \tag{9.5}$$

where the residual J^{-1} inhibits the invariance of the entropy. Although Shannon's definition lacks invariance and although it is not always positive, it generally performs very well. Moreover, its fundamental character is emphasized by Jaynes's maximum entropy principle, which permits the construction of statistical physics, based on the concept of information [16, 17]. In the last decade several investigations of the Shannon entropy in a quantum chemical context have been reported. Those relevant to our research are discussed in more detail below.

9.3 Kullback-Leibler Missing Information

Kullback-Leibler's information deficiency was introduced in 1951 as a generalization of Shannon's information entropy [18]. For a continuous probability distribution $P(x)$, relative to the reference distribution $P_0(x)$, it is given by

$$\Delta S[P(x)|P_0(x)] = \int P(x) \log \frac{P(x)}{P_0(x)}\, dx. \tag{9.6}$$

As can easily be seen from expression (9.5), the introduction of a reference probability distribution $P_0(x)$ yields a measure independent of the choice of the coordinate system. The Kullback-Leibler functional quantifies the amount of information which discriminates $P(x)$ from $P_0(x)$. In other words, it quantifies the distinguishability of the two probability distributions. Sometimes it can be useful to see $\Delta S[P(x)|P_0(x)]$ as the distance in information from P_0 to P, although strictly speaking the lack of symmetry under exchange of $P(x)$ and $P_0(x)$ makes it a directed divergence.

Kullback-Leibler's measure is an attractive quantity from a conceptual and formal point of view. It satisfies the important properties positivity, additivity, invariance, respectively:

1. $\Delta S[P(x)|P_0(x)] \geq 0$;
2. $\Delta S[P(x,y)|P_0(x,y)] = \Delta S[P(x)|P_0(x)] + \Delta S[P(y)|P_0(y)]$ for independent events i.e. $P(x,y) = P(x)P(y)$;
3. $\Delta S[P(y)|P_0(y)] = \Delta S[P(x)|P_0(x)]$ if $y = f(x)$.

Besides the lack of symmetry, the Kullback-Leibler functional has other formal limitations e.g. it is not bound, nor is it always well defined. In [19] the lack of these properties was addressed and the Jensen-Shannon divergence was introduced as a symmetrized version of Kullback-Leibler's functional. In [20] the Jensen-Shannon distribution was first proposed as a measure of distinguishability of two quantum states. Chatzisavvas et al. investigated the quantity for atomic density functions [21].

For our investigation of atomic and molecular density functions, as carrier of physical and chemical information, we constructed functionals based on the definition of information measures. In the Sects. 9.5–9.11 below, the research is discussed in depth.

9.4 Fisher Information

The Fisher information was originally introduced in statistical estimation theory [22]. Fisher's functional was constructed to express the amount of information a data sample, with a given parametric distribution model, contains about the parameters. We limit ourselves to the relevant so-called diagonal form of Fisher's measure and write for an univariate probability distribution $P(x)$

$$I[P(x)] = \int \frac{[\nabla P(x)]^2}{P(x)} dx. \tag{9.7}$$

This functional is also known as the "intrinsic accuracy" and it measures the "narrowness" of the probability distribution.

As stated in [5], Fisher's measure for accuracy or narrowness (9.7) and Shannon's measure for randomness or spread (9.2), can be considered as two sides of the same coin. Since the first application by Parr, Sears and Dinur [5], where the quantum mechanical kinetic energy was expressed as an information measure of the electron density function, numerous studies involving Fisher's functional have been reported. Whilst a thorough mathematical treatment of the Fisher information measure lies beyond the scope of this thesis, the fact that the measure provides a functional which can be interpreted and linked with physical properties is of great interest. The functional, like others in this chapter, provides a way of analyzing the density function.

9.5 Reading Chemical Information from the Atomic Density Functions

9.5.1 Introduction

This section contains a detailed description of our research on the recovery of the periodicity of Mendeleev's Table. The novelty in this study is that we managed to generate the chemical periodicity of Mendeleev's table in a natural way, by constructing and evaluating a density functional. As reported earlier in [23], the comparison of atomic density functions on the basis of a quantum similarity index (using the $\delta(\mathbf{r}_1 - \mathbf{r}_2)$ operator), masks the periodic patterns in Mendeleev's table. On the other hand, the importance of the periodicity, as one of the workhorses in chemistry, can hardly be underestimated. Due to the Hohenberg-Kohn theorems, the electron density can be considered as the basic carrier of information, although, for many properties it is unknown how to extract the relevant information from the density function. This prompted us to investigate whether the information measures, which gained a widespread attention by the quantum chemical community, could be used to help extract chemical information from atomic density functions in general and help to regain chemical periodicity in particular.

Gadre and Sears for example studied the near Hartree-Fock information entropy $S = -\int \rho(\mathbf{r}) \log \rho(\mathbf{r})\, d\mathbf{r}$ and its analogue in momentum space [24]. Fazal presented an overview of computations on the Shannon entropy of the 1-normalized density $S = -\int \sigma(\mathbf{r}) \log \sigma(\mathbf{r})\, d\mathbf{r}$ [25]. Nagy and Parr [26] argued that information entropy gives a measure for the quality of an approximate electron wave function. Finally the important work of Nalewajski and Parr [27] should be mentioned, pointing out that the Kullback-Leibler entropy deficiency promotes Hirshfeld's [28] stockholder procedure as a natural way to generate atomic charges.

In this study we used ground state density functions obtained from the numerical LS-dependent Hartree-Fock scheme For a thorough discussion on atomic density functions and a code to calculate them, we refer to [29].

9.5.2 Construction and Evaluation of the Functional

A point of major interest in recent literature on the information theory approach is the information discrimination (9.6). Use of this functional necessitates the choice of a reference density $P_0(x)$. Still concentrating on the electron density as an information carrier we chose $P_0(x)$ in a way which bears analogy to the way Sanderson formulated his electronegativity scale [30]. Sanderson based his reasoning on the average electron density of a given atom A

$$ED_A = \frac{Z_A}{\frac{4}{3}\pi r_A^3}, \tag{9.8}$$

where Z_A is the atomic number and r_A the covalent radius. The corresponding value for a (hypothetical) noble gas atom with the same number of electrons, obtained by linear interpolation between the noble gases, is denoted as ED_0. The ratio

$$S_A = \frac{ED_A}{ED_0} \tag{9.9}$$

was called the stability ratio for atom A and put in relation with the electronegativity of atom A on the basis of the compactness of the electron cloud, reflecting, in a present-day terminology, its shape. In the same vein we considered as reference density in (9.6) the renormalized noble gas density of the row preceding the considered atom. The quantity to be evaluated then becomes

$$\Delta S_A^\rho \equiv \Delta S[\rho_A(\mathbf{r})|\rho_0(\mathbf{r})] = \int \rho_A(\mathbf{r}) \log \frac{\rho_A(\mathbf{r})}{\frac{N_A}{N_0}\rho_0(\mathbf{r})}\, d\mathbf{r}, \tag{9.10}$$

with $\rho_0(\mathbf{r})$ the density of the reference noble gas, scaled by the factor $\frac{N_A}{N_0}$ to yield upon integration the same number of electrons as in atom A, characterised by $\rho_A(\mathbf{r})$. It is easily seen that the integrand in (9.10) can be rewritten in a way that the shape functions $\sigma_A(\mathbf{r})$ and $\sigma_0(\mathbf{r})$, appear in ΔS_A^ρ

$$\Delta S_A^\rho \equiv \Delta S[\rho_A(\mathbf{r})|\rho_0(\mathbf{r})] = \int \rho_A(\mathbf{r}) \log \frac{\sigma_A(\mathbf{r})}{\sigma_0(\mathbf{r})}\, d\mathbf{r}. \tag{9.11}$$

Figure 9.1 indicates that a ΔS_A^ρ versus Z plot reveals periodicity to some extent. Intuitively we can consider the data as reflecting the difference between the information content of an atom when the information of the inner (completely filled) shells has been removed, by putting it in the reference density. So essentially information on the valence electrons is displayed, the noble gas considered having the same core as the atom under consideration.

Note however that in (9.11) an explicit dependency of ΔS_A^ρ on N_A is observed as $\rho_A(\mathbf{r}) = \sigma_A(\mathbf{r})N_A$. This behavior might be at the origin of the similar steepness of the curves belonging to first, second, third and fourth row atoms.

It is tempting to eliminate the N_A dependency, easily justified by building the theory directly in terms of shape functions, yielding

$$\Delta S_A^\sigma \equiv \Delta S[\sigma_A(\mathbf{r})|\sigma_0(\mathbf{r})] = \int \sigma_A(\mathbf{r}) \log \frac{\sigma_A(\mathbf{r})}{\sigma_0(\mathbf{r})}\, d\mathbf{r}. \tag{9.12}$$

Figure 9.2 now shows that periodicity is more pronounced coupled to the fact that the distance between points in a given period is decreasing gradually from first

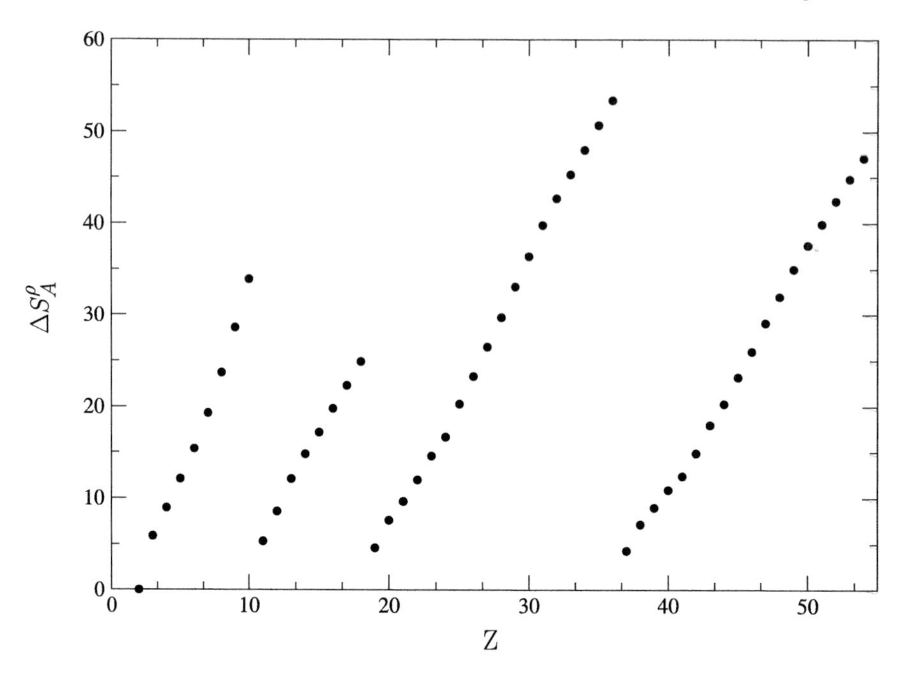

Fig. 9.1 Kullback-Leibler information (9.11) versus Z for atomic densities with the noble gas of the previous row as reference

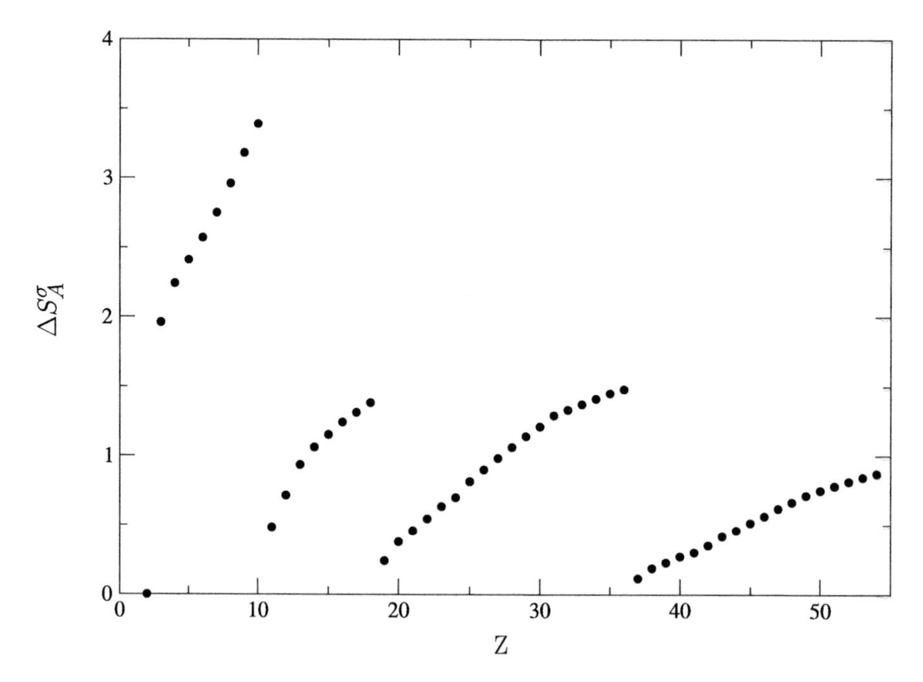

Fig. 9.2 Kullback-Leibler information (9.12) versus Z for atomic shape functions with the noble gas of the previous row as reference

to fourth row. One hereby regains one of the basic characteristics of the Periodic Table namely that the evolution in (many) properties through a given period slows down when going down in the Table. The decrease in slope of the four curves is a further illustration.

9.5.3 Conclusion

As reported earlier in [23], LS-dependent numerical Hartree-Fock densities for atoms H-Xe combined in a similarity index with a Dirac-delta function separation operator yield a nearest neighbour dominated similarity, masking periodicity. Introduction of the information discrimination concept with reference to the noble gas atom of the previous row leads to periodicity, with more pronounced results when densities are replaced by shape functions throughout. The present study also highlights the importance of the choice of the reference used to discriminate information in (9.12), which in our work has been fixed as the noble gas density of the previous row in the periodic table. As shown above the chemically intuitive choice of the reference density plays a key role in unmasking the periodicity in the table of Mendeleev.

The final result can be translated in the gain of information that valence electrons bear in their shape as compared to that of the noble gas with the same core. This result indicates the importance of the shape function as carrier of physical information.

9.6 Information Theoretical QSI

9.6.1 Introduction

Continuing the search for periodic patterns based on similarity measures motivated by the results obtained in an information theoretical frame work in Sect. 9.5, we will now combine the ideas from quantum similarity and information theory to construct an information theoretical similarity measure.

9.6.2 Construction of the Functional

In Sect. 9.5 we reported on the information entropy of atomic density and shape functions respectively defined as (9.11) and (9.12). As motivated above we set the density function of the reference equal to the density of the noble gas preceding the atom under investigation in the periodic table, scaled by the factor $\frac{N_A}{N_0}$, where N_A

and N_0 are the number of electrons, respectively of atom A and its reference. In this way the reference density $\rho_0(\mathbf{r})$ and $\rho_A(\mathbf{r})$, the density of atom A, yield the same number of electrons upon integration. In the previous section we demonstrated that these quantities reflect the periodic evolution of chemical properties in the Periodic Table and that Kullback's interpretation can be formulated in terms of chemical information stored in the density functions when we make this particular choice for the prior densities.

Following the conclusions in the previous section, one can see that it would be interesting to compare the information entropy, evaluated locally as

$$\Delta S_A^{\rho}(\mathbf{r}) \equiv \rho_A(\mathbf{r}) \log \frac{\rho_A(\mathbf{r})}{\frac{N_A}{N_0}\rho_0(\mathbf{r})}, \tag{9.13}$$

for two atoms by use of a quantum similarity measure (QSM). To that purpose the integrand in expression (9.12) is considered as a function, which gives the information entropy locally—at a given point $\mathbf{r}$. The construction of the corresponding QSM becomes straightforward by considering the overlap integral (with Dirac-δ as separation operator) of the local information entropies of two atoms A and B

$$Z_{AB}(\delta) = \int \rho_A(\mathbf{r}) \log \frac{\rho_A(\mathbf{r})}{\frac{N_A}{N_0}\rho_0(\mathbf{r})} \rho_B(\mathbf{r}) \log \frac{\rho_B(\mathbf{r})}{\frac{N_B}{N_{0'}}\rho_{0'}(\mathbf{r})}\, d\mathbf{r}. \tag{9.14}$$

Now a quantum similarity index (QSI) can be defined by normalizing the QSM, i.e. $\mathrm{QSI} = \frac{Z_{AB}}{\sqrt{Z_{AA}}\sqrt{Z_{BB}}}$. The QSM and the normalized QSI give a quantitative way of studying the resemblance in the information carried by the valence electrons of two atoms. The obtained QSI trivially simplifies to a shape based expression

$$SI_{(\delta)} = \frac{\int \Delta S_A^{\sigma}(\mathbf{r}) \Delta S_B^{\sigma}(\mathbf{r}) d\mathbf{r}}{\sqrt{\int \Delta S_A^{\sigma}(\mathbf{r}) \Delta S_A^{\sigma}(\mathbf{r}) d\mathbf{r}}\sqrt{\int \Delta S_B^{\sigma}(\mathbf{r}) \Delta S_B^{\sigma}(\mathbf{r}) d\mathbf{r}}}. \tag{9.15}$$

The simplification can be generalized from the local information discrimination $\Delta S_A^{\rho}(\mathbf{r})$ to any expression $F^{\rho}(\mathbf{r})$, which is linear in ρ (thus satisfying $F^{\rho}(\mathbf{r}) = N F^{\sigma}(\mathbf{r})$), as follows:

$$\frac{\int F_A^{\rho}(\mathbf{r}) F_B^{\rho}(\mathbf{r}) d\mathbf{r}}{\sqrt{\int F_A^{\rho}(\mathbf{r}) F_A^{\rho}(\mathbf{r}) d\mathbf{r}}\sqrt{\int F_B^{\rho}(\mathbf{r}) F_B^{\rho}(\mathbf{r}) d\mathbf{r}}} = \frac{\int F_A^{\sigma}(\mathbf{r}) F_B^{\sigma}(\mathbf{r}) d\mathbf{r}}{\sqrt{\int F_A^{\sigma}(\mathbf{r}) F_A^{\sigma}(\mathbf{r}) d\mathbf{r}}\sqrt{\int F_B^{\sigma}(\mathbf{r}) F_B^{\sigma}(\mathbf{r}) d\mathbf{r}}}. \tag{9.16}$$

In agreement with the fact that the shape function completely determines the properties of a system, the relevance of the QSI as a tool to compare physical properties of atomic electron density functions is confirmed. This characteristic distinguishes the QSI above, together with the Carbó QSI from other similarity measures. (For a recent overview about other similarity indices we refer to [31] and Hodgkin-Richards [32].)

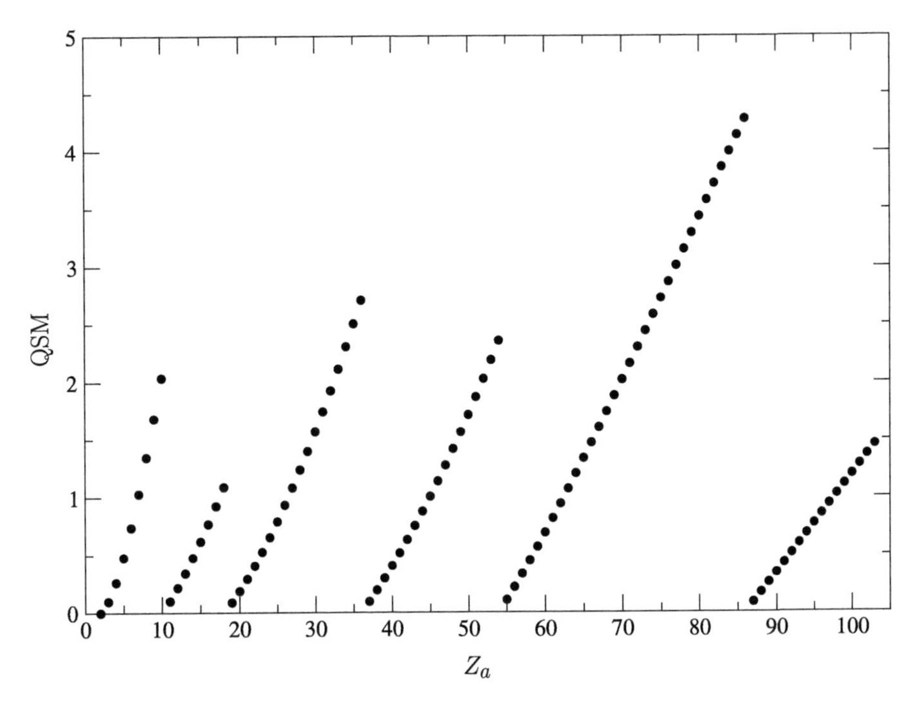

Fig. 9.3 A selection of the results for $Z_B = 82$ of the Information entropy QSI. The change of reference atom is still visible, but the periodicity is not regained

9.6.3 Evaluation of the Functional

In this section the results of the QSM and QSI, evaluated for shape functions of all pairs of atoms in the periodic table are discussed.

The evaluation of the information theory based QSM reveals a picture corresponding to the periodicity of Mendeleev's Table, which can be distinguished by for example selecting the results involving the density of Pb in Fig. 9.3. The results correspond to the evolution of chemical properties first of all in the sense that for each period the QSM increases gradually from the first column to the last. Ionization energy and hardness are properties which reveal a similar evolution throughout [33]. Secondly in the sense that neighbouring atoms with large nuclear charge differ less than neighbouring light atoms, e.g. the difference between the QSM values of two atoms in the first period is large in comparison to the difference in QSM between two neighbouring Lanthanides. The periodicity is regained throughout by the choice of the reference atoms, as it yields low QSM values for atoms similar to the chosen prior. One notes however that the QSM does not reveal results, which reach maxima when a given atom is compared with another atom of the same group. Moving to the QSI, the periodicity of the QSM is lost due to the normalization as illustrated by selection of results involving Pb in Fig. 9.4. The change of prior is still visible (gaps) at the positions where the prior changes, but the normaliza-

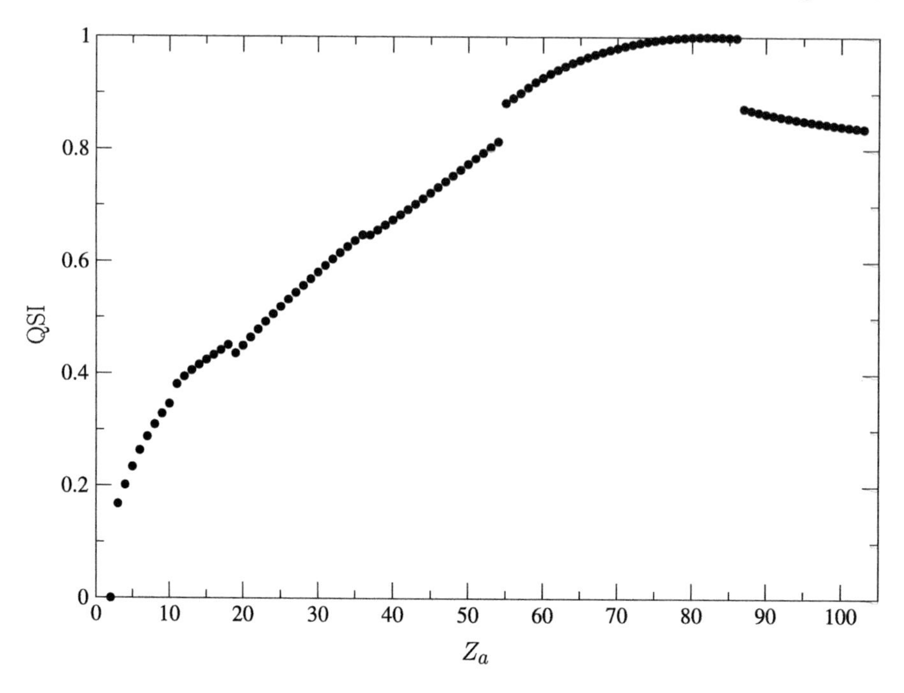

Fig. 9.4 A selection of the results of the information entropy QSI for $Z_b = 82$. The change of reference atom is still visible, but the periodicity is not regained

tion blends out the clear periodic evolution of the QSM in graph 9.3. This leads to the conclusion that the normalization, which yielded the nearest neighbour effect for the Carbó, which were reported in [23], can overwhelm the characteristics of a QSM.

Changing the point of view, we can opt to investigate which atom of a given period of the table belongs to a certain column and in which way the atoms should be ordered within the period. This can be done by investigating the QSI with the top atoms of each column as prior. Formulated in terms of Kullback-Leibler information discrimination the following is evaluated. For instance, when we want to investigate the distance of the atoms Al, Si, S and Cl from the N-column (group Va), we consider the information theory based QSI in expression (9.14), where the reference densities ρ_0 and $\rho_{0'}$ are set to ρ_N, ρ_A to ρ_{Al}, ρ_{Si}, ρ_P, etc. respectively and ρ_B to ρ_P, i.e. we compare the information contained in the shape function of N to determine that of P, with its information on the shape function of Al, Si, S, Cl. Due to the construction a 1. is yielded for the element P and the other values for the elements to the left and to the right of the N-column decrease, as shown in Fig. 9.5. This pattern is followed for the periods 3 up to 6, taking As, Sb and Bi as reference, with decreasing difference along a given period in accordance with the results above. Note that the difference from 1. remains small, due to the effect of the renormalization used to obtain the QSI.

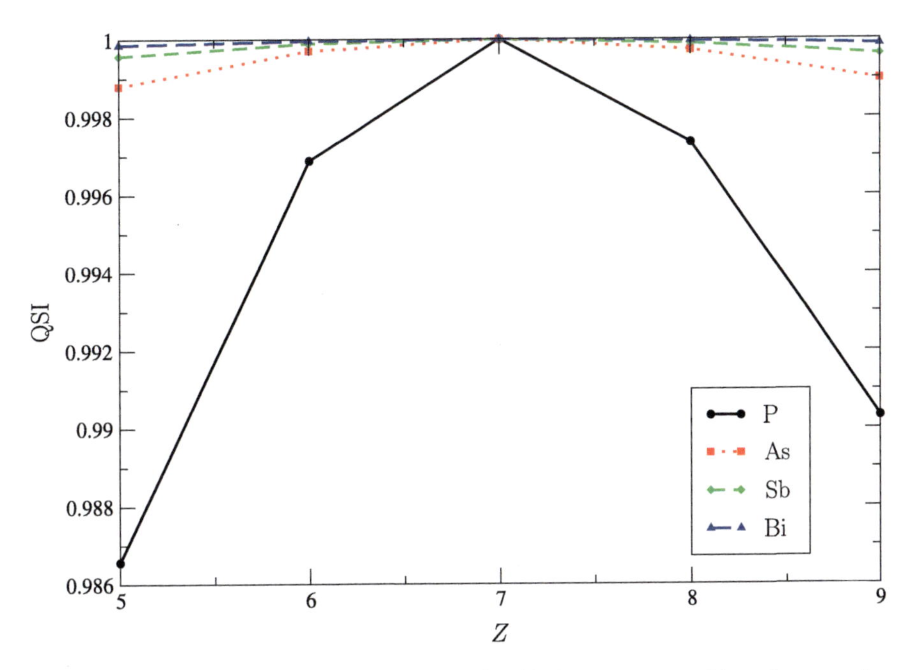

Fig. 9.5 Results of the information theory based QSI with the atom on top of the column as prior. The symbol in the legend indicates the period of the investigated atom and the nuclear charge Z-axis indicates the column of the investigated atom. (For example Ga can be found as a square $Z = 5$)

9.6.4 Conclusion

In this study we reported on the development and calculation of a new information theory based quantum similarity measure (QSM) and the corresponding quantum similarity index (QSI) for atoms, using their density functions and shape functions. We showed that a QSM constructed with the Kullback-Leibler information deficiency looses its periodic character upon normalization. One might say that the normalisation renders the QSI insensitive to certain characteristics, clearly present in the QSM. To regain the periodicity with the information theory based QSM, the choice of the reference for each atom as the density of the noble gas of the previous row, normalized to the same number of electrons in the atom under investigation, is crucial. The results of the QSM are in agreement with chemical intuition in the sense that the difference in QSM of two successive light atoms is large in comparison to the difference in QSM of two successive heavy atoms, which reflects that light atoms next to each other in the table differ more than neighbouring heavy atoms. In particular, when looking at the results of Lanthanides and Actinides we find high similarities indeed. This interpretation is not regained by looking at the QSI, with the prior set to the noble gas atoms. It is rewarding that the comparison of information content of the shape function of a given top atom in a column with the atoms of the subsequent period(s) reveals another periodicity pattern.

9.7 Relativistic Effects on the Fisher Information

9.7.1 Introduction

The Shannon information entropy (9.2) and Fisher information measure (9.7) of a probability distribution are two of the most investigated measures of information. In this section they will be used to investigate the shape function of atoms. We shall thus refer all the information measures to the shape function. The Shannon information is then defined as

$$S = -\int \sigma(\mathbf{r}) \log \sigma(\mathbf{r})\, d\mathbf{r}. \tag{9.17}$$

We note here that S is a global measure of the spread of the probability density $\sigma(\mathbf{r})$. We shall denote the integrand in (9.17) as $s(\mathbf{r})$. The Fisher information (intrinsic accuracy) in the position space is defined as

$$I = \int \frac{[\nabla \sigma(\mathbf{r})]^2}{\sigma(\mathbf{r})} d\mathbf{r}. \tag{9.18}$$

The quantity I measures the 'narrowness' of the electron distribution and presents itself as a local measure of the oscillations in $\sigma(\mathbf{r})$. The integrand in (9.18) will be denoted by $i(\mathbf{r})$. An equivalent form of I has been proposed for atomic and molecular systems [34], which is given by

$$I' = -\int [\nabla^2 \sigma(\mathbf{r})] \log[\sigma(\mathbf{r})]\, d\mathbf{r}. \tag{9.19}$$

The integrand in (9.19) will be denoted by $i'(\mathbf{r})$, distinguishing it from $i(\mathbf{r})$. Further, using an identity based on Green's theorem [35, 36], it has been shown [34] that (under the assumption that $-\sigma(\mathbf{r}) \log \sigma(\mathbf{r})$ is strongly decaying) an interesting relationship holds between the Shannon entropy density $s(\mathbf{r})$ and the "potential" derived from the Fisher information integrands $i(\mathbf{r})$ in (9.18) and $i'(\mathbf{r})$ in (9.19). More specifically,

$$-\sigma(\mathbf{r}) \log[\sigma(\mathbf{r})] = -\sigma(\mathbf{r}) + \frac{1}{4\pi} \int \frac{i(\mathbf{r}')}{|(\mathbf{r}-\mathbf{r}')|} d\mathbf{r}' - \frac{1}{4\pi} \int \frac{i'(\mathbf{r}')}{|(\mathbf{r}-\mathbf{r}')|} d\mathbf{r}' \tag{9.20}$$

and

$$S = -1 + \frac{1}{4\pi} \int\int \frac{i(\mathbf{r}')}{|\mathbf{r}-\mathbf{r}'|} d\mathbf{r}\, d\mathbf{r}' - \frac{1}{4\pi} \int\int \frac{i'(\mathbf{r}')}{|\mathbf{r}-\mathbf{r}'|} d\mathbf{r}\, d\mathbf{r}', \tag{9.21}$$

have been derived. The local function $-\sigma(\mathbf{r}) \log \sigma(\mathbf{r})$ is said to be *strongly decaying* if it falls faster than $1/r$ and its derivative decays faster than $1/r^2$. Numerical tests for the ground state atoms with atomic number $Z = 2$–18 have been presented in [34], employing numerical Hartree-Fock densities, confirming the validity of (9.21).

In this study, we present the first calculations of expressions (9.18) and (9.19) for atoms $Z = 1$–103, using the relativistic Dirac-Fock (DF) wave functions [37]. In addition to testing (9.21), (9.20) has been studied, focusing on the evaluation at the nucleus $\mathbf{r} = \mathbf{0}$. In particular, we write

$$-\sigma(\mathbf{0}) \log[\sigma(\mathbf{0})] = -\sigma(\mathbf{0}) + \frac{1}{4\pi} \int \frac{i(\mathbf{r}')}{\mathbf{r}'} d\mathbf{r}' - \frac{1}{4\pi} \int \frac{i'(\mathbf{r}')}{\mathbf{r}'} d\mathbf{r}'. \tag{9.22}$$

This equality can be used to evaluate the difference between the two Fisher information potentials (second and third terms) at $\mathbf{r} = \mathbf{0}$ directly as $\sigma(\mathbf{0})[1 - \log(\sigma(\mathbf{0})]$. Thus, the effective potential difference at the nucleus due to the density $(i(\mathbf{r}) - i'(\mathbf{r}))$ is completely defined in terms of the shape function at the nucleus. This result assumes further significance in the light of the importance of Fermi-contact interaction in the measurements of hyperfine interactions in atoms, molecules and solids [38, 39]. The Fermi-contact term, which is directly proportional to $\sigma(\mathbf{0})$, is particularly sensitive to relativistic effects as a result of orbital and bond length contractions.

9.7.2 Calculations

In the present work, we have used the MDF/GME program of Desclaux and Indelicato [37] including both the magnetic and retardation part of the Breit interaction in the self-consistent process, but not the vacuum polarization under the option of a point nucleus. For the details about the DF density functions we refer to [40]. We employed both the finite and point nucleus options to generate the DF densities. We used the Z-dependent mesh with the "RHOMIN" parameter of the logarithmic mesh [22] as -16.0. The non-relativistic HF densities are similar to those employed earlier [41] and are generated using Koga-Roothaan-Hartree-Fock wave functions [42, 43]. All the necessary integrals have been computed using numerical quadrature on the spherically averaged electron density.

9.7.3 Results and Conclusions

In Fig. 9.6, we displayed the results of our calculations of the equivalent I and I' for neutral atoms with nuclear charge $Z = 1$–103. These numbers are based on DF density functions using a finite nucleus. The two sets of results are found to be in excellent numerical agreement. This numerical test confirms the equivalence of the two forms of the Fisher information measures in (9.18) and (9.19) for all atoms. In Fig. 9.7, we compared the I values determined from the point nucleus DF densities as well as the finite nucleus ones with those of the HF (point nucleus) densities for neutral atoms. From this plot it is clear that the relativistic effects lead to a significant non-linear increase in I for atoms with $Z \geq 40$. A good linear fit to the non-relativistic HF estimates was obtained as $I = 8.3088Z - 12.454$, with

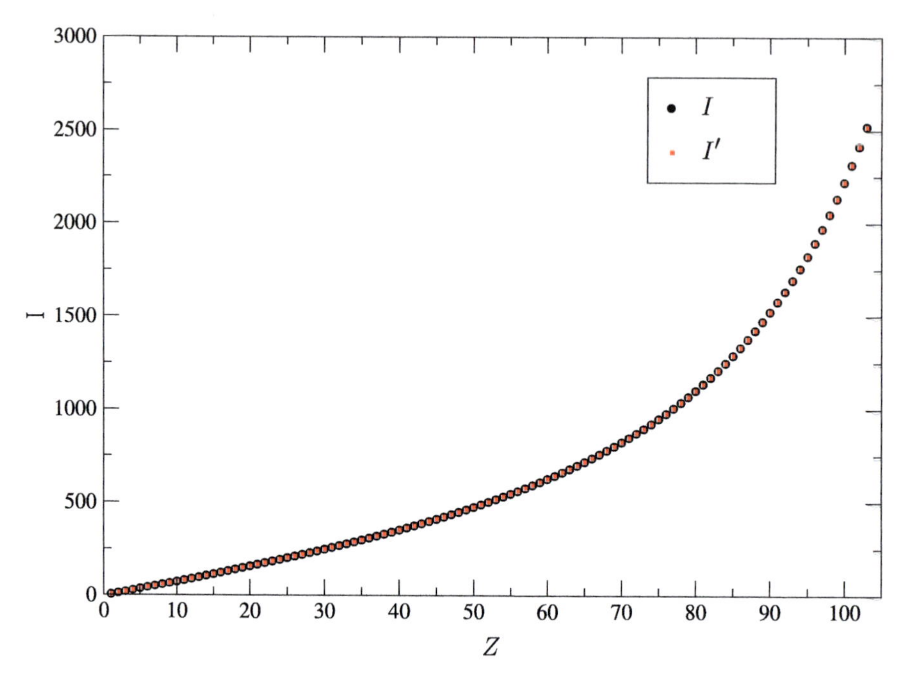

Fig. 9.6 A plot of I (x) and I' (o) vs. Z for neutral atoms using the finite nucleus DF densities

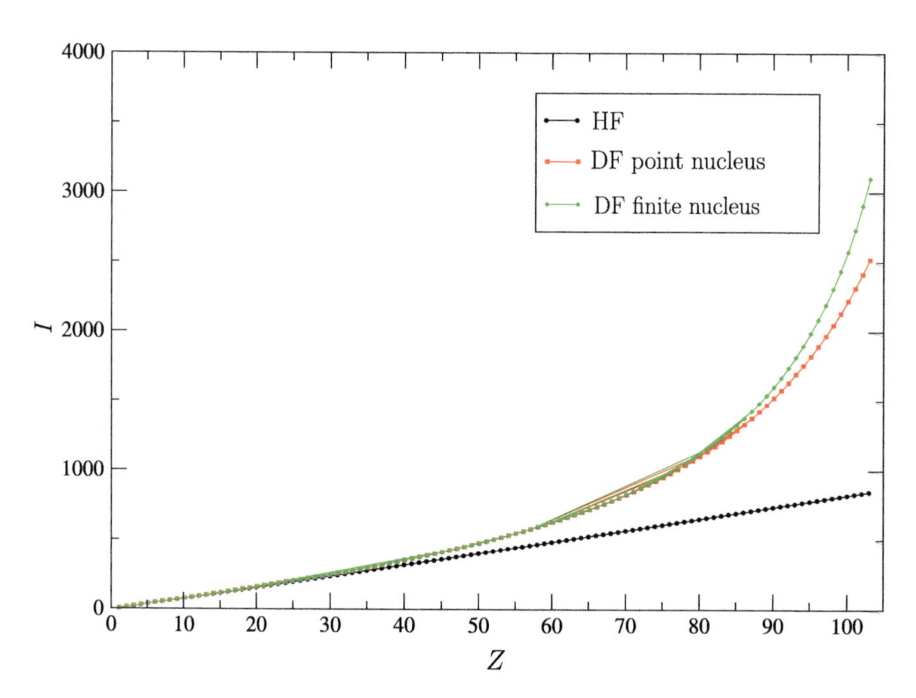

Fig. 9.7 A plot showing comparison of I for all neutral atoms using the finite nucleus DF densities (x) and point nucleus DF densities (o) with those derived from HF density (+)

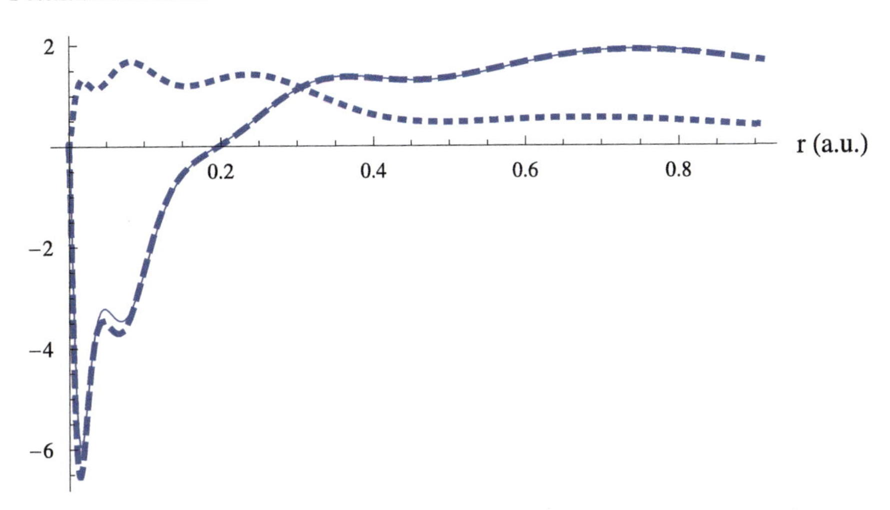

Fig. 9.8 A plot of the radial distribution of the potential difference, $\frac{1}{4\pi}\int \frac{i(\mathbf{r}')}{|\mathbf{r}-\mathbf{r}'|}d\mathbf{r}' - \frac{1}{4\pi}\int \frac{i'(\mathbf{r}')}{|\mathbf{r}-\mathbf{r}'|}d\mathbf{r}'$ for Xe comparing the HF (*Thin/Solid*) and the finite nucleus DF (*Thick/Dashed*) estimate to show the local effects due to the relativistic interactions. The standard radial distribution function $4\pi r^2[\sigma(\mathbf{r})]$ (*Thick/Dotted*) for the finite nucleus DF density has been plotted alongside to show that the shell boundaries are represented as the maxima in the difference of Fisher potential curves

correlation coefficient of 0.9999. The observed trend is attributed to the nearly linear dependence on Z of the Weizsäcker kinetic energy, which is a functional of the shape function $\sigma(\mathbf{r})$ and is proportional to the Fisher measure [5] in (9.18). For heavy atoms, the relativistic effects appear to break this linearity. A comparison of Fisher information using point nuclei with those using finite nuclei within the DF model shows that the finite nucleus densities lead to slightly reduced values of the Fisher information. This is probably related to the presence of more rapidly changing behavior of the electron density near the nucleus in case of the point nucleus DF model (see below). In Fig. 9.8, we present a case study for Xe of the local variation of the radial distribution of the difference in the two Fisher information potentials, given by the second and third terms in the right hand side of (9.20). Here, the HF estimates have been compared with those derived from the finite nucleus DF density. The DF densities lead to an enhancement of the corresponding integrand at smaller radial distances. It is striking how the detailed features of the shell effects are clearly revealed by the radial distribution function of the difference in Fisher potential. Moreover, comparison with the radial distribution function of the shape function, $4\pi r^2\sigma(\mathbf{r})$, displayed in Fig. 9.8 shows that the shell boundaries are marked by the locations of maxima in the radial density distribution of the difference in Fisher potentials. In Fig. 9.9, we have plotted the quantity $-4\pi r^2[\sigma(\mathbf{r})\log\sigma(\mathbf{r}) - \sigma(\mathbf{r})]$ for Xe. As expected this curve reproduces quantitatively the radial density distribution of Fisher potential difference plotted in Fig. 9.8. This provides the numerical verification of the identity in (9.20) at

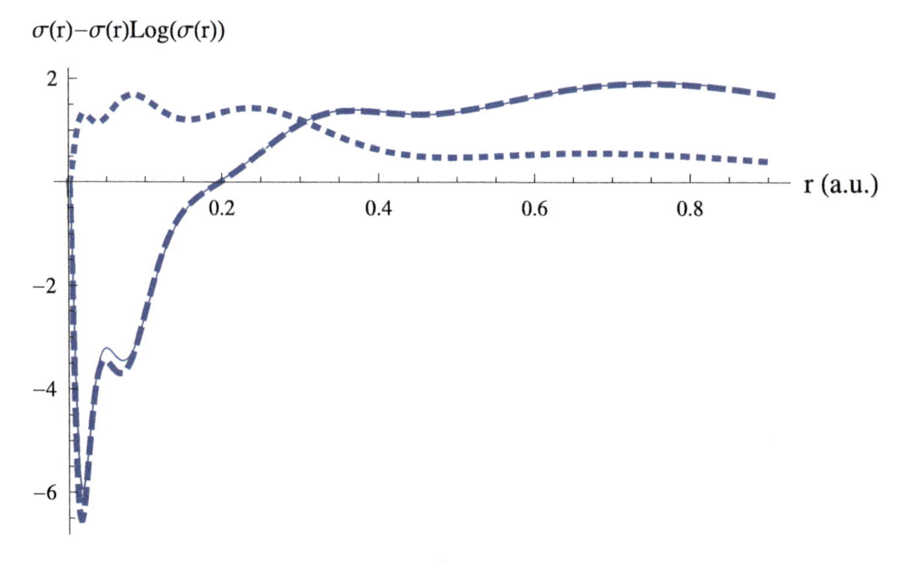

Fig. 9.9 A plot of the density function, $-4\pi r^2[\sigma(\mathbf{r})\log\sigma(\mathbf{r})-\sigma(\mathbf{r})]$ for Xe, comparing the HF (*Thin/Solid*) and finite nucleus DF (*Thick/Dashed*) results. The results are quantitatively the same as those displayed in Fig. 9.8

Table 9.1 Equivalence of (9.22) of the HF estimates of the difference in Fisher potentials due to ($i(\mathbf{0})$ and $i'(\mathbf{0})$) with $\sigma(0)[1-\ln(\sigma(0))]$ for a representative set of atoms [He, Ne, Ar, Xe, Rn, Li, Na, K, Rb, I]

Atom	$\sigma(\mathbf{0})$	$\sigma(\mathbf{0})-\log\sigma(\mathbf{0})$	$i'(\mathbf{0})$	$i(\mathbf{0})$	$i-i'(\mathbf{0})$
He	1.79795	0.743187	0.960499	1.70369	0.743187
Ne	61.9919	−193.849	251.368	57.5189	−193.849
Ar	213.32	−930.673	1122.73	192.057	−930.673
Kr	895.538	−5191.81	5977.51	785.694	−5191.81
Xe	2059.77	−13657	15440.6	1783.57	−13657
Li	4.6052	−2.4278	6.85062	4.42281	−2.4278
Na	75.7962	−252.253	322.195	69.9421	−252.253
K	238.931	−1069.5	1284.05	214.549	−1069.5
Rb	947.59	−5547.12	6377.67	830.557	−5547.12
I	1982.35	−13067.7	14785.3	1717.55	−13067.7

all radial points. In Table 9.1, we have presented the numerical results of the difference of the Fisher potentials due to $i(\mathbf{r})$ and $i'(\mathbf{r})$ in (9.20) evaluated at the nuclear position, $r=0$, with the corresponding estimates of $\sigma(\mathbf{0})[1-\log\sigma(\mathbf{0})]$ for a representative set of atoms. Here we only provide numerical results corresponding to HF densities. This is due to the fact that the Dirac $j=1/2$ wave functions diverge as $\sqrt{(1-Z\alpha^2)}$; therefore there is no total electron density at the DF point nucleus. The equivalence of $-4\pi r^2[\sigma(\mathbf{r})\log\sigma(\mathbf{r})-\sigma(\mathbf{r})]$ and the difference between the radial distribution of the Fisher potentials at all other radial points has already established in Figs. 9.8 and 9.9. An equivalence at the nucleus similar to that of the HF results in Table 9.1 can be obtained from DF density functions by integrating the density over the volume of the finite nucleus. We have not carried out such calculations in the present work as there exist several possibilities of modelling the finite nuclear volume. Our observations in Figs. 9.8 and 9.9 and in Table 9.1, suggest that the probability distribution $-4\pi r^2[\sigma(\mathbf{r})\log\sigma(\mathbf{r}) - \sigma(\mathbf{r})]$ can be used equivalently to represent the radial distribution of the difference of the two Fisher potentials $\frac{i(\mathbf{r})}{|\mathbf{r}-\mathbf{r}'|} - \frac{i'(\mathbf{r})}{|\mathbf{r}-\mathbf{r}'|}$ in (9.20). More interestingly, for any electronic system the difference in the Fisher potential at the nucleus is completely defined in terms of the value of the shape function at this point. The important effects on the effective electrostatic property given by the difference of the potential due to the Fisher densities $i(\mathbf{r})$ and $i'(\mathbf{r})$ can therefore be probed using the quantity $-\sigma(\mathbf{r})\log[\sigma(\mathbf{r})] + \sigma(\mathbf{r})$, which is much simpler to calculate. In this manner, the effect of the overall bonding interactions, including the long range effects, can be readily calculated by studying the corresponding changes in the potential difference. The results reported in this section provide an information theoretical interpretation of the observation [44] that the electron density (shape function) at the nucleus, through Kato's cusp condition [45], enables the ground state electron density function to carry information on all the other properties [1].

9.8 Two Complexity Measures

Until present no general definition of complexity is available and one has to use a definition which is workable for the situation at hand. Here we will introduce two complexity measures, which we employed for an investigation of atomic density functions.

9.8.1 LMC Complexity Measures

Most measures from information theory are constructed with the idea to quantify the randomness of a probability distribution. For example the outcome of a process described by a uniform probability distribution is harder to predict compared to one determined by a sharp normal distribution, where the expected outcome would be indicated by the peak of the distribution. In a conceptually related approach to the analysis of systems described by probability distributions another type of measure has been developed. These new measures serve to quantify structure, where structure most generally refers to "the relationship between a system's components" [46]. In this picture more correlation between constituents implies more structure. In the literature the quantities that have been constructed to quantify structure are

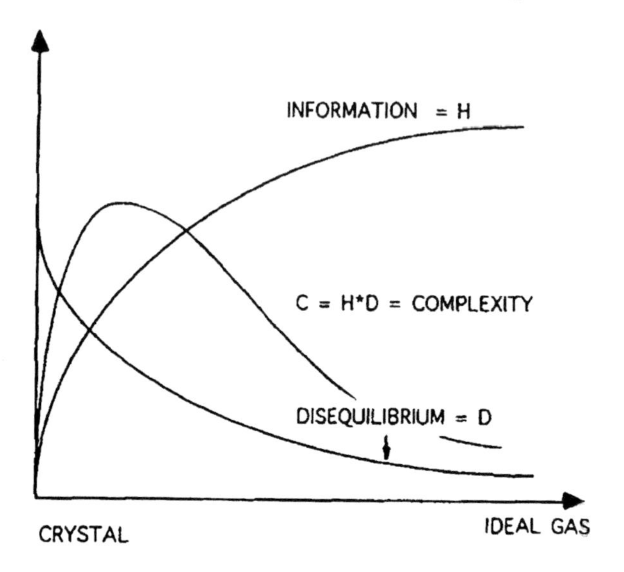

Fig. 9.10 Intuitive qualitative plot of the behavior of the LMC measure. (Reproduced from [48])

often referred to as complexity measures or *statistical* complexity measures. Unlike some other concepts like entropy (or information) in information theory, there is no universal applicable definition of complexity and in the literature there are numerous different definitions available which are built to be adequate for specific types of structure. Different approaches can be found e.g. in the domains of information theory and computation theory. What is important for this thesis is that complexity measures provide functionals which are applicable to investigate density functions. Some complexity measures recently appeared in the context of quantum chemistry [21, 47].

One way to construct a new complexity measure, is to build a functional, which satisfies certain exact requirements. This is the approach adopted by Lopéz-Ruiz, Mancini and Calbet for their measure C_{LMC}. In their paper [48], the authors argue that a statistical complexity measure should yield zero in both the case of perfect order and in a situation of maximal randomness. In the original LMC work an intuitive plot is given of the complexity measure is supposed to evolve between perfect order and maximal randomness. In that picture, here reproduced in Fig. 9.10, the authors modeled perfect order by a perfect crystal and maximal randomness with an ideal gas. The plot shows the expected asymptotic behavior for the extremes.

LMC start by constructing a measure for a discrete probability distribution (i.e. with a finite number of possible outcomes of the random variable and corresponding probabilities p_i) and they proposed to achieve the desired behavior by writing the complexity as a product of an information measure H and a measure disequilibrium D.

$$C_{LMC} = H.D. \tag{9.23}$$

The former, modeled by Shannon's entropy $H = -\sum p_i \log p_i$, yields zero in the case of a probability distribution with a unique outcome for the random variable (i.e.

perfect order), whilst reaching its maximum for a uniform distribution (i.e. maximal randomness), whereas the latter, modeled by $D = \sum(p_i - 1/N)^2$, expresses the "departure from equilibrium". At maximum randomness, clearly $D = 0$. The clever choice of H and D do indeed give a vanishing complexity for the extremes of order and randomness.

The generalization to continuous probability distributions is made by replacing the Shannon entropy expression by its continuous counterpart $-\int \rho(\mathbf{r}) \log \rho(\mathbf{r})\, d\mathbf{r}$ and by taking $D = \int \rho^2(\mathbf{r})\, d\mathbf{r}$ for the disequilibrium [7, 48]. In our work where the complexity was evaluated for atomic density functions $\rho(\mathbf{r})$ (see Sect. 9.9), we used the "exponential power Shannon entropy" [7, 12]

$$H = \frac{1}{2\pi e} e^{\frac{2}{3}S}, \tag{9.24}$$

where the Shannon information entropy in position space S is given by expression (9.2). The obtained form of the LMC complexity measure satisfies particular properties [7].

The disequilibrium for continuous distributions, on a finite interval $[-L, L]$, can be found by first considering the straightforward generalization of the discrete version

$$D' = \int_{-L}^{L} \left(p(x) - \frac{1}{2L}\right)^2 dx = \int_{-L}^{L} p^2(x)\, dx - \frac{1}{2L} \tag{9.25}$$

and then omitting the constant $-\frac{1}{2L}$, giving

$$D = \int_{-L}^{L} p^2(x)\, dx. \tag{9.26}$$

This yields a $D > 0$, which attains its minimum for the constant rectangular distribution i.e. corresponding to maximum randomness. Note that this quantity also appears as the self similarity in the context of quantum similarity.

9.8.2 SDL Complexity Measures

In a different effort to construct a complexity measure, Landsberg [49] defined the parameters order Ω and disorder Δ as

$$\Omega = 1 - \Delta = 1 - \frac{S}{S_{\max}}, \tag{9.27}$$

where the maximum Shannon entropy value $S_{\max}$ is given by [50]

$$S_{\max} = \frac{3}{2}(1 + \log \pi) + \frac{3}{2} \log\left(\frac{2}{3}\langle r^2 \rangle\right). \tag{9.28}$$

According to this definition $\Omega = 1$ corresponds to perfect order and predictability, whereas complete disorder and randomness yields $\Omega = 0$.

In [51] a measure of complexity $\Gamma_{\alpha,\beta}$ was defined as

$$\Gamma_{\alpha,\beta} = \Delta^{\alpha}\Omega^{\beta} = \Delta^{\alpha}(1-\Delta)^{\beta} = \Omega^{\beta}(1-\Omega)^{\alpha}, \tag{9.29}$$

which is referred to as the "simple complexity of disorder strength α and order strength β". Three categories of complexity measures are distinguished according to the values of α and β. If $\beta = 0$ and $\alpha > 0$ the measure belongs to category I. When $\alpha > 0$ and $\beta > 0$ it belongs to category II and it belongs to category III when $\alpha = 0$ and $\beta > 0$. In the first category, complexity is an increasing function of disorder whereas in the third category it is an increasing function of order. In category II complexity vanishes both at zero order and zero disorder, respectively, and has a maximum given by

$$(\Gamma_{\alpha,\beta})_{\max} = \frac{\alpha^{\alpha}\beta^{\beta}}{(\alpha+\beta)^{(\alpha+\beta)}} \quad \text{at } \Delta = \frac{\alpha}{\alpha+\beta} \text{ and } \Omega = \frac{\beta}{\alpha+\beta}. \tag{9.30}$$

The SDL and LMC measures have been critically analyzed in terms of their general applicability. We refer the readers to the published literature on this debate [46, 52, 53]. Recently both the LMC and the SDL complexity measures were applied in a quantum chemical context. In the next section we employ both to analyze atomic density functions as probability distributions.

9.9 Complexity of Atomic Densities

9.9.1 Introduction

Surprisingly, inspite of its simplicity the complexity measures introduced in Sect. 9.8 have been applied, only very recently [21, 47], in a quantum chemical frame work for the analysis of the electronic structural complexity of atoms using the non-relativistic Hartree-Fock (HF) wave functions [54] for atoms with atomic number $Z = 2$–54. It is interesting to study the trends in the complexity measures with specific reference to the nature of interaction, e.g. the relativistic effects in atoms. In this section, we present the first calculations of C_{LMC} and $\Gamma_{\alpha,\beta}$ for atoms with atomic number $Z = 1$–103 using the relativistic Dirac-Fock (DF) wave functions [37] in the position space. We shall compare these results with the corresponding complexity measures for all atoms derived from the non-relativistic HF wave functions [42, 43] in order to ascertain the influence of relativistic interactions on complexity. The present work thus also extends the previous non-relativistic HF calculations [21, 47] of the two complexity measures [7, 48, 49, 51] for atoms with $Z = 1$–103, which, in our case, are derived from the non-relativistic Koga–Roothaan–HF wave functions [42, 43, 55]. We shall be particularly concerned with the complexity measures of the heavy atoms ($Z > 54$) wherein the relativistic effects are well known to be dominating over the electron-electron correlation effects [56–58]. We note here that electron correlation effects have been neglected in the present work.

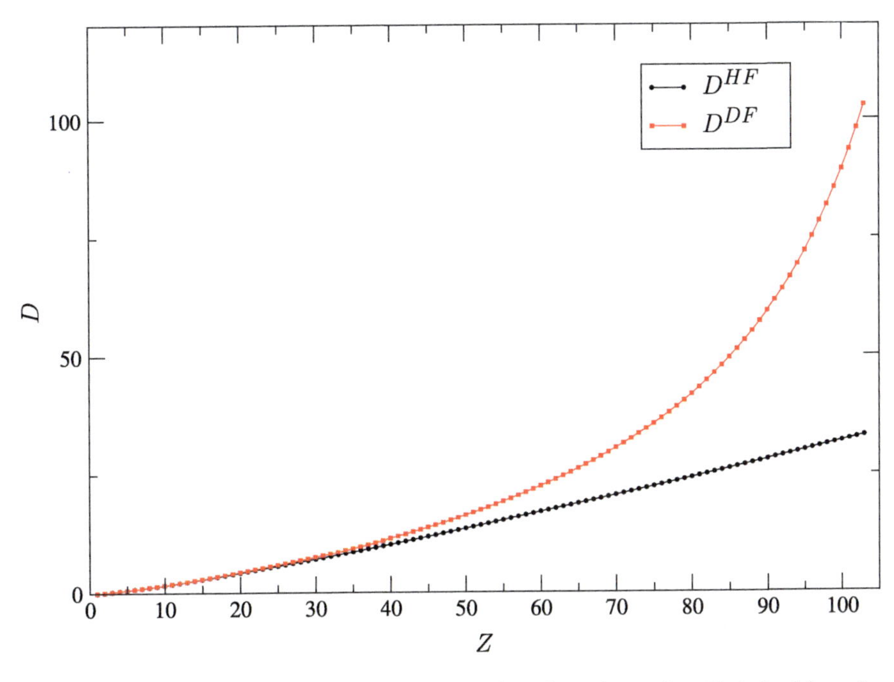

Fig. 9.11 A plot of the disequilibrium, D as a function of atomic number, Z, derived from the non-relativistic Hartree-Fock (*squares*) and relativistic Dirac-Fock densities (*circles*). The densities are each normalized to unity

9.9.2 Construction of the Functional

The functionals evaluated in this study, are defined in the expressions (9.23) and (9.30), where the atomic shape function $\sigma(\mathbf{r})$ acts is the relevant probability distribution.

9.9.3 Results and Conclusions

In the present work, the relativistic DF densities used are similar to those used in [40]. In Fig. 9.11, we have compared the variation of D from (9.26), which appears in the LMC definition (9.23), as a function of Z resulting from the DF and HF wave functions for the neutral atoms ($Z = 1$–103). The quantity D represents the quantum self similarity [40, 59], information energy [60], or linear entropy [61, 62]. Most significantly, it is also an experimentally measurable quantity [63]. For a given atom it expresses the extent of charge concentration in it. As shown in Fig. 9.11, while both HF and DF densities show increase in D as Z is varied, the relativistic effects make a significant non-linear departure as compared to the linear behavior, for atoms with $Z \geq 20$. In the context of disequilibrium this amounts to a significantly larger rate of change of D with Z for the DF atoms at large Z values. In

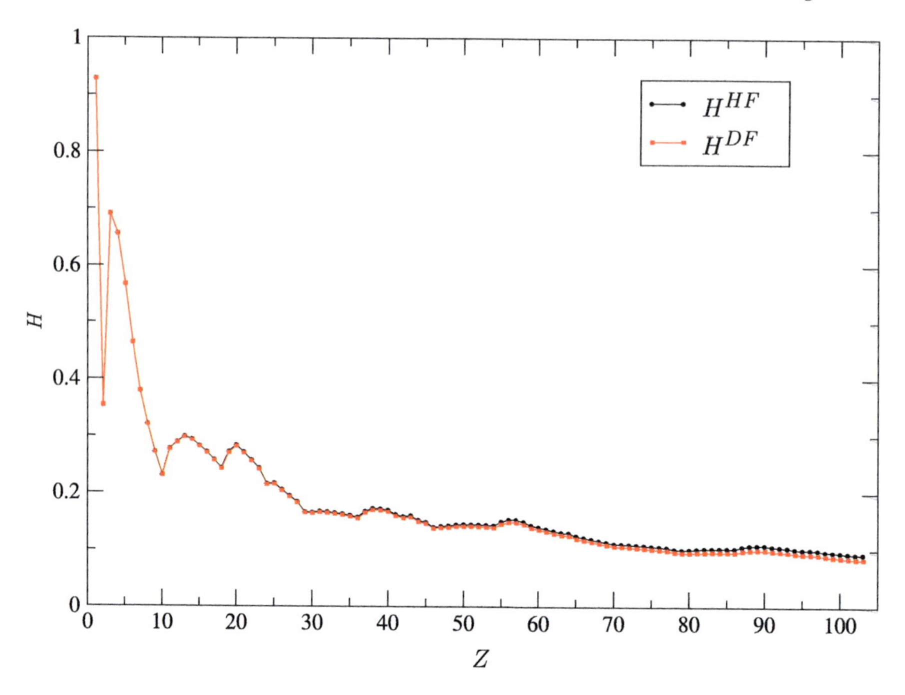

Fig. 9.12 A plot of the exponential power Shannon entropy, H, as a function of atomic number, Z, derived from the non-relativistic Hartree-Fock (*squares*) and relativistic Dirac-Fock densities (*circles*). The densities are each normalized to unity

Fig. 9.12, we have displayed H as defined in (9.24) in terms of the exponential power entropy. As discussed in our earlier work [64], the shell structure is revealed in both the cases with the relativistic DF estimates (smaller) showing stronger localization as Z increases. In Fig. 9.13, the LMC complexity, C_{LMC}, as calculated using DF densities have been compared with the corresponding HF estimates. The shell structure is more clearly revealed as regions of comparable complexity. The relativistic values are found to be more sensitive to the changes due to the occupation of the outer valency sub-shells. This is indicated by sudden change in the slope of C_{LMC} as Z increases. While the complexity within the lanthanide series is found to slowly increase with Z, the actinide atoms show a sharper rise in complexity along the series. This is attributed to the more readily accessible (larger D) 5f6d7s orbitals which are relatively closer in energy than 4f5d6s orbitals. In fact, for the non-relativistic atom the complexity is found to even decrease in the region $Z = 60$–80, after which it remains nearly constant. The plot in Fig. 9.13 of C_{LMC} derived from the non-relativistic HF densities suggests that atoms can not grow in complexity as Z increases. On the other hand, C_{LMC} estimated from the DF densities clearly indicate that C_{LMC} increases sharply at large Z suggesting that the relativistic effects dominantly contribute to the complexity of atoms growing with Z.

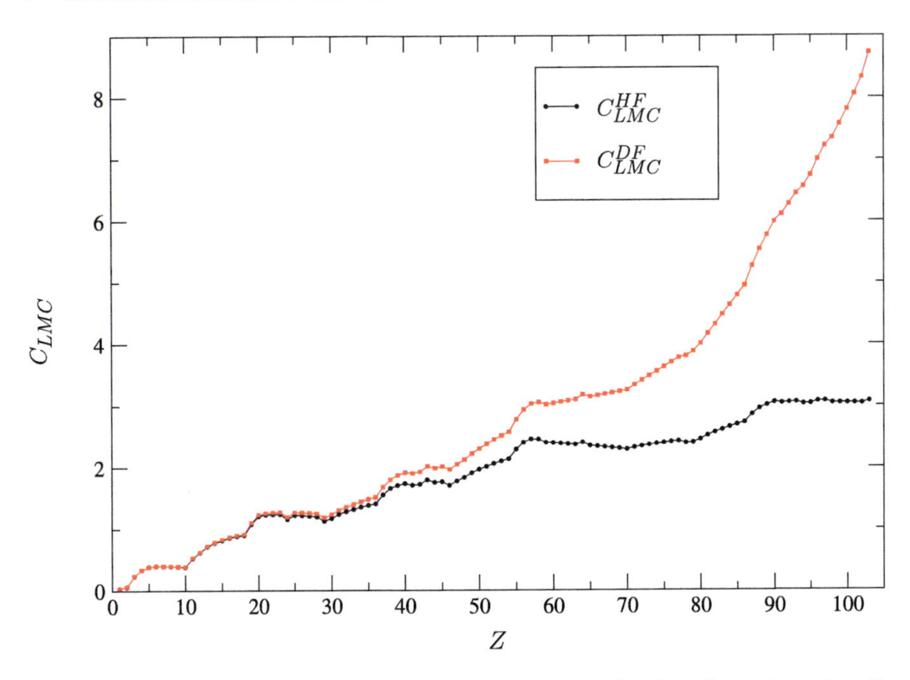

Fig. 9.13 A plot of the LMC complexity measure, C_{LMC}, as a function of atomic number, Z, derived from the non-relativistic Hartree-Fock (*squares*) and relativistic Dirac-Fock densities (*circles*)

In Fig. 9.14, we have plotted the simple SDL complexity measures, $\Gamma_{\alpha,\beta}$ in expression (9.30) defined by the four sets of (α, β) values given by $(1, 1)$, $(1, 1/4)$, $(1/4, 0)$, and $(0, 4)$, respectively. Among them, the SDL measure $\Gamma_{0,4}$ derived from the HF as well as the DF densities lead to an increase in complexity under category III [51], i.e. the complexity is an increasing function of order. However, relative to the HF data the DF results display a significantly sharper increase in $\Gamma_{0,4}$ as Z increases. This particular measure of SDL complexity behaves very similar to the LMC measure obtained in Fig. 9.13. Thus, our results on DF atoms lead to the C_{LMC} as well as $\Gamma_{0,4}$ showing the similar trend of increasing complexity with Z.

In conclusion the two measures show dissimilar trends for the non-relativistic HF results when all atoms in the periodic table are included in the test set. More specifically, the non-relativistic HF densities lead to decreasing LMC complexity with increasing Z beyond $Z \geq 54$, a trend which is clearly reversed when the relativistic effects are included to calculate C_{LMC} and $\Gamma_{\alpha,\beta}$ using the DF densities. The DF densities are found to dramatically increase the values of disequilibrium D for large Z in a sharp nonlinear manner in comparison to their non-relativistic estimates which follow a linear trend. The use of DF density is essential to reconcile the results derived from the two complexity measures in the region of large Z values $(Z \geq 54)$. The only assumption made in arriving at the this conclusion is that both the measures, C_{LMC} and $\Gamma_{\alpha,\beta}$, vary similarly with Z, which is reasonable. The result that the complexity of Dirac-Fock atom increases with Z as supported by two

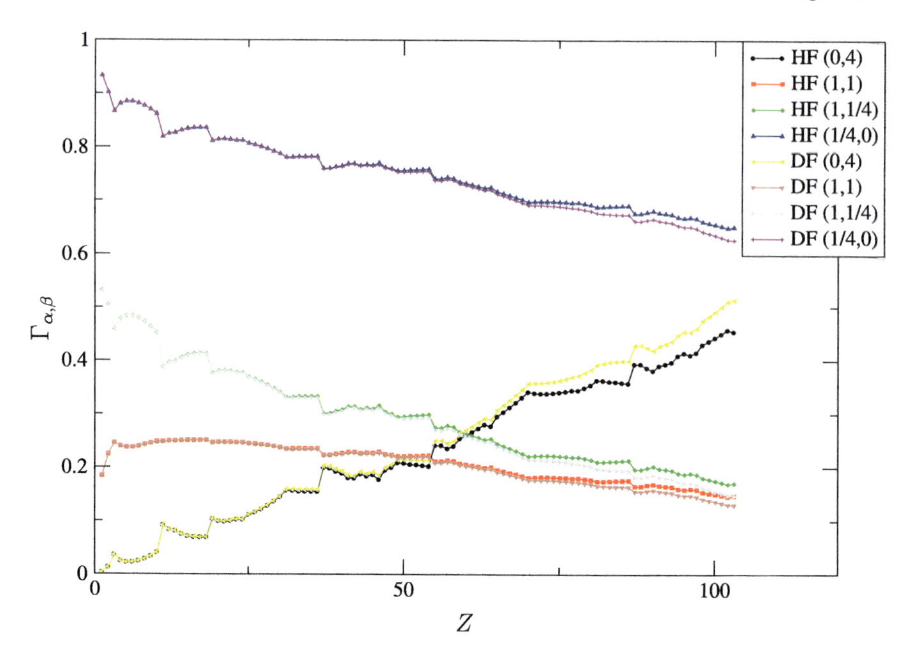

Fig. 9.14 A plot of the SDL complexity measures, $\Gamma_{\alpha,\beta}$, as a function of atomic number, Z, derived from the non-relativistic Hartree-Fock (*squares*) and relativistic Dirac-Fock densities (*circles*). The four SDL complexity measures are defined by the four sets of (α, β) values given by $(1, 1)$, $(1, 1/4)$, $(1/4, 0)$, and $(0, 4)$, have been displayed. The $\Gamma_{0,4}$ follows the trends similar to C_{LMC} plotted in Fig. 9.13 as a function of Z

intrinsically different measures of complexity measures C_{LMC} and $\Gamma_{0,4}$ is the main result of this study.

9.10 Relative Complexity Measures

It is interesting to extend the definition of statistical complexity of an atom A with respect to atom B and thus introduce the definition of relative complexity according to

$$C^{Rel} = H^R.D^R, \tag{9.31}$$

where H^R and D^R define the relative information content and the relative disequilibrium, respectively. In particular,

$$H^R = \frac{1}{2\pi e} e^{\frac{2}{3}S^R}, \tag{9.32}$$

where the relative entropy is defined, as earlier,

$$S^R = \int \sigma_A(\mathbf{r}) \log \frac{\sigma_A(\mathbf{r})}{\sigma_B(\mathbf{r})} \, d\mathbf{r}, \tag{9.33}$$

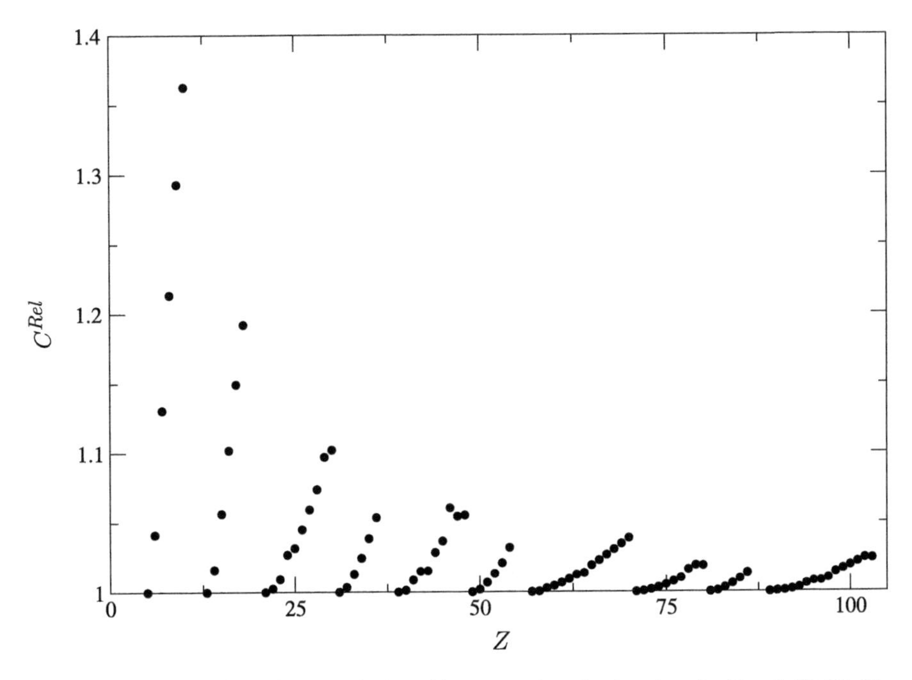

Fig. 9.15 Variation of relative complexity with a pre-assigned prior given by $Z = 5$, 13, 21, 31, 39, 49, 57, 71, 81, and 89 as atom B. In each curve, the atom A is defined by Z, $Z + 1$, $Z + 2$, until the sub-shell is completely filled

and the relative disequilibrium is evaluated as

$$\frac{D^R[A, B]}{\sqrt{D^R[A, A]}\sqrt{D^R[B, B]}}, \tag{9.34}$$

where

$$D^R[A, B] = \int \sigma_A(\mathbf{r})\sigma_B(\mathbf{r})\, d\mathbf{r}. \tag{9.35}$$

For a given set of structurally related atoms, the C^{Rel} values with respect to a common prior shape density, σ_B could be treated as measuring the *diversity* within the group. In Fig. 9.15, we have plotted the C^{Rel} values by choosing the prior shape density corresponding to the atoms with $Z = 5$, 13, 21, 31, 39, 49, 57, 71, 81, and 89 as atom B. For a given choice of B, the shape density corresponding to the atom A is given by $A = B(Z)$, $Z + 1$, $Z + 2, \ldots$. In all 10 sets of atoms A are chosen as a given sub-shell in them gets filled up successively beyond the ground state electronic structure of the prior atom B. For example, in the first set, the 2p sub-shell gets filled up successively along the series $2p^1 \rightarrow 2p^6$. All calculations have been performed using the relativistic DF densities. It is found that the relative complexity increases in a linear fashion within the group. The rate of increase within a group decreases as the prior atom becomes heavier, indicating that a constant diversity, i.e. a relatively flat C^{Rel} is gradually attained in moving towards the heavier end of the periodic table.

9.11 Information Theoretical Investigation of Reaction Paths

9.11.1 Introduction

From the construction of the functional which generated the chemical periodicity in Mendeleev's table (see Sect. 9.5), we obtained a generic method for constructing a functional which can be used to distinguish density functions. In the present application we will use the advantage of the reference choice to construct a simple functional which reveals reaction profiles. The aim is to obtain profiles which reflect the progress of the reaction, in a similar way as is done by energy profiles.

In this study the Kullback-Leibler information deficiency will be evaluated as a functional of the one-normalized density function for several processes: the internal rotation and resulting rotational isomerization in nitrous acid (HO–NO) and hydrogen peroxide (HO–OH), symmetric and asymmetric vibrational modes of the water molecule, and an intramolecular (HO-N=S $\rightarrow$ O=N-SH) and intermolecular (HO-N=S $\cdots$ OH_2 $\rightarrow$ O=N-SH $\cdots$ OH_2) proton transfer reaction. To analyze the specific role of the atoms during these processes, Hirshfeld's stockholder partitioning [28] of the electron density, which permits the evaluation of atomic contributions to the Kullback-Leibler information deficiency, has been introduced. Finally we report the information theoretical analysis of a more complex chemical reaction, the bimolecular nucleophilic substitution (S_N2) reaction. Some of these processes have been used before, for benchmarking the concepts and principles of conceptual DFT [2].

9.11.2 Construction of the Functionals

For the purpose of investigating chemical processes we introduce a single coordinate (bond length, bond angle, or dihedral angle) or the intrinsic reaction coordinate (IRC) characterizing the minimum energy path. We write ΔS_ξ for a given coordinate value ξ as:

$$\Delta S_{KL}[\sigma_\xi(\mathbf{r})|\sigma_0(\mathbf{r})] = \int \sigma_\xi(\mathbf{r}) \log \frac{\sigma_\xi(\mathbf{r})}{\sigma_0(\mathbf{r})} \, d\mathbf{r}, \tag{9.36}$$

where $\sigma_\xi(\mathbf{r})$ is the shape function for a given value of the reaction coordinate ξ and $\sigma_0(\mathbf{r})$ corresponds to the selected reference. In the case of the rotational and the vibrational motions, the shape function of the equilibrium configuration has been taken as reference. For the chemical reactions the shape function of the transition state (TS) has been chosen. Now the result of (9.36) can be interpreted as a measure of the information which distinguishes the shape function at position ξ from the TS shape function or alternatively as the distance in information between the TS and the system for an arbitrary value of ξ.

Hirshfeld's stockholder partitioning [28] for atoms in molecules (AIM) is closely linked with information theory. Nalewajski showed that [27] the minimum entropy deficiency principle of Kullback and Leibler applied to the problem of defining AIM yields the Hirshfeld definition—by relating the molecular density function to that of the promolecule and by assuming that the AIM densities preserve as much information as possible about the isolated atoms. In this context, the contribution of atom A to ΔS_ξ is given by

$$\Delta S_{KL}^{A}[\sigma_\xi^A(\mathbf{r})|\sigma_0^A(\mathbf{r})] = \int \sigma_\xi^A(\mathbf{r}) \log \frac{\sigma_\xi^A(\mathbf{r})}{\sigma_0^A(\mathbf{r})}\, d\mathbf{r}, \tag{9.37}$$

where $\sigma_\xi^A(\mathbf{r})$ and $\sigma_0^A(\mathbf{r})$ are the atom A's Hirshfeld contributions to the total shape function and to shape function for atom A in the TS. To obtain proper probability distributions the atomic Hirshfeld density functions were normalized to unity. The confrontation of the different atomic information deficiencies indicate which atomic shape function differs in information from its TS counterpart. As discussed below, by identifying the atomic regions where the local contributions increase or decrease one gets an idea of the role the atoms play as the rotational, vibrational or chemical process advances. Note that the sum of the atomic information deficiencies differs from the global one, by a term which can be interpreted as an entropy of mixing [65].

9.11.3 Methodology and Implementation

The geometry optimization and the electron densities were calculated using the hybrid exchange–correlation functional B3LYP [66, 67] combined with a standard 6-311G(d, p) basis set [68] available in the Gaussian03 program [69]. In order to discriminate between a minimum and a saddle points on the potential energy surface a vibrational analysis was carried out. The minimum energy paths in going from reactants to products were calculated through the Intrinsic Reaction Coordinate (IRC) procedure [70, 71] using a gradient step size of 0.10 amu$^{1/2}$ bohr.

The information theory concepts were implemented in the program STOCK, part of BRABO package [72, 73]. For a detailed discussion of the implemented numerical techniques we refer to [74]. To evaluate the integrals in (9.36) and (9.37) the structure, at a given ξ needs to be aligned with the reference structure. This procedure was carried out aligning the center of mass (CM) of both structures, and fixing the last degree of freedom by minimizing the mean root square deviation of the distances between the corresponding nuclei. Other alignments were tested but revealed the same tendencies in the resulting profiles.

9.11.4 Results and Discussion

9.11.4.1 Internal Rotations: HO–NO and HO–OH

The rotational isomerization *cis* ↔ *trans* of HO–NO is graphically represented in Fig. 9.16(a). This process is a typical example of a double-well energy profile, as

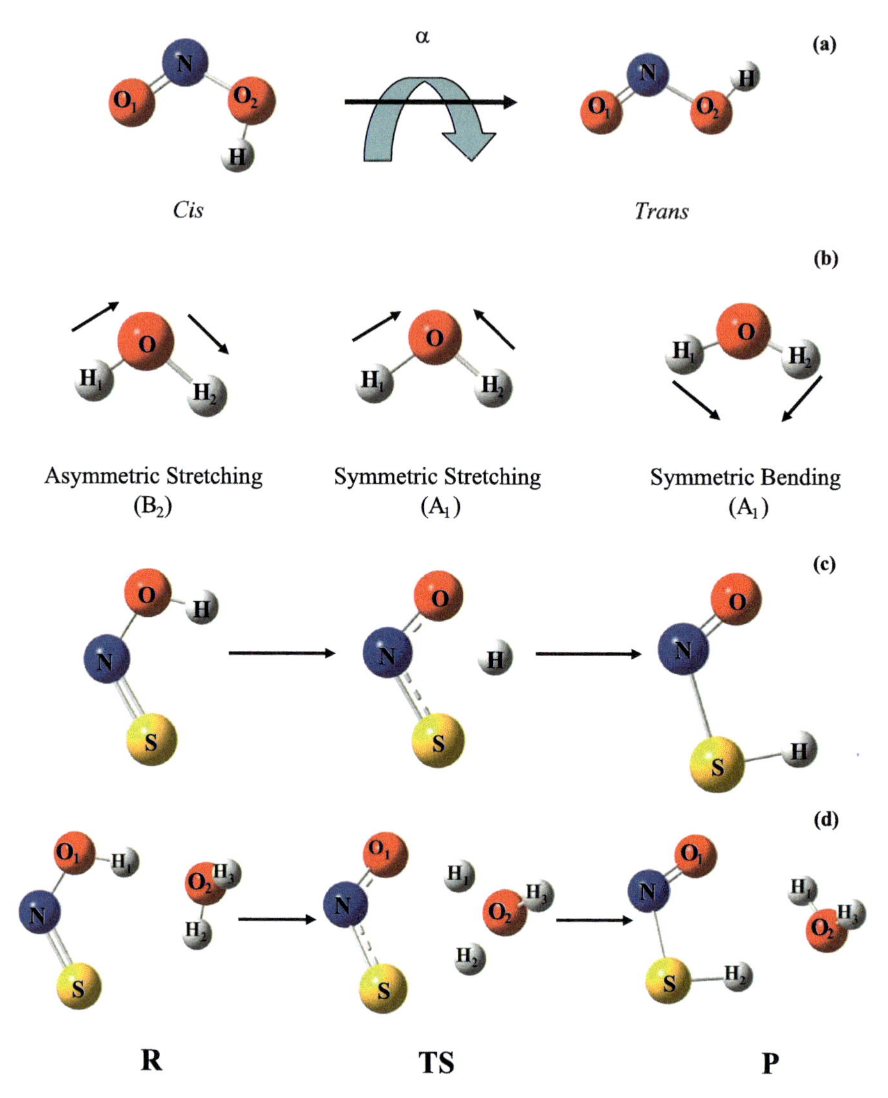

Fig. 9.16 Schemes of the studied processes: (**a**) internal rotation, (**b**) vibrational modes in H_2O, (**c**) intramolecular, and (**d**) intermolecular proton transfer reactions

displayed in Fig. 9.17(a). The planar *cis* and *trans* are stable conformers (minima) whereas the non-planar *gauche* conformation is a first-order saddle point or TS. The shape function of this *gauche* conformation will serve as our reference for the evaluation of (9.36). For the construction of the profiles we evaluated the energy and information deficiency in (9.36) using a step of 10° along the torsional angle within the range $0°$ (*cis*) $\leq \alpha \leq 180°$ (*trans*). The energy profile in Fig. 9.17(a) shows that the energetically unstable *gauche* conformation is positioned midway between the planar isomers ($\alpha = 90°$).

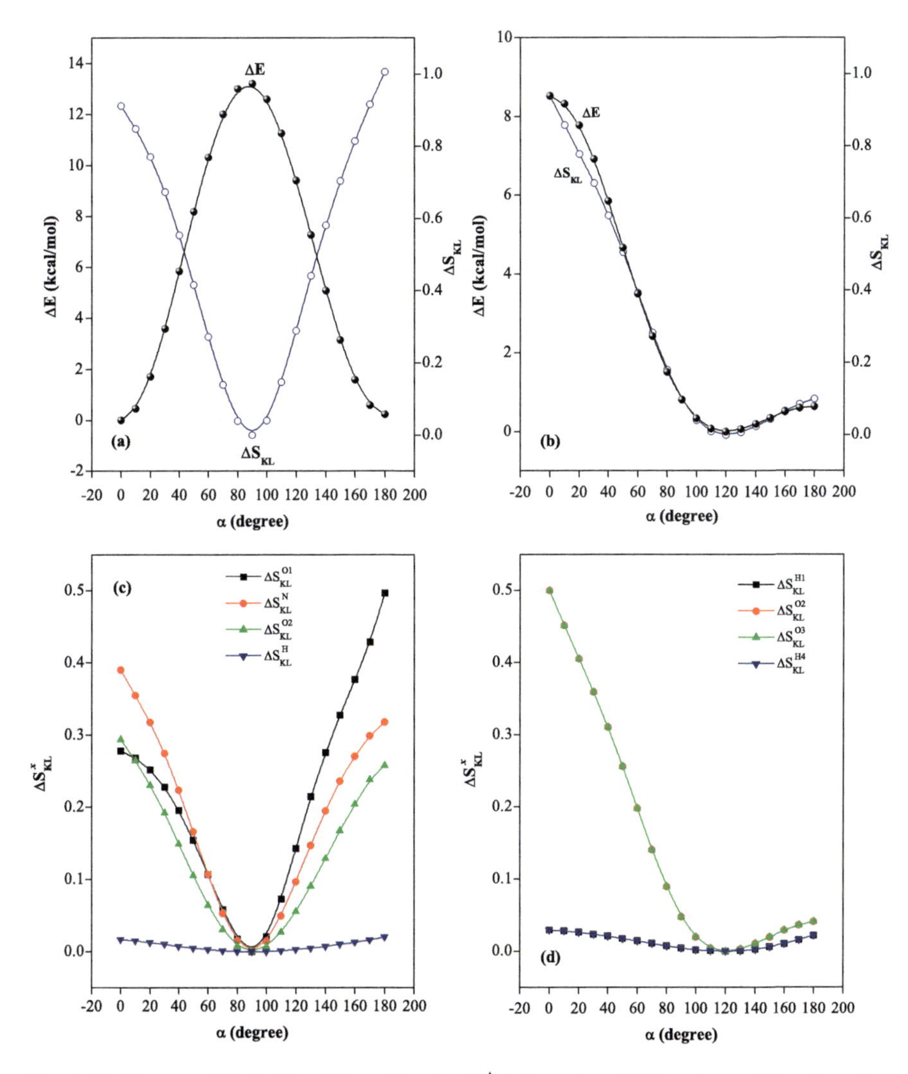

Fig. 9.17 Energy, global $\Delta S_{KL}(\xi)$ and local ΔS^A information deficiencies profiles along the torsional coordinate of HO–NO ((**a**) and (**c**)) and HO–OH ((**b**) and (**d**))

From Fig. 9.17(a) it is clear that the choice of the *gauche* conformer as the reference implies a minimum when the energy profile reaches its maximum. We also see that the small stability difference between *cis* and *trans* is reflected in ΔS_{KL}.

As a second example we investigate the internal rotation process of HO–OH, which is characterized by a non-planar *gauche* stable conformer, connected to planar *cis* and *trans* conformations through two barriers. This process is a typical example of a double-barrier energy profile. Choosing the *gauche* conformation as reference reveals a profile with a close resemblance to the energy profile, as can be seen in

Fig. 9.17(b). One notices that the large energy difference between the planar conformers is recovered in the ΔS_{KL} profile.

The atomic information deficiency of Hirshfeld's density for atom A, $\Delta S_{KL}^{A}(\xi)$, is studied for both rotational isomerization processes. This quantity can be separated in two contributions: a delocalization and a localization component [75], as given by

$$\Delta S_{KL}^{A}(\xi) = \int \sigma_{\xi}^{A}(\mathbf{r}) \log \sigma_{\xi}^{A}(\mathbf{r})\, d\mathbf{r} - \int \sigma_{\xi}^{A}(\mathbf{r}) \log \sigma_{0}^{A}(\mathbf{r})\, d\mathbf{r}, \qquad (9.38)$$

where $\sigma_{\xi}^{A}(\mathbf{r})$ and $\sigma_{0}^{A}(\mathbf{r})$ are the shape functions of the subsystem A at IRC coordinate value ξ and the TS, respectively. The first term in (9.38) is a reference independent term, which can be identified as the Shannon information entropy (cf. Sect. 9.2) of the subsystem's shape function, whereas the second term is a reference dependent term which gives information about the localization of the subsystem's density at IRC value ξ, with respect to the density of the reference. Starting with HO–NO, in Fig. 9.17(c) one observes the following: (i) the shape function of the O_1 atom is more delocalized in the *trans* isomer than in the *cis* isomer whereas the opposite tendency is observed for the O_2 atom; (ii) the information deficiency of the N atom is higher in the *cis* conformer than in the other planar structure; (iii) the atomic contribution corresponding to the hydrogen atom does not change significantly along the torsional coordinate. These results together with the global ΔS_{ξ} indicate that the process where the unstable isomer reaches any of the stable isomers is driven by an increase of the delocalization. Moreover, it allows to understand the similarity of the energy barriers of the forward and the backward processes. Moving to HO–OH, in Fig. 9.17(d) for the local contributions, the equivalence of both H atoms and both O atoms is clearly reflected. The large difference between the H and the O atoms can be linked to the alignment of the CM. Indeed, since the CM of both structures are aligned, the density rotates around it. The fact that the O atoms are closer to the CM and have more diffuse electron distributions (due to the presence of the lone pairs electron) can then explain the large difference between the H and the O profiles. The small differences in global and local information entropy for the stable *gauche* and the unstable *trans* isomers are a consequence of the similarity in the shape functions, whereas these are different for *gauche* and *cis* conformers.

9.11.4.2 Molecular Vibrations: Stretching and Bending of a Water Molecule

In this section we analyze the information entropy along the coordinates for the vibrational modes for H_2O as given in Fig. 9.16(b). Figure 9.18(a), (c), and (e) show the energy and the global information deficiency profiles for the three vibration modes of H_2O. As before the choice of the equilibrium structure as reference ensures that the minimum in information deficiency corresponds to the minimum of the energy profile.

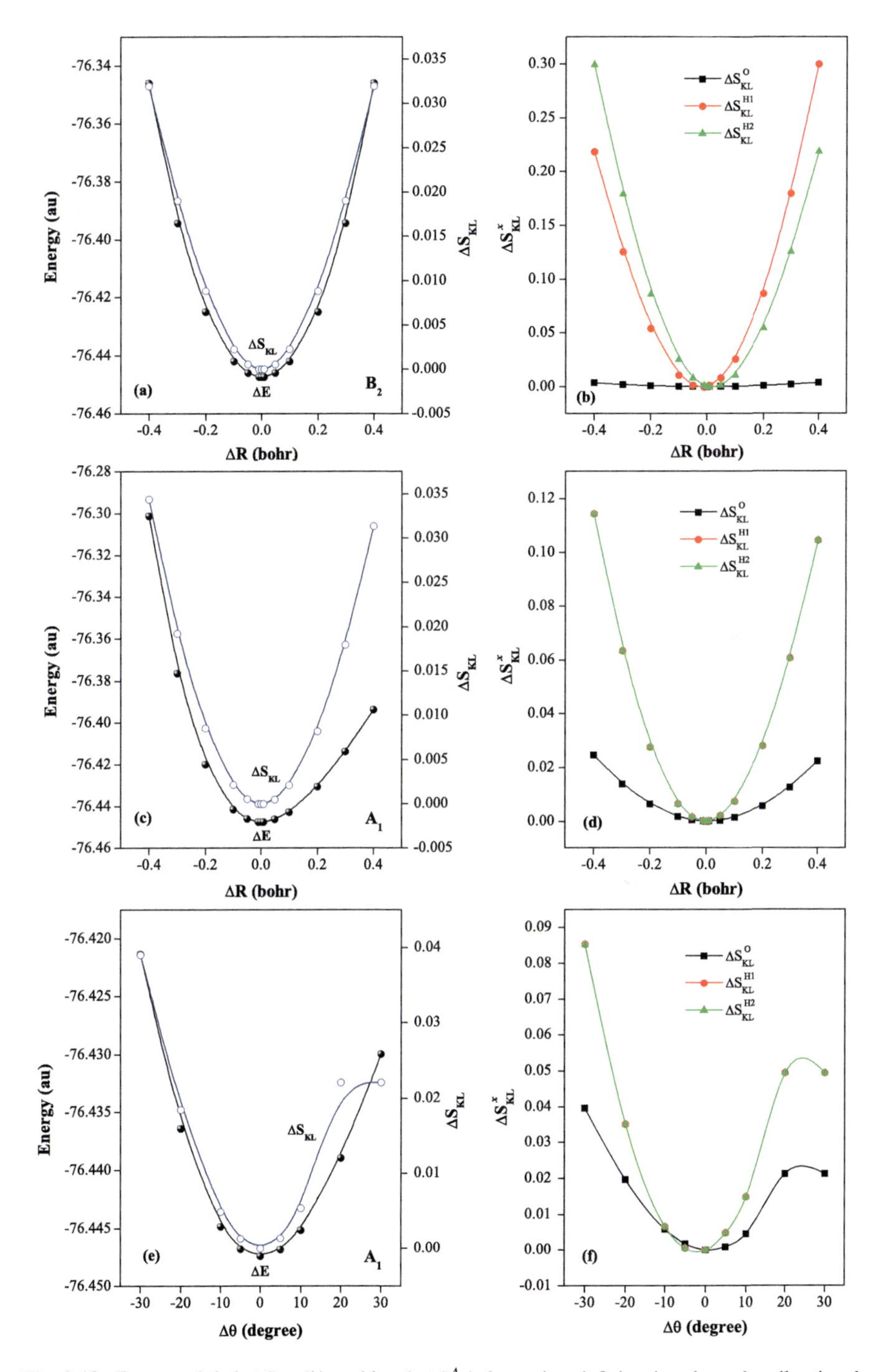

Fig. 9.18 Energy, global $\Delta S_{KL}(\xi)$ and local ΔS^A information deficiencies along the vibrational coordinates of H_2O: (**a**) and (**b**) asymmetric stretching; (**c**) and (**d**) symmetric stretching; (**e**) and (**f**) symmetric bending

An analysis of the information deficiency profiles of the Hirshfeld atoms in molecules (AIM) helps to understand the role atoms play along the vibrational coordinate. In Fig. 9.18(b), (d), and (f) we give the profiles of the atomic information deficiency for the three modes. In the asymmetric mode we notice that the ΔS for the O atom remains small throughout the process, whereas the corresponding results in the symmetric process are significantly larger (approximately a factor of 2). This can be linked with the comparatively small changes in the external potential of the asymmetric stretching [76, 77] and the fact that the CM is located close to the O atom. By comparing the profiles of the H atoms it is clear that the results reflect the symmetry of the considered process.

After investigating two elementary processes, we can conclude that the global ΔS_ξ in (9.36) reflects the behavior of the energy profile. We now consider that it may be a useful complementary tool for the description of chemical reactions. The results at the local level could indicate that the analysis of the information profiles of the Hirshfeld AIM help to understand the role the atoms play within a chemical reaction.

9.11.5 Chemical Reactions

9.11.5.1 Intramolecular Proton Transfer: HO-N=S → O=N-SH

First we investigate the intramolecular proton transfer reaction HO-N=S → O=N-SH, sketched in Fig. 9.16(c), to test whether the information deficiency profile reveals known properties of this chemical reaction. By choosing the TS as the reference we obtain a correspondence of the energy and information deficiency extrema, as seen in Fig. 9.19(a). The confrontation of the thermodynamic driving force $\Delta E^\circ = E(P) - E(R)$ (the energy difference between product and reactant) and the energy barrier $\Delta E^{\neq} = E(TS) - E(R)$ (the energy difference between TS and reactant) with the information distance between the TS and the reactants, $\Delta S_\xi[\sigma_R|\sigma_{TS}]$, and the information distance between the TS and the products, $\Delta S[\sigma_P|\sigma_{TS}]$, quoted in Table 9.2, indicates an interesting link with the renowned Hammond postulate (HP) [78]. The HP is a classic concept in physical organic chemistry, which asserts that the transition state of an elementary reaction step will be localized closer to the reactant or product, depending on which has the highest energy [78]. As can be seen from the energetic data given in Table 9.2, O=N-SH is more stable by about 7 kcal/mol than HO-N=S and the energy barrier for the forward and reverse processes are about 27 and 34 kcal/mol respectively. This, on one hand, suggests that the O atom is a better proton donor than the S atom and that the N=O double bond is stronger than the double bond formed between N and S atoms. On the other hand, an early transition state is expected according to the HP. Consistent with the energetic data and the HP, the $\Delta S_\xi[\sigma_R|\sigma_{TS}]$ and $\Delta S[\sigma_P|\sigma_{TS}]$ show that the TS contains more information about the reactants than about the products.

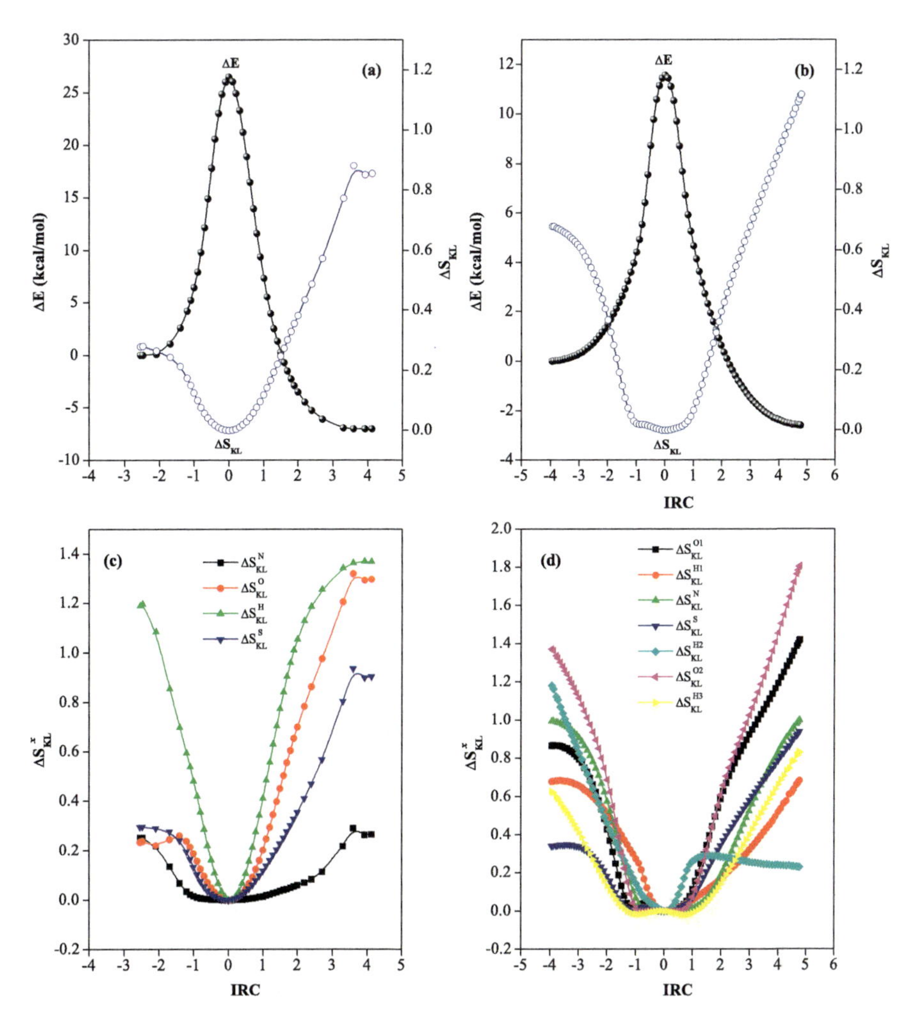

Fig. 9.19 Energy, global $\Delta S_{KL}(\xi)$ and local ΔS^A information deficiencies along the intrinsic reaction coordinate of an intramolecular (HO-N=S→O=N-SH) ((**a**) and (**c**)) and intermolecular (HO-N=S···OH_2 → O=N-SH ··· OH_2) ((**b**) and (**d**)) proton transfer reactions

Analyzing the atomic profiles, given in Fig. 9.19(c), we see that the ΔS_ξ values of the N atom remains small in both the reactant and the product regions. This can be expected since the density around the nucleus, closest to the CM, undergoes the smallest influence, although the N atom switches single and double bonding between O and S atoms, respectively. For both the O and S atoms one notices a difference between the left and the right side of the TS, which can be understood since the electron reorganization which takes place on the left is less significant than for the process on the right, indicating that the through-bond interactions are more important in the product side than in the reactant side as pointed out in reference [79, 80].

Table 9.2 Reaction energy (ΔE°), energy barrier for the forward ($\Delta E^{\neq}_{forward}$) and reverse ($\Delta E^{\neq}_{reverse}$) processes. Information deficiency with the transition state (TS) as reference of the reactant ($\Delta S[\sigma_R, \sigma_{TS}]$) and the product ($\Delta S[\sigma_P, \sigma_{TS}]$). Energy values are given in kcal/mol. Reaction A is HO-N=S → O=N-SH, reaction B is HO-N=S···OH_2 → O=N-SH···OH_2 and reaction C is OH^- + CH_3-F → HO-CH_3 + F^-

Reaction	ΔE°	$\Delta E^{\neq}_{forward}$	$\Delta E^{\neq}_{reverse}$	$\Delta S_{\xi}[\sigma_R, \sigma_{TS}]$	$\Delta S_{\xi}[\sigma_P, \sigma_{TS}]$
A	−7.00	26.46	33.46	0.28	0.86
B	−2.59	11.56	14.15	0.68	1.12
C	−36.75	7.85	44.60	1.83	3.44

9.11.5.2 Intermolecular Proton Transfer: HO-N=S ··· OH_2 → O=N-SH ··· OH_2

The process we consider here is obtained by the assistance of a water molecule in the reaction analyzed above, as displayed Fig. 9.16(d). In Fig. 9.19(b), one observes that the choice of the TS as reference assures the minimum of the profile to coincide with the maximum in the energy profile. By comparing the energetic data reported in Table 9.2 for this reaction with the intramolecular proton transfer discussed in the previous section, we see that the presence of a water molecule decreases the energy barrier (for the forward reaction) by 14.9 kcal/mol and the thermodynamic driving force by 4.4 kcal/mol. Again, the ΔS_{ξ} values, quoted in Table 9.2, indicate that the TS contains more information about the reactants than about the products, which is in agreement with the HP.

After investigating the profiles of the local information deficiencies, corresponding to the Hirshfeld AIM partitions, depicted in Fig. 9.19(d), we report two observations. First, the profiles indicate that the electron reorganization taking place in both the forward and the reverse processes are equally important, certainly when compared to the difference noticed in the intramolecular proton transfer process. In other words, the through bond interactions drive both the forward and the reverse processes. Second, by comparing the results of the HO-N=S backbone with those of the intramolecular process, we can identify a delocalization effect on the backbone, induced by the H_2O molecule. This favors the proton transfer from a kinetic point of view.

9.11.5.3 Nucleophilic Substitution at Carbon Center (S_N2) Reaction: OH^- + CH_3-F → HO-CH_3 + F^-

In the final example we analyze the S_N2 reaction displayed in Fig. 9.20(a). The reaction starts with a [F-CH_3 ··· OH]$^-$ ion–molecule complex which arises from the backside attack of the OH^- group on the C atom. After reaching the TS, the system relaxes toward the products by repositioning the F atom, driven by the formation of a hydrogen bond.

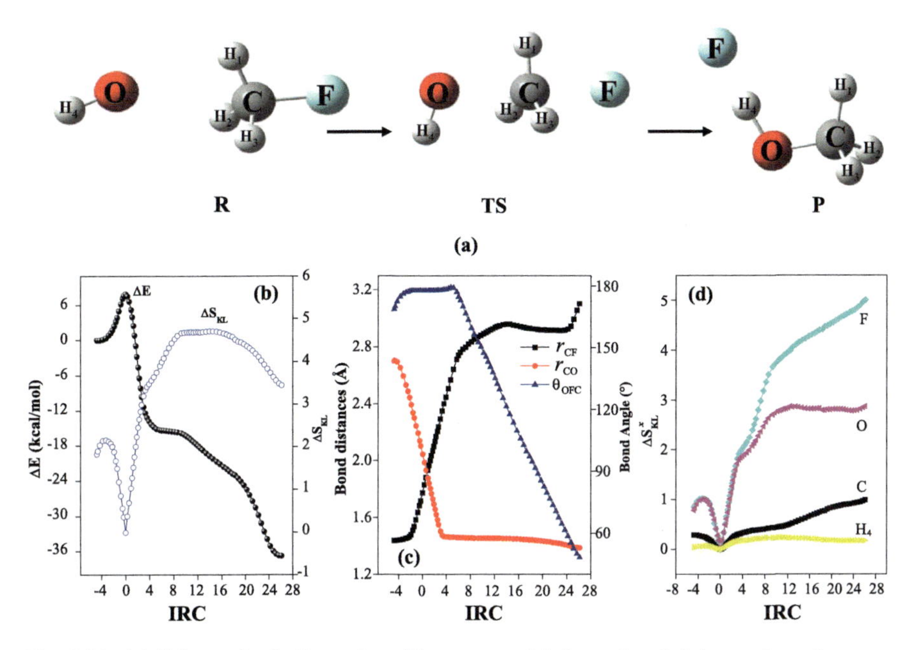

Fig. 9.20 (**a**) Scheme for S_N2 reaction; (**b**) energy and information deficiency along the reaction path; (**c**) bond distances and angle profiles, and (**d**) atomic contributions for the information deficiency associated to the S_N2 reaction

For the electronic structure calculations of this reaction we follow the study by Gonzales et al. [81], where it is stated that the hybrid exchange–correlation functional B3LYP [66, 67] performs better than other functionals for describing the structure of the stationary points and it gives the best energetic values for the S_N2 reaction. The authors also stressed the necessity to include diffuse functions in the basis set. We used the aug-cc-pVTZ basis set [82, 83]. We start the discussion of the results by comparing the energy and the information deficiency profiles, as presented in Fig. 9.20(b). One can distinguish the correspondence of the minimum of ΔS_ξ and the maximum of the energy, which is guaranteed by choosing the TS as reference in (9.36). The data collected in Table 9.2 show that the TS contains more information about the reactants ($\Delta S_\xi = 1.8276$), than about the products ($\Delta S_\xi = 3.4442$). We found an energy barrier of 7.85 kcal/mol for the forward reaction and a barrier of 44.60 kcal/mol for the reverse reaction. This indicates that the Hammond's behavior of the reaction is again reflected in the information profile.

The steep slopes around the minimum in the information deficiency profile can be linked with the behavior of the geometrical parameters, plotted in function of the IRC in Fig. 9.20(c). One notices indeed that, in the vicinity of the TS ($IRC = 0$), the distances between the nuclei vary significantly, while the angle remains practically constant ($\theta_{FCO} \approx 178°$), which indicates that the behavior of the profile around the TS is governed by the bond-formation and bond-breaking processes. Further analyzing the information profile in Fig. 9.20(b), one notices a small bump close to the reactant side and a plateau after $IRC = 9$. The first feature can be linked to the rapid variation of the FCO angle in the beginning of the process. The second feature seems to correspond with the sudden drop in energy for $IRC > 9$.

In Fig. 9.20(d), we plotted the results of the atomic information deficiency values along the IRC. Note that we left the results for the three H atoms (H1, H2, H3) bonded to the C atom out of the graph, since they only revealed minor values which is in agreement with the fact that they do not directly participate in the reaction. The local contributions reflect the importance of the electron reorganization in the product side, which permits the stabilization of the final structure. On the other hand, it is interesting to note that the global behavior is mainly determined by the contributions of the F and O atoms. These centers follow the same tendency until F leaves the quasi linear FCO framework to reach the hydrogen bonding structure (at $IRC = 3$), which can be deduced from the behavior of the geometrical parameters given in Fig. 9.20(c).

9.11.6 Conclusion

In this section we demonstrated the value of the Kullback-Leibler information deficiency by analyzing profiles for several benchmark processes and chemical reactions. To obtain the profiles we evaluated the Kullback-Leibler information deficiency for a significant set of IRC values. The choice of the transition state as reference in the Kullback-Leibler's information deficiency ensures that the information deficiency and the energy profiles simultaneously reach an extremum. We have been able to identify important chemical properties in the information deficiency profiles. In particular the recovery of results consistent with Hammond's postulate strongly indicates that the information deficiency carries valuable chemical information. Evaluating Kullback-Leibler's measure for Hirshfeld's atoms-in-molecules shape functions, revealed that, when choosing the corresponding atomic shape function of the transition state as reference, the results indicate the most electron reorganizing sites, which can help to understand the role different atomic regions play during a chemical reaction.

9.12 Concluding Remarks

Vested in the spirit of the Hohenberg-Kohn theorems of DFT, we investigated atomic and molecular density functions to try and devise a way to extract chemical information from them. In our search several functionals from information theory have proven to be particularly successful. In particular the functional form of the Kullback-Leibler measure permitted the construction of density functionals, which help us to read chemical information from the density function.

The main success of our results based on the Kullback-Leibler information discrimination, is based on the choice a good reference. For example, it is the specific choice of the reference as noble gas elements, which helped to reveal the periodicity in Mendeleev's table and it is the choice of the transition state as reference, which revealed chemically interesting reaction profiles.

The investigation of complexity measures, which came to quantum chemistry more recently, throughout the periodic table showed that these functionals too have revealed features of the atomic electronic structure, indicating their potential to investigate the chemical information carried in density functions and shape functions. It is shown that a relative complexity measure can be defined which reflects the diversity of electron density with respect to a prior atom as a given sub-shell is filled across the periodic table.

References

1. Hohenberg P, Kohn W (1964) Phys Rev B 136(3B):B864
2. Geerlings P, De Proft F, Langenaeker W (2003) Chem Rev 103(5):1793
3. Geerlings P, De Proft F (2008) Phys Chem Chem Phys 10(21):3028
4. Levine RD (1978) Annu Rev Phys Chem 29:59
5. Sears SB, Parr RG, Dinur U (1980) Isr J Chem 19(1–4):165
6. Gadre S (2000) Information theoretical approaches to quantum chemistry, vol 1. World Scientific, Singapore, pp 108–147
7. Catalan RG, Garay J, Lopez-Ruiz R (2002) Phys Rev E 66(1):011102
8. Geerlings P, Borgoo A (2010) Phys Chem Chem Phys 13:911
9. Borgoo A, Jaque P, Toro-Labbé A, Van Alsenoy C, Geerlings P (2009) Phys Chem Chem Phys 11(3):76
10. Borgoo A, Geerlings P, Sen KD (2008) Phys Lett A 372(31):5106
11. Borgoo A, De Proft F, Geerlings P, Sen KD (2007) Chem Phys Lett 444:186
12. Shannon S (1948) Bell Syst Tech J 27:379
13. Mathai A (1975) Basic concepts in information theory and statistics axiomatic foundations and applications. Wiley Eastern, New Delhi
14. Ash R (1967) Information theory. Interscience Publishers, New York
15. Kinchin A (1957) Mathematical foundations of information theory. Dover, New York
16. Jaynes ET (1957) Phys Rev 106(4):620
17. Jaynes ET (1957) Phys Rev 108(2):171
18. Kullback S, Leibler RA (1951) Ann Math Stat 22(1):79. URL http://www.jstor.org/stable/2236703
19. Lin J (1991) IEEE Trans Inf Theory 37(1):145
20. Majtey A, Lamberti PW, Martin MT, Plastino A (2005) Eur Phys J D 32(3):413
21. Chatzisavvas KC, Moustakidis CC, Panos CP (2005) J Chem Phys 123(17):174111
22. Fisher R (1925) Proc Camb Philos Soc 22:700
23. Borgoo A, Godefroid M, Sen KD, De Proft F, Geerlings P (2004) Chem Phys Lett 399(4–6):363
24. Gadre SR, Sears SB, Chakravorty SJ, Bendale RD (1985) Phys Rev A 32(5):2602
25. Fazal SP, Sen KD, Gutierrez G, Fuentealba P (2000) Indian J Chem, Sect A 39(1–3):48
26. Nagy A, Parr RG (1996) Int J Quant Chem 58(4):323
27. Nalewajski RF, Parr RG (2000) Proc Natl Acad Sci USA 97(16):8879
28. Hirshfeld FL (1977) Theor Chim Acta 44(2):129
29. Borgoo A, Scharf O, Gaigalas G, Godefroid M (2010) Comput Phys Commun 181(2):426. URL http://www.sciencedirect.com/science/article/B6TJ5-4XG3SF0-1/2/d040eb496c97b1d109b779bede692437
30. Sanderson RT (1951) Science 114:670
31. Maggiora GM, Petke JD, Mestres J (2002) J Math Chem 31(3):251

32. Hodgkins EE, Richards WG (1987) Int J Quant Chem S14:105
33. Liu SB, De Proft F, Parr RG (1997) J Phys Chem A 101(37):6991
34. Liu SB (2007) J Chem Phys 126(19):191107
35. Jackson J (1975) Classical electrodynamics. Wiley, New York
36. Liu SB, Parr RG, Nagy A (1995) Phys Rev A 52(4):2645
37. Desclaux JP (1975) Comput Phys Commun 9(1):31
38. Kutzelnigg W (1988) Theor Chim Acta 73(2–3):173
39. Hada M, Ishikawa Y, Nakatani J, Nakatsuji H (1999) Chem Phys Lett 310(3–4):342
40. Borgoo A, Godefroid M, Indelicato P, De Proft F, Geerlings P (2007) J Chem Phys 126(4):044102
41. Sen KD, Panos CP, Chatzisavvas KC, Moustakidis CC (2007) Phys Lett A 364(3–4):286
42. Koga T, Kanayama K, Watanabe S, Thakkar AJ (1999) Int J Quant Chem 71(6):491
43. Koga T, Kanayama K, Watanabe T, Imai T, Thakkar AJ (2000) Theor Chem Acc 104(5):411
44. Handy NC (1996) Quantum mechanical simulation methods for studying biological systems. Springer, Berlin
45. Kato T (1957) Commun Pure Appl Math 10(2):151
46. Feldman DP, Crutchfield JP (1998) Phys Lett A 238(4–5):244
47. Panos CP, Chatzisavvas KC, Moustakidis CC, Kyrkou EG (2007) Phys Lett A 363(1–2):78
48. LopezRuiz R, Mancini HL, Calbet X (1995) Phys Lett A 209(5–6):321
49. Landsberg PT (1984) Phys Lett A 102(4):171
50. Gadre SR, Bendale RD (1987) Phys Rev A 36(4):1932
51. Shiner JS, Davison M, Landsberg PT (1999) Phys Rev E 59(2):1459
52. Crutchfield JP, Feldman DP, Shalizi CR (2000) Phys Rev E 62(2):2996
53. Stoop R, Stoop N, Kern A, Steeb WH (2005) J Stat Mech-Theory Exp P11009
54. Bunge CF, Barrientos JA, Bunge AV (1993) At Data Nucl Data Tables 53(1):113
55. Angulo JC, Romera E, Dehesa JS (2000) J Math Phys 41(12):7906
56. Das BP, Latha KVP, Sahoo BK, Sur C, Chaudhuri RK, Mukherjee D (2005) J Theor Comput Chem 4(1):1
57. Lindgren I (1996) Int J Quant Chem 57(4):683
58. Engel E, Bonetti AF, Keller S, Andrejkovics I, Dreizler RM (1998) Phys Rev A 58(2):964
59. Carbo R, Leyda L, Ardau M (1980) Int J Quant Chem 17(6):1185
60. Onicescu O (1966) C R Hebd Seances Acad Sci, Ser A 263(22):841
61. Hall MJW (1999) Phys Rev A 59(4):2602
62. Pennini F, Plastino A (2007) Phys Lett A 365(4):263
63. Hyman AS, Yaniger SI, Liebman JF (1978) Int J Quant Chem 14(6):757
64. Sen KD, De Proft F, Borgoo A, Geerlings P (2005) Chem Phys Lett 410(1–3):70
65. Parr RG, Ayers PW, Nalewajski RF (2005) J Phys Chem A 109(17):3957
66. Becke AD (1993) J Chem Phys 98(7):5648
67. Lee CT, Yang WT, Parr RG (1988) Phys Rev B 37(2):785
68. Krishnan R, Bnkley JS, Seeger R, Pople JA (1980) J Chem Phys 72(1):650
69. Frisch MJ, Trucks GW, Schlegel HB, Scuseria GE, Robb MA, Cheeseman JR, Montgomery JA Jr, Vreven T, Kudin KN, Burant JC, Millam JM, Iyengar SS, Tomasi J, Barone V, Mennucci B, Cossi M, Scalmani G, Rega N, Petersson GA, Nakatsuji H, Hada M, Ehara M, Toyota K, Fukuda R, Hasegawa J, Ishida M, Nakajima T, Honda Y, Kitao O, Nakai H, Klene M, Li X, Knox JE, Hratchian HP, Cross JB, Bakken V, Adamo C, Jaramillo J, Gomperts R, Stratmann RE, Yazyev O, Austin AJ, Cammi R, Pomelli C, Ochterski JW, Ayala PY, Morokuma K, Voth GA, Salvador P, Dannenberg JJ, Zakrzewski VG, Dapprich S, Daniels AD, Strain MC, Farkas O, Malick DK, Rabuck AD, Raghavachari K, Foresman JB, Ortiz JV, Cui Q, Baboul AG, Clifford S, Cioslowski J, Stefanov BB, Liu G, Liashenko A, Piskorz P, Komaromi I, Martin RL, Fox DJ, Keith T, Al-Laham MA, Peng CY, Nanayakkara A, Challacombe M, Gill PMW, Johnson B, Chen W, Wong MW, Gonzalez C, Pople JA (2004) Gaussian 03, Revision C.02. Gaussian Inc, Wallingford
70. Fukui K (1981) Acc Chem Res 14(12):363
71. Gonzalez C, Schlegel H (1990) Reaction-path following in mass-weighted internal coordinates

72. Van Alsenoy C, Peeters A (1993) Theochem 286:19
73. Rousseau B, Peeters A, Van Alsenoy C (2000) Chem Phys Lett 324:189
74. Mandado M, Van Alsenoy C, Geerlings P, De Proft F, Mosquera RA (2006) Chem Phys Phys Chem 7:1294
75. Nalewajski RF, Parr RG (2001) J Phys Chem A 105(31):7391
76. Pearson RG, Palke WE (1992) J Phys Chem 96(8):3283
77. Chattaraj PK, Gutierrez-Oliva S, Jaque P, Toro-Labbé A (2003) Mol Phys 101(18):2841
78. Hammond GS (1955) J Am Chem Soc 77(2):334
79. Morell C, Grand A, Toro-Labbé A (2005) J Phys Chem A 109(1):205
80. Jaque P, Toro-Labbé A, Politzer P, Geerlings P (2008) Chem Phys Lett 456(4–6):135
81. Gonzales JM, Cox RS, Brown ST, Allen WD, Schaefer HF (2001) J Phys Chem A 105(50):11327
82. Kendall RA, Dunning TH, Harrison RJ (1992) J Chem Phys 96(9):6796
83. Woon DE, Dunning TH (1993) J Chem Phys 98(2):1358

Index

K.D. Sen (ed.), *Statistical Complexity*,
DOI 10.1007/978-90-481-3890-6, © Springer Science+Business Media B.V. 2011

V

W

Y

9789048138890